The Centrosome

150 Years of Research

nova
science publishers
www.novapublishers.com

Cell Biology Research Progress

More information about this series can be found at
https://novapublishers.com/product-category/series/cell-biology-research-progress/

Life Sciences Research and Development

More information about this series can be found at
https://novapublishers.com/product-category/series/life-sciences-research-and-development/

Rustem E. Uzbekov
Editor

The Centrosome

150 Years of Research

nova
science publishers
www.novapublishers.com

DOI: https://doi.org/10.52305/YCLY4083

Library of Congress Cataloging-in-Publication Data

ISBN: 979-8-89530-815-8 (Hardcover)
ISBN: 979-8-89530-900-1 (eBook)

Published by Nova Science Publishers, Inc. † New York

The front cover presents illustrations showing four main stages in the history of centrosome research.

1. Study of the centrosome at the light-optical level - a drawing of the mitotic spindle by the author of the terms centrosome and centriole, Theodore Boveri (Boveri, 1887).
2. Study of the centrosome ultrastructure using an electron microscope - a photograph of a cross-section of a centriole by Romano Dallai (Dallai et al., 2017).
3. Study of the protein composition of the centrosome using immunofluorescence methods - a cell in anaphase of mitosis, MTs are shown in green, actin in red, with DNA staining (DAPI) in blue, and gamma-tubulin staining revealing the centrosome in purple, photograph kindly provided by Aleksey L. Khodjakov.
4. Study of the fine structure of centrioles using fluorescently labeled proteins and high-resolution microscopy methods – schematic drawing of a centriole from a Paul Guichard paper showing the precise localization of centriolar proteins (Laporte et al., 2024)

Contents

Editor's Introduction

The centrosome and its constituent centrioles are probably among the most remarkable components of eukaryotic cells. During evolution, centrioles first appeared in ancient single-celled flagellates as the basal bodies of flagella. Their striking geometric organization - combining ninefold symmetry in cross-sectional structure and longitudinal polarity - renders centrioles unique in their orderly macromolecular arrangement, inviting comparison to viruses.

While their role in forming flagella and cilia persists in multicellular organisms (e.g., ciliated epithelia and sperm), centrioles have acquired additional critical cellular functions. This adaptation arose because microtubule (MT) nucleation centers on centrioles began organizing not only ciliary MTs but also cytoplasmic MTs. The evolutionary advantage of compartmentalizing multicomponent biochemical reactions in two-dimensional planes over three-dimensional cellular volumes likely drove the development of membranous organelles. However, confining reactions to a single point proved even more advantageous, positioning the centrosome as this specialized hub. Within the centrosome, regulatory molecules concentrate and interact.

Prior to mitotic chromosome segregation and cytokinesis, centrosomes duplicate, forming two centers from one. This process initiates with centriole duplication in late G1 phase, marking the first morphological sign of cell preparation for mitosis. In mitosis, two centrosomes (each containing two centrioles) orchestrate spindle apparatus formation. Microtubules - dynamic polymers ubiquitous in eukaryotes - and their prokaryotic homolog FtsZ filaments (Bi and Lutkenhaus, 1991) form the spindle's structural backbone, akin to their role in flagella.

Centrosomal functions universally involve MT organization: flagellar MTs (excluding the central pair) extend directly from centriolar MTs, while MTs of the spindle are nucleated by the mitotic halo surrounding centrioles or transported by motor proteins. Regulatory molecules are similarly trafficked to the centrosome via the interphase MT network. Thus, the term "microtubule-organizing center" (MTOC) aptly describes a key centrosomal role. However, MTOCs and centrosomes are not synonymous; for example, ciliary basal bodies are MTOCs but not centrosomes. Conversely, centrosomes lose MTOC function in many differentiated cells. Despite taxonomic variation in tubulin sequences, MTs in centrioles, flagella, spindles, and interphase cytoplasm exhibit nearly identical structures.

The centriole's role in cell division remains less clear than its function in ciliogenesis. Spindle MTs are not direct extensions of centriolar MTs but instead grow from nucleation centers in the mitotic halo. Notably, centrioles are absent at spindle poles in higher plants, yeasts, and meiotic spindles of many organisms. Nevertheless, these acentriolar poles retain MT-nucleating proteins. Spindle assembly relies heavily on motor proteins that cluster MT minus ends and on NuMA, which bundles them. Although acentrosomal spindles can form *in vitro*, centrosomes serve as the dominant MTOCs *in vivo*, ensuring bipolar spindle formation and faithful genome segregation.

This book explores the history of centrosome research, modern insights into its morphology and duplication mechanisms, functional analyses of centrosomal proteins, and the biochemistry of MTs - which constitute the centriole's structural framework.

Rustem E. Uzbekov

Chapter 1

The History of Centrosome Discovery and Its Description at the Light Microscopy Level

Rustem E. Uzbekov[*], **PhD, HDR, IR**
Laboratory of Cell Biology and Electron Microscopy, Faculty of Medicine, University of Tours, Boulevard Tonnellé, France;
Faculty of Bioengineering and Bioinformatics, Lomonosov Moscow State University, Moscow, Russia

"Omnis cellula e cellula" – a fundamental postulate formulated by Rudolf Virchow (Virchow, 1855), which, on one hand, summarizes research on living matter by his predecessors, and on the other, established the foundation of a science now called "Cell Biology." Somewhat earlier, Matthias Jakob Schleiden (Schleiden, 1838) and Theodor Schwann (Schwann, 1839) published the fundamentals of their cell theory, according to which all living organisms consist of cells. However, the authors of the cell theory erred in understanding how cells are formed – they assumed that cells could spontaneously generate from acellular protoplasm. A logical conclusion from R. Virchow's statement can be drawn: at a certain point in its existence, the mother cell needs to produce two daughter cells, i.e., divide. Research in the second half of the XIX century broadly described the general process of cell division (Wilson, 1900). The elucidation of the role of chromosomes was essential for understanding how hereditary features are transferred from cell to cell. But chromosomes themselves appeared as passive participants in mitosis, which led one of the classics in cytology – D. Mazia – to compare their role to that of "a dead man at a funeral" during a scientific conference:

[*] Corresponding Author's Email: rustem.uzbekov@univ-tours.fr

In: The Centrosome
Editor: Rustem E. Uzbekov
ISBN: 979-8-89530-815-8

"Everyone gathered for him, but he himself doesn't actively participate in the process."

Even when examining cells under a light microscope, it was observed that spindle fibers attached to chromosomes at their centromeres and pulled them toward opposite poles. These fibers, initially identified by histochemical staining and later observed in living cells using specialized microscopy techniques, were termed spindle fibers, and the structure they formed was named the spindle apparatus due to its resemblance to a textile spindle.

But these spindle fibers do not just pull the chromosomes apart; they pull them strictly into specific parts of the cytoplasm – the spindle's poles. So, what are the poles of the spindle?

In the 70s of the XIX century, two symmetrically located structures were described. They had the form of a "radiant glow" and were called centrospheres (Flemming, 1875; Hertwig, 1875; van Beneden, 1876) (Figure 1.1.). Granules, initially called "polar corpuscles," were observed in the focus of the centrosphere (van Beneden, 1876).

Van Beneden and Neyt (1887) and, independently, Theodor Boveri (1887a) discovered that the polar corpuscles do not disappear at the end of mitosis but are preserved in the interphase, while frequently positioning near the geometrical center of the cell (Figure 1.2.). Van Beneden offered to rename the "polar corpuscles" to "central corpuscles" or "central bodies," but Boveri (1888) proposed to call this structure a "centrosome" and later – a "centriole" (Boveri, 1895).

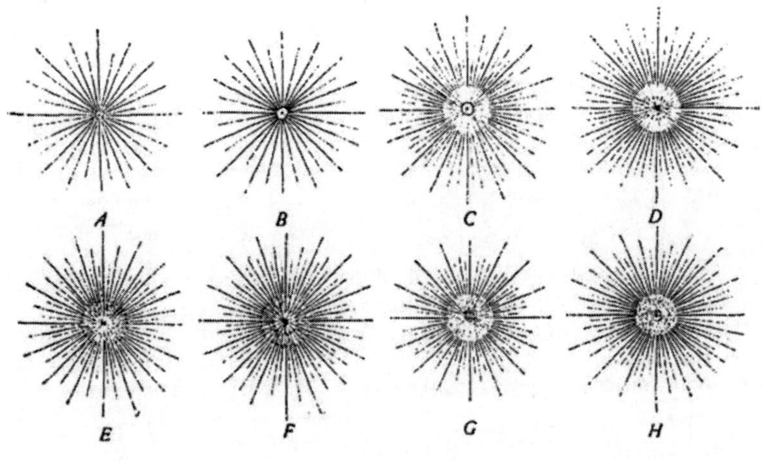

Figure 1.1. Different types of centrosome structure (Wilson, 1900).

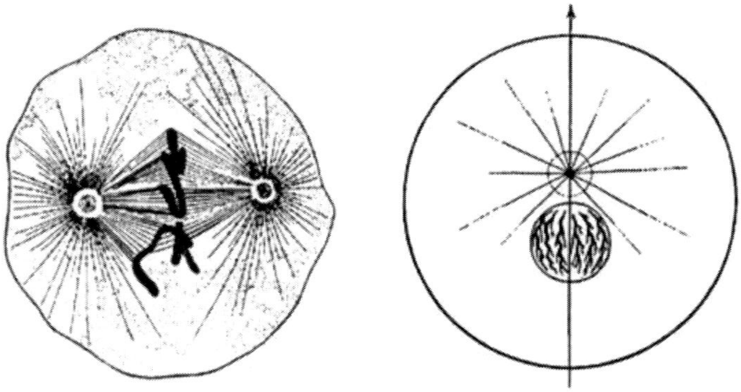

Figure 1.2. Left – centrosomes at the poles of the mitotic spindle (Boveri, 1887a); right – a centrosome in an interphase cell (Van Beneden, 1887).

In some specimens, where centrioles are large, it was stated that at both poles of the spindle there may be visible two of these organelles, and they may take the form of rods (Schreiner and Schreiner, 1905). Not only was the shape of the centrioles described, but the mechanism of their division and the existence of a ligament between the two centrioles were demonstrated in the cited work (Figure 1.3.).

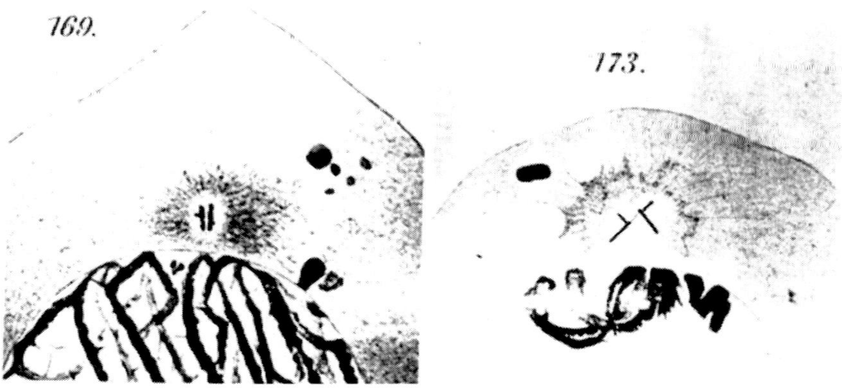

Figure 1.3. Centrosome duplication in spermatocytes of Myxine (Schreiner A. and Schreiner K. E., 1905). Left – a cell in pachytene and right - a cell in diplotene of meiosis. Reproduction from Gall (1961).

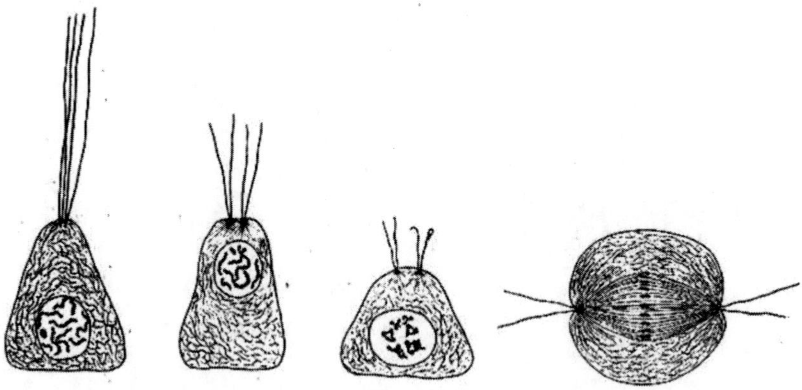

Figure 1.4. Formation of spindle poles from basal bodies in spermatocytes of Bombyx mori (Henneguy, 1898).

After receiving data about the ultrastructure of these organelles by electron microscopy, the centriolar cylinders were directly called centrioles (de Harven and Bernard, 1956; Bessis and Breton-Gorius, 1958), and the name "centrosome" was proposed by Mazia (Mazia, 1984) as a general collective term to designate the centrioles with a plurality of structures associated with them.

For several decades after the first description of the centrioles, it was shown that these organelles are essential components of many animal cells, which was mostly possible due to the fact that centrioles can be intensely stained with iron hematoxylin, crystal violet, and some other histochemical dyes introduced into practice at that time (Heidenhain, 1907; Wilson, 1925).

Somewhat after the centrosomes, at the end of the XIX century, organelles were spotted at the base of flagella and cilia. They were named kinetosomes, or basal bodies (Henneguy, 1898; von Lenhossek, 1898). Henneguy and von Lenhossek observed a mutual transition of basal bodies and centrosomes (Figure 1.4.) and in the year 1898 proposed a hypothesis concerning the homology of these organelles.

Chapter 2

Ultrastructure of the Centrioles and Basal Bodies

Rustem E. Uzbekov*, **PhD, HDR, IR**
Laboratory of Cell Biology and Electron Microscopy, Faculty of Medicine, University of Tours, Boulevard Tonnellé, France;
Faculty of Bioengineering and Bioinformatics, Lomonosov Moscow State University, Moscow, Russia

2.1. Ultrastructure of Centriole

Since the sizes of centrioles and basal bodies are near the resolution limit of light microscopy, detailed investigation of their ultrastructure only became possible with the advancement of electron microscopy in the 1950s. Notably, Brinkley and Stubblefield (1970) already described the literature on centrioles as "voluminous," and since then, the number of studies on these organelles has increased a hundredfold.

The hypothesis of Henneguy and von Lenhossek, which proposed the homology of basal bodies and centrioles, was definitively confirmed upon comparing their ultrastructure. Both organelles are based on cylindrical structures with a diameter of ~0.2 μm and a length of 0.3–0.7 μm. Insect centrioles share a similar diameter (~0.2 μm) but are shorter (0.15–0.3 μm), with their diameter occasionally exceeding their length (Uzbekov et al., 2024a). Exceptions to these dimensions include giant centrioles, reaching several micrometers in length (Schreiner andand Schreiner, 1905; Friedlander and Wahrman, 1966), though such cases are rare. Furthermore, numerous

* Corresponding Author's Email: rustem.uzbekov@univ-tours.fr

In: The Centrosome
Editor: Rustem E. Uzbekov
ISBN: 979-8-89530-815-8

deviations from the canonical centriolar structure have since been documented (Gupta and Kitagawa, 2018).

The ultrastructure of basal bodies was first described in ciliary epithelium (Fawcett and Porter, 1954) and spermatozoa (Burgos and Fawcett, 1955), while mitotic and interphase centrioles were characterized in vertebrate cells (de Harven and Bernhard, 1956). Interestingly, in an earlier study (Selby, 1953), centrioles were visible in electron micrographs of mitotic cells, but the author failed to identify them, instead suggesting that osmophilic granules near the poles might correspond to centrioles. Only after the defining morphological features of centriolar cylinders were established through studies of basal bodies were centrioles reliably detected in mitotic and interphase cells.

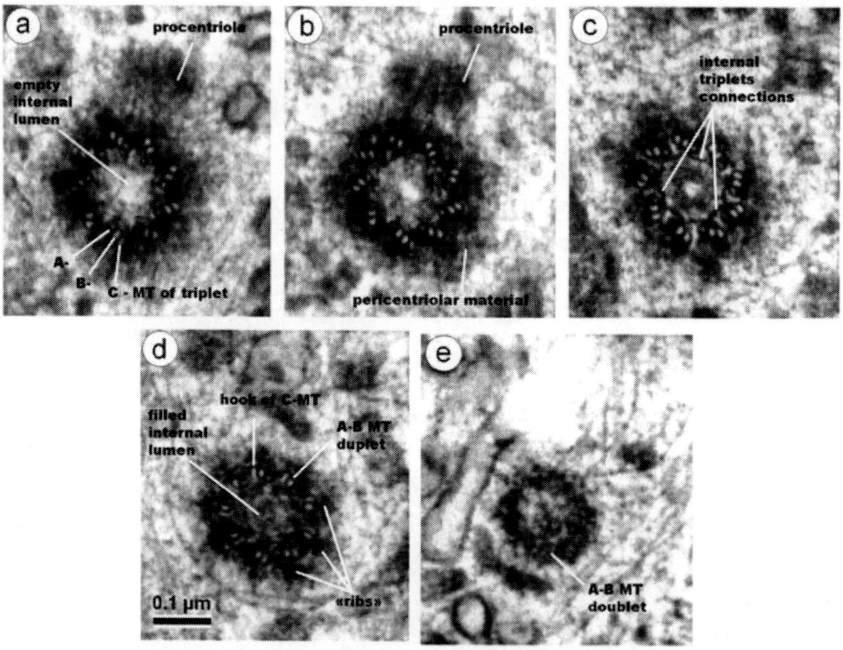

Figure 2.1. Ultrastructure of a centriole in an interphase mammalian cell (pig kidney embryonic cell line SPEV). Panels (a)–(e) show five serial ultrathin sections of a daughter centriole. A short procentriole is visible in panels (a) and (b). Adapted from Uzbekov and Prigent (2007).

Early research revealed that the walls of centriolar cylinders are not uniform but instead consist of nine cords (Fawcett and Porter, 1954), each composed of two or three interconnected tubules (~20 nm in diameter) (de Harven and Bernhard, 1956; Yamada, 1956). These structures were later termed triplets (Afzelius, 1959; Gibbons and Grimstone, 1960), with their constituent tubules designated as the A-, B-, and C-tubules (innermost to outermost, respectively; Gibbons and Grimstone, 1960). Although some authors used numerical labeling (tubules "1," "2," and "3"; de Harven, 1968), the letter-based nomenclature ultimately became standard.

Microtubule (MT) triplets are oriented at an angle to the centriolar cylinder's radius (Figures 2.1., 2.2.), a configuration likened to turbine blades (Bernhard and de Harven, 1960). Although cytoplasmic microtubules were not yet characterized at the time, early researchers noted that the tubular structures (termed *canaliculi*) forming centriolar triplets resembled 20–25 nm filaments found in mitotic spindle fibers and interphase (then called *interkinetic*) cytoplasm (de Harven and Bernhard, 1956). Subsequent work confirmed this observation, and Slauterback (1963) formally proposed the term microtubules (MT) for these filaments.

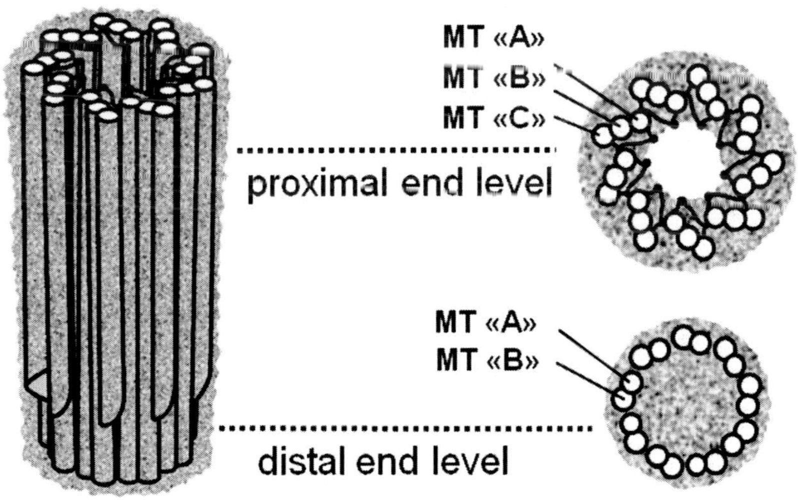

Figure 2.2. Schematic representation of centriole structure.

MT triplets are embedded in amorphous *pericentriolar material.* Each 'C-tubule' connects to the 'A-tubule' of the adjacent triplet (Gibbons and Grimstone, 1960; André, 1964; Szöllősi, 1964; de Harven, 1968). Additionally, electron-dense filaments project inward from each 'A-tubule,' linking it to the 'A-C' bridge (de Harven, 1968).

Centrioles are intrinsically *polar structures.* Bernhard and de Harven (1960) first demonstrated this polarity by showing that the ends of the centriolar cylinder are functionally distinct: the *distal end* (from which the cilium extends toward the cell surface) opposes the *proximal end* (oriented inward, toward the nucleus).

Polarity is further reflected in the orientation of MT triplets. Viewed from the proximal end, a vector drawn from the 'A-' to the 'C-tubule' (radially outward) points counterclockwise (Gibbons and Grimstone, 1960; Gall, 1961; Brinkley and Stubblefield, 1970; Fulton, 1971; Albrecht-Bühler, 1990; Paintrand et al., 1992; Chrétien et al., 1997). This orientation was initially described in basal bodies and later confirmed in centrosomal centrioles (O'Hara, 1970).

To date, no experimental evidence supports the existence of centriolar cylinders with reverse-twist triplets. Thus, depictions of opposite triplet orientation in centrosome diagrams (e.g., Krstić, 1976; Vorobjev and Chentsov, 1980, 1982; Vorobjev and Nadezhdina, 1987; Doxsey et al., 1994) are likely erroneous, as they conflict with established data. For a comprehensive critique of such inaccuracies in centrosome models, see Uzbekov and Prigent (2007).

The centriole exhibits structural polarity, characterized by differences in the inclination angles of MT triplets relative to the centriolar cylinder's radius at its two ends. At the distal end, the line connecting the centers of the MT triplets is nearly perpendicular to the radius (inclination angle $\approx 85°$), whereas at the proximal end, the inclination angle is approximately 55° (Vorobjev and Chentsov, 1982) (Figure 2.2.). This angular disparity results in a slightly larger diameter of the centriolar cylinder at the proximal end compared to the distal end, a feature clearly observable in cryo-electron microscopy images (Chrétien et al., 1997).

MT, which are inherently polar biopolymers, exhibit uniform orientation within centriolar triplets: their minus-ends are positioned at the proximal end of the centriolar cylinder, while their plus-ends are located at the distal end (Snell et al., 1974; Gould and Borisy, 1977).

In transverse cross-section, each MT comprises multiple protofilaments (Figure 2.3.). These protofilaments correspond to 4-nanometer globular subunits of α- and β-tubulin, the proteins that form the structural basis of MTs (Mandelkow et al., 1986).

In the distal region of the centriolar cylinder, MT "C" is typically shorter than MTs "A" and "B," resulting in the appearance of doublets rather than complete triplets in cross-sections (Figure 2.1.).

Initially, MT "C" loses its inner protofilaments adjacent to MT "B," leading to an inwardly open structure. Only MT "A" exhibits a closed circular profile in transverse sections and consists of 13 protofilaments, consistent with most cytoplasmic MTs (Ringo, 1967; Tilney et al., 1973). MT "B" partially overlaps with MT "A," while MT "C" overlaps with MT "B" (Gibbons and Grimstone, 1960; Andre, 1964).

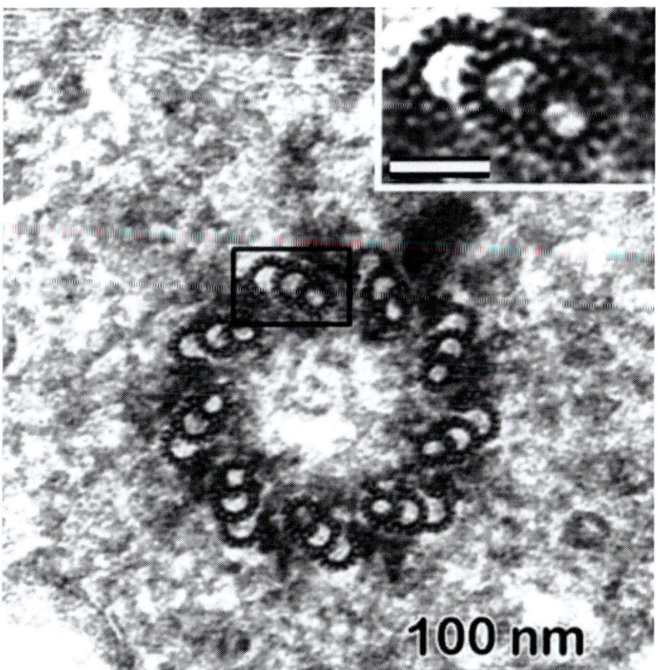

Figure 2.3. MT triplets with clearly visible protofilaments in a cross-section of a centriole from the early spermatid of Adalia decempunctata (Dallai et al., 2017). The scale bar in the inset represents 25 nm.

Thus, MT "B" forms an arc adjacent to MT "A," and MT "C" forms an arc adjacent to MT "B." The number of protofilaments in MTs "B" and "C" varies between 10 and 11, resulting in a total of 33–35 protofilaments per triplet in cross-section (Tilney et al., 1973; Wheatley, 1982; Ou and Rattner, 2004). Notably, the 11th protofilament may exhibit distinct biochemical properties (Tilney et al., 1973). These ultrastructural variations could reflect differences among studied specimens. Additionally, the number of protofilaments shared between adjacent triplets ranges from 3 to 5, as reported by various authors (Fulton, 1971; Tilney et al., 1973; Wheatley, 1982).

MTs within centriolar cylinders exhibit significantly greater stability than spindle or interphase MTs. They resist disassembly upon treatment with microtubule-depolymerizing agents or exposure to cold (De Harven and Bernhard, 1956; Osborn and Weber, 1976; Jensen et al., 1979). Triplet MTs can only be disrupted by high salt concentrations (1–2 M NaCl or KCl) in isolated centrioles (Fais et al., 1986). This stability likely arises from associated proteins, such as tektin (Steffen and Linck, 1988; Stephens and Lemieux, 1998), and post-translational polyglutamylation of centriolar tubulin (Plessmann and Weber, 1997; Bobinnec et al., 1998b).

Even after MT triplet removal, the centriole retains its structural integrity, forming a "centriolar rim" (Fais et al., 1986). This observation suggests that the centriolar cylinder's structural framework is primarily maintained by the surrounding matrix rather than the MT triplets themselves. Intracentriolar structures, particularly the inter-triplet linkage system, also play a crucial role (Figures 2.1., 2.2.). Andre (1964) first reported that centriolar triplet MTs are not strictly parallel to the cylinder's longitudinal axis but may exhibit slight twisting. Subsequent studies confirmed this helical arrangement in both centrosomal centrioles (Stubblefield and Brinkley, 1967; Stubblefield, 1968) and basal bodies (Anderson, 1972; Aist and Williams, 1972). However, other studies suggest that triplet MTs align nearly parallel in certain cell types (Vorobjev and Chentsov, 1980; Paintrand et al., 1992). Thus, triplet twisting appears characteristic of basal bodies and centrosomes in specific cell types.

2.2. Ultrastructure of the Procentriole

Gibbons and Grimstone (1960) first described an internal structure within the basal body's centriolar cylinder, consisting of a central strand (often containing a hub) with nine radiating spokes. These spokes connected distally to filaments adjacent to MT "A" (Figure 2.4.). Gall (1961) later identified an identical structure in centrosomal centrioles. In both studies, this assembly was termed the "*cartwheel structure*" or "cartwheel with an indistinct hub and radial spokes." Gall (1961) also noted that while some centrioles exhibited this feature, others displayed an "empty" hub. Subsequent studies demonstrated that the cartwheel structure is characteristic of immature procentrioles in vertebrates (Figure 2.4a) but absent in mature centrioles (Vorobiev and Nadezhdina, 1986).

The cartwheel is localized asymmetrically within the lumen of the centriolar cylinder, restricted to the proximal end (Gall, 1961). In mature vertebrate centrioles, this structure is absent, leaving the interior lumen appearing electron-lucent in longitudinal sections (Murray et al., 1965). Conversely, the distal end often contains amorphous, electron-dense material (Murray et al., 1965). In insects, the cartwheel structure is retained even in fully mature mother centrioles (Friedlander and Wahrman, 1966, 1970) (Figure 2.4b).

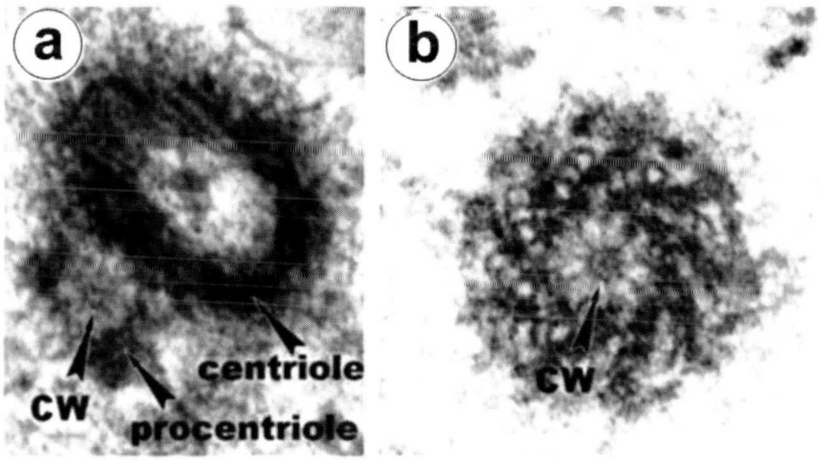

Figure 2.4. Ultrastructure of the cartwheel in procentrioles and mature centrioles. (a) Procentriole adjacent to a mother centriole in SPEV porcine kidney epithelial cells (Uzbekov and Prigent, 2007). (b) Mature centriole from the wasp Nasonia vitripennis (Uzbekov et al., 2018a). CW indicates cartwheel structure. Scale bars: 100 nm (A), 200 nm (B).

In basal bodies, the cartwheel is typically more elongated and may extend beyond the proximal end into the mid-region of the centriolar cylinder. As in centrosomes, it is often absent in mature basal bodies (Wolfe, 1972). However, certain protists (e.g., ciliates) retain the cartwheel in fully differentiated basal bodies (Allen, 1969; Tamm, 1972).

Chapter 3

Structures Associated with the Centriolar Cylinder

Rustem E. Uzbekov*, **PhD, HDR, IR**
Laboratory of Cell Biology and Electron Microscopy, Faculty of Medicine, University of Tours, Boulevard Tonnellé, France;
Faculty of Bioengineering and Bioinformatics, Lomonosov Moscow State University, Moscow, Russia

3.1. Subdistal Appendages (Pericentriolar Satellites)

Early electron microscopy studies, facilitated by improved specimen fixation and staining methods (Watson, 1958; Sabatini et al., 1963; Reynolds, 1963), enabled the identification of important ultrastructural details of centrioles and basal bodies.

Bessis and Breton-Gorius (1958) discovered that the surface of leukocyte centrioles was often covered with outgrowths (Figure 3.1.), consisting of a stem with a round cap, which these authors termed *"massules."* Bernhard and de Harven (1960) proposed naming these centriole-associated outgrowths *"pericentriolar satellites."* These structures measure approximately 0.14 µm in length, with caps approximately 0.04 µm in diameter.

Subsequently, these structures were alternatively named *"subdistal appendages"* (Paintrand et al., 1992; Tournier and Bornens, 1994) - a designation that may be less accurate, as pericentriolar satellites in some cell types are distributed along the entire length of the centriolar cylinder (Vorobjev and Chentsov, 1977). Nevertheless, this terminology has become

* Corresponding Author's Email: rustem.uzbekov@univ-tours.fr

In: The Centrosome
Editor: Rustem E. Uzbekov
ISBN: 979-8-89530-815-8

the common usage (Uzbekov and Alieva, 2018), and will be employed hereafter.

Early ultrastructural analyses revealed that the two centrioles in a cell exhibit distinct morphological features. These ultrastructural differences persist after the daughter centriole reaches full length. Murray et al., (1965) were the first to note that subdistal appendages are present on only one of the paired centrioles, an observation subsequently confirmed in most cell types.

The appendage-bearing centriole has been variously designated as mature (Gall, 1961; Rattner and Phillips, 1973), primary (Murray et al., 1965), parental (Sorokin, 1968; Brinkley and Stubblefield, 1970), or mother centriole (Vorobjev and Chentsov, 1982; Paoletti and Bornens, 1997). The accompanying centriole is typically called immature or daughter (Bernhard and de Harven, 1960; Szollosi, 1964; Rattner and Phillips, 1973; Vorobjev and Chentsov, 1982). In French literature, these were historically termed "centriole paternal" and "centriole filial," reflecting the masculine gender of 'centriole' in French.

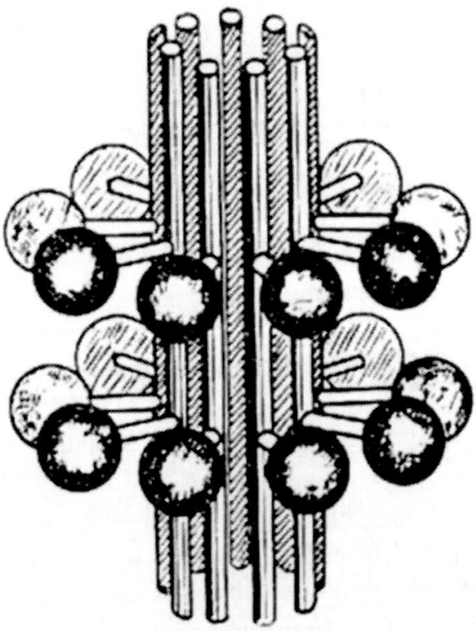

Figure 3.1. Centriole structure diagram adapted from Bessis and Breton-Gorius (1958). Two rows of nine outgrowths (termed "massules" by the authors) associated with the centriole's lateral surface. Note that the stems connecting the heads to the centriole appear significantly thinner in this schematic representation than in subsequent studies.

De Harven and Dustin (1960) first reported that subdistal appendages are exclusively present on interphase centrioles and absent during mitosis. Subsequent ultrastructural analysis revealed that these appendages disassemble during the G2-phase preceding mitosis and reassemble during early G1-phase (Vorobiev and Chentsov, 1982).

Initial reports regarding appendage number were contradictory. While Bessis and Breton-Gorius (1958) proposed a symmetrical arrangement of two rows containing nine appendages each (Figure 3.1.), later studies challenged this model. Image analysis suggests their schematic reconstruction, based on limited-quality oblique sections, may have conflated *subdistal appendages* with *distal appendages* (which possess thinner stems, see below). Bernhard and de Harven (1960) subsequently demonstrated that fewer than nine subdistal appendages may be present, a finding corroborated by later work. For instance, thymocyte centrosomes contain no more than six such structures (termed "arms" in Komesli et al., 1989). Moreover, some cell types completely lack these appendages (Stubblefield and Brinkley, 1967; Wheatley, 1969, 1974; Onishchenko and Chentsov, 1986), while others exhibit more than nine per centriole (Vorobiev and Chentsov, 1977).

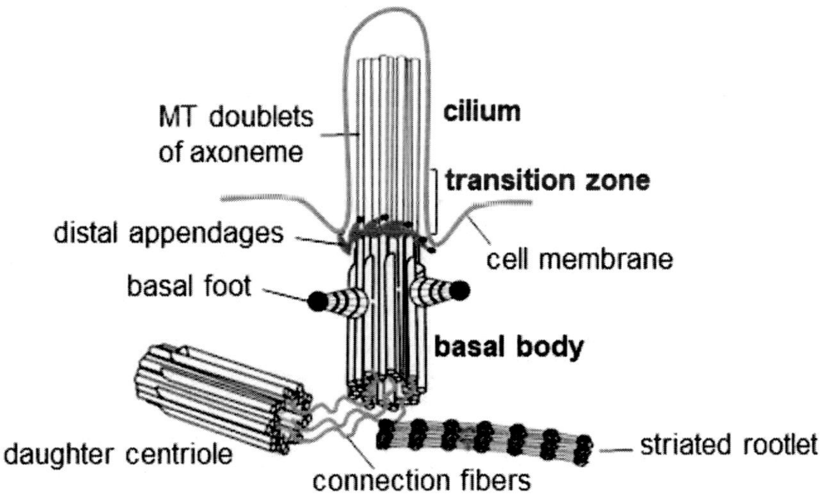

Figure 3.2. Schematic structure of a basal body. Structural details of the cilium are not depicted.

A centrally symmetric arrangement of nine appendages - each stem linked to a centriolar triplet - was described in KE37 cells (Paintrand et al., 1992). However, the authors themselves noted considerable variability in appendage number and distribution, even within single experiments. Subsequent studies often overlooked this variability, mistakenly generalizing the KE37 pattern. Furthermore, isolation artifacts (including cytochalasin/nocodazole pretreatment, hypotonic treatment and ultracentrifugation) may alter appendage morphology. Ultrastructural evidence suggests native appendages comprise strands extending from 2-3 triplets, converging at a head to form conical structures. Isolation procedures may disrupt these connections, leaving single strands attached to individual triplets - as frequently observed in isolated KE37 centrioles

Collectively, these observations demonstrate that both the number and spatial organization of subdistal appendages exhibit remarkable cell-type specificity and functional dependence (Uzbekov and Alieva, 2018). Ultrastructural studies consistently show these appendage heads serving as anchoring sites for centrosomal MTs, with their architecture reflecting the cell's physiological state.

The evolutionarily conserved basal foot structure represents a functional homolog of subdistal appendages in basal bodies (Figure 3.2.). Typically numbering 1-2 per centriolar cylinder (Wolfe, 1972) and positioned near the distal end, basal feet in motile cilia show strict planar polarization aligned with the ciliary beating direction (Gibbons, 1961). This organization reveals an intriguing inverse correlation: cell types rich in subdistal appendages (e.g., hematopoietic lineages) never develop primary cilia, whereas ciliated cells exhibit appendage numbers matching basal foot quantities in basal bodies.

In most cell types, subdistal appendages are exclusively associated with mature mother centrioles (Vorobiev and Nadezhdina, 1987). Although observed on both centrioles in neutrophils, the mother centriole invariably possesses a greater number of these appendages compared to its daughter counterpart (Gudima et al., 1984).

3.2. Distal Appendages

Unlike subdistal appendages, distal appendages persist throughout the cell cycle, maintaining their presence during the transition from interphase to mitosis (Figure 3.3j, k). These structures serve as reliable markers for identifying mature mother centrioles, as they develop only after the centriole has completed more than one cell cycle.

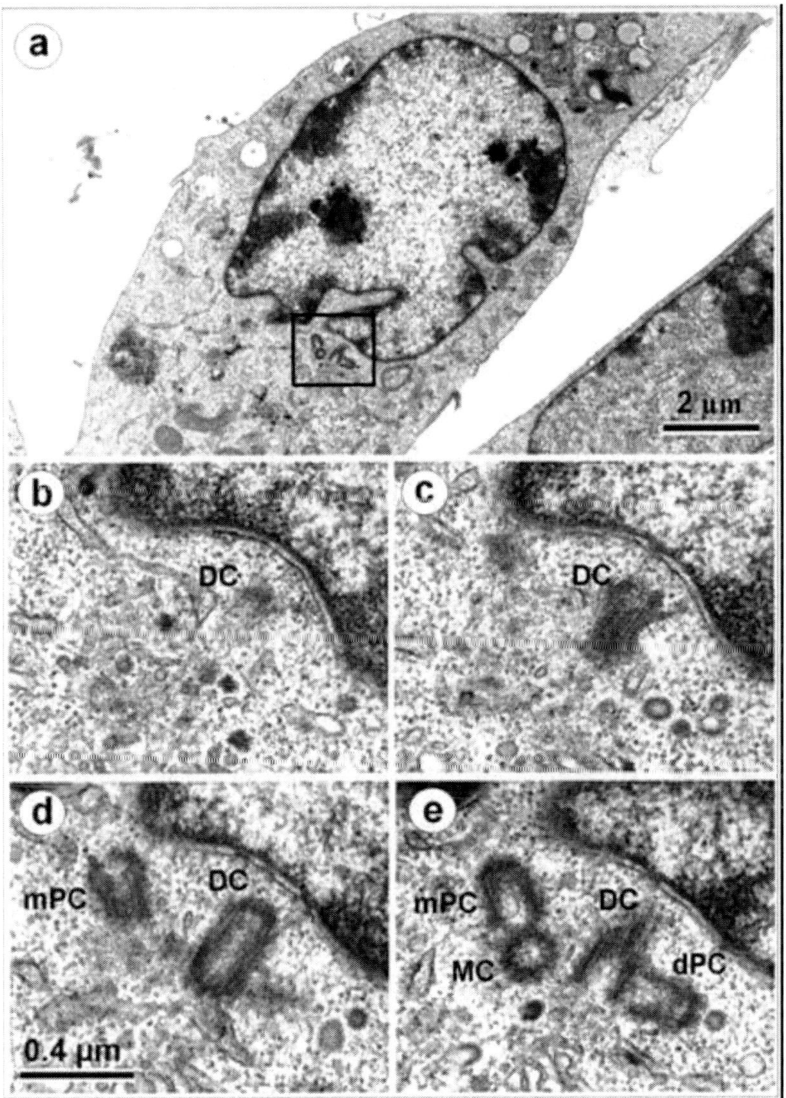

Figure 3.3. (part 1). Centrosomal region in BHK-21 cells during G2 phase (a) Low-magnification overview (centrosomal region marked by rectangle). (b-e) Serial ultrathin sections through the centrosomal region. MC - mother centriole, DC - daughter centriole, mPC - mother procentriole, dPC - daughter procentriole (Images: R. Uzbekov, from Uzbekov and Alieva, 2013, modified).

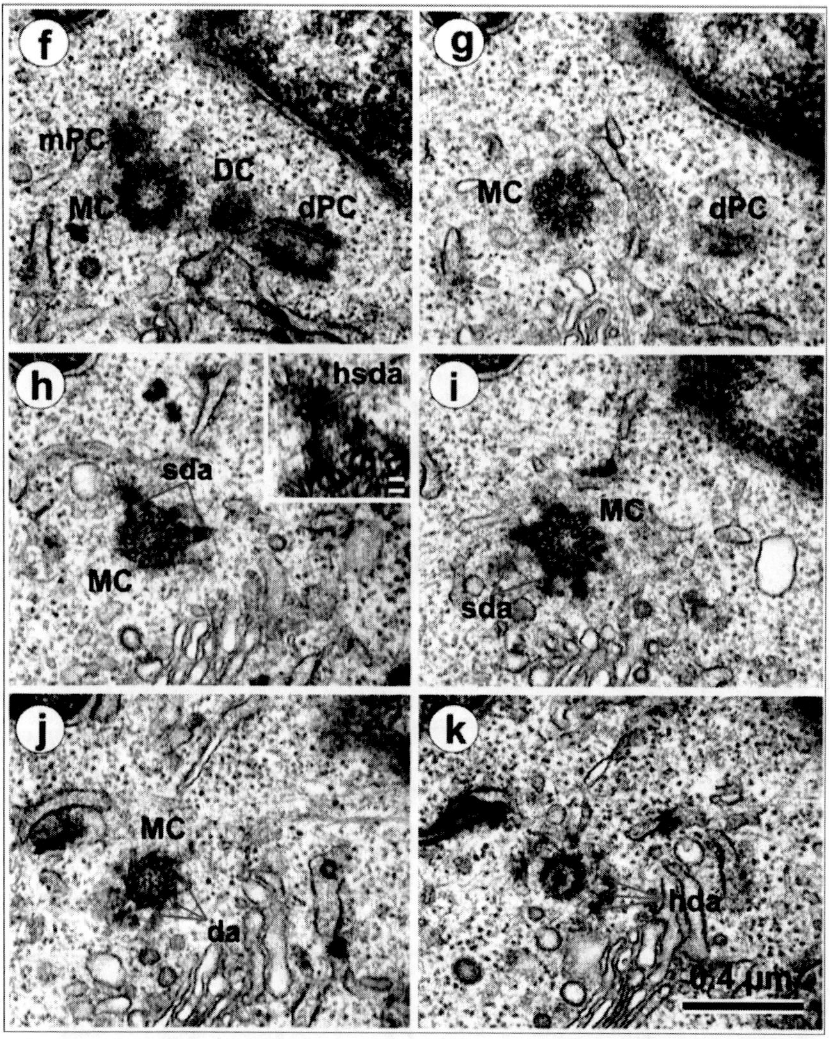

Figure 3.3. (part 2). Centrosomal ultrastructure in BHK-21 cells during G2 phase. (f-k) Serial ultrathin sections. MC - mother centriole, DC - daughter centriole, mPC - mother procentriole, dPC - daughter procentriole, pcs - subdistal appendages, ap - distal appendages, hda - distal appendage heads, hsda - subdistal appendage heads. Note exclusive presence of appendages on mother centriole (Images: R. Uzbekov, from Uzbekov and Alieva, 2013, modified). Scale bar in panel h inset: 25 nm.

Distal appendages were first described in centrioles by Gachet and Thiery (1964) and in basal bodies by Szollosi (1964). However, retrospective analysis suggests that the ninefold radial crown structures reported earlier by Bessis and Breton-Gorius (1958) likely represented distal appendages. Both distal appendages and their homologous structures - termed alar sheets (Anderson, 1972; Figure 3.2.) or satellites (Szollosi, 1964) - consistently number nine per basal body (Szollosi, 1964; Anderson, 1972). Notably, in jellyfish spermatids, these alar sheets (referred to as satellites in the original publication) exhibit terminal bifurcation into three distinct projections (Szollosi, 1964).

Distal appendages project at approximately 50° relative to the centriolar radius. When viewed from the proximal end, they exhibit a clockwise helical arrangement, angled at about 45° to the longitudinal axis. This configuration creates the appearance of the centriole standing on supporting "legs" that frequently extend beyond the triplet/doublet MTs longitudinally.

Consequently, the head domains of distal appendages often remain discernible in the immediately adjacent serial section even after the distal extremity of the centriolar cylinder is no longer present in the micrograph.

In summary, three distinct centriolar variants can coexist within cells: procentrioles, daughter centrioles, and mother centrioles (Figure 3.3.). The sequential transformation from procentriole to daughter centriole and ultimately to mother centriole is schematically illustrated in Figure 3.4.

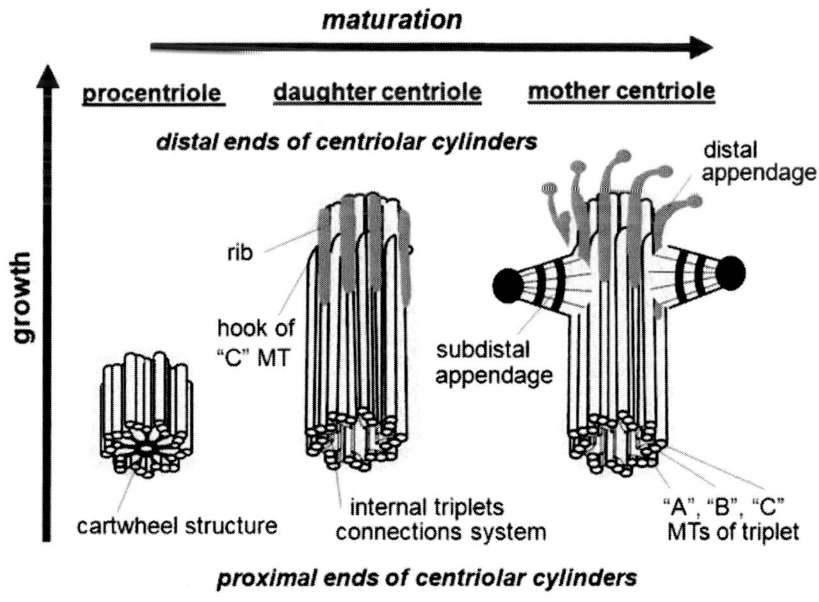

Figure 3.4. Centriolar cylinder growth and maturation cycle (Prigent and Uzbekov, 2022).

During the first cell cycle, when the developing daughter centriole remains attached to and significantly shorter than the mother centriole, it is referred to as a daughter procentriole (Gall, 1961). In interphase cells exhibiting radial microtubule organization (e.g., fibroblasts), the mother centriole serves as the primary MT-organizing center, hence its designation as the "active" centriole, contrasting with the "inactive" daughter centriole.

During G2 phase, when four centriolar structures are present (mother centriole, daughter centriole, and two nearly formed procentrioles), only the mother centriole displays complete morphological and functional maturation, evidenced by the presence of both subdistal and distal appendages (Figure 3.3.).

Stubblefield (1968) first demonstrated that procentriole elongation follows linear kinetics. Our studies reveal that procentriole growth, initiated during G1 phase, enters a paused state until S-phase onset, with DNA replication initiation appearing to license further centriolar maturation (Uzbekov, 2007). This regulatory checkpoint may exhibit cell-type specific timing.

Chapter 4

Centrosome Structure in Mitotic Cells

Rustem E. Uzbekov*, PhD, HDR, IR

Laboratory of Cell Biology and Electron Microscopy, Faculty of Medicine, University of Tours, Boulevard Tonnellé, France;
Faculty of Bioengineering and Bioinformatics, Lomonosov Moscow State University, Moscow, Russia

At the onset of mitosis, the daughter centriole remains shorter than the mother centriole (Stubblefield, 1968; Rattner and Phillips, 1973; Chretien et al., 1997). In various cell types, the mother centriole reaches its full length during prometaphase (Rattner and Phillips, 1973; Vorobjev and Chentsov, 1982) or metaphase (Robbins et al., 1968). Notably, Chretien and colleagues demonstrated that even after mitosis, the daughter centrioles in the lymphoblastoid cell line KE37 may still be slightly shorter than the mother centriole (Chretien et al., 1997).

Until the end of telophase of mitosis, the young daughter centriole (formerly the procentriole) maintains a perpendicular orientation relative to the mother centriole, forming a diplosome (Andre, 1964; Robbins et al., 1968; Vorobjev and Chentsov, 1982). The distance between the two centrioles typically does not exceed 0.2 micrometers.

The mother centriole is positioned at the focus of the mitotic spindle and is oriented perpendicular to its axis (Murray et al., 1965). It is surrounded by a "cloud" of fine fibrillar material, referred to as the *mitotic halo*. The daughter centriole is partially immersed in this halo, with only its proximal end embedded within it (Figure 4.1.). The diameter of the mitotic halo reaches approximately 1 micrometer in prophase cells and decreases to about 0.5

* Corresponding Author's Email: rustem.uzbekov@univ-tours.fr

In: The Centrosome
Editor: Rustem E. Uzbekov
ISBN: 979-8-89530-815-8

micrometers in metaphase cells (Robbins et al., 1968). Following the completion of mitosis, the centrioles lose their mutually perpendicular orientation, and the mitotic halo dissipates.

During mitosis, the centrosome structure exhibits less variability across different organisms compared to interphase. At the onset of mitosis, the primary cilium (if present) is resorbed (Rieder et al., 1979), and the subdistal appendages disappear (De Harven and Dustin, 1960). Toward the end of interphase or during prophase, the two pairs of centrioles begin to separate and form two equivalent microtubule (MT) polymerization centers, known as prophase "asters," which were previously referred to as centrospheres. The process of centrosome separation is inhibited by MT depolymerization (Hyman and White, 1987; Schatten et al., 1988).

The interphase microtubule (MT) network undergoes complete disassembly during prophase. However, in the early stages of prophase, as asters begin to form, remnants of interphase MTs may persist along the cell periphery (Vandre et al., 1984b). Each spindle pole in mitosis comprises a diplosome, consisting of two mutually perpendicular centrioles (Robbins and Gonatas, 1964; Roos, 1973). The mother centriole is readily distinguishable from the daughter centriole by its two "free ends." Additionally, the more mature centriole in mitosis can be identified morphologically by the presence of distal appendages (Vorobjev and Chentsov, 1982). The proximal end of the daughter centriole is positioned adjacent to the surface of the centriolar cylinder of the mother centriole. Typically, only this end is embedded within the mitotic halo, which serves as the focal point for the convergence of MTs of the mitotic spindle.

The centrosome's capacity to nucleate MTs remains relatively stable throughout interphase but significantly increases during the transition from interphase to prophase, peaking in prometaphase and subsequently declining after the onset of anaphase (Snyder and McIntosh, 1975; Brinkley et al., 1981; Kuriyama and Borisy, 1981a, 1981b). This enhanced activity is accompanied by a 4-7-fold increase in the amount of gamma-tubulin within the mitotic centrosome (Dictenberg et al., 1998; Vorobjev et al., 2000). These findings align well with *in vitro* MT polymerization experiments, which demonstrated that the mitotic centrosome can nucleate up to five times more MTs than its interphase counterpart (Snyder and McIntosh, 1975; Kuriyama and Borisy, 1981b).

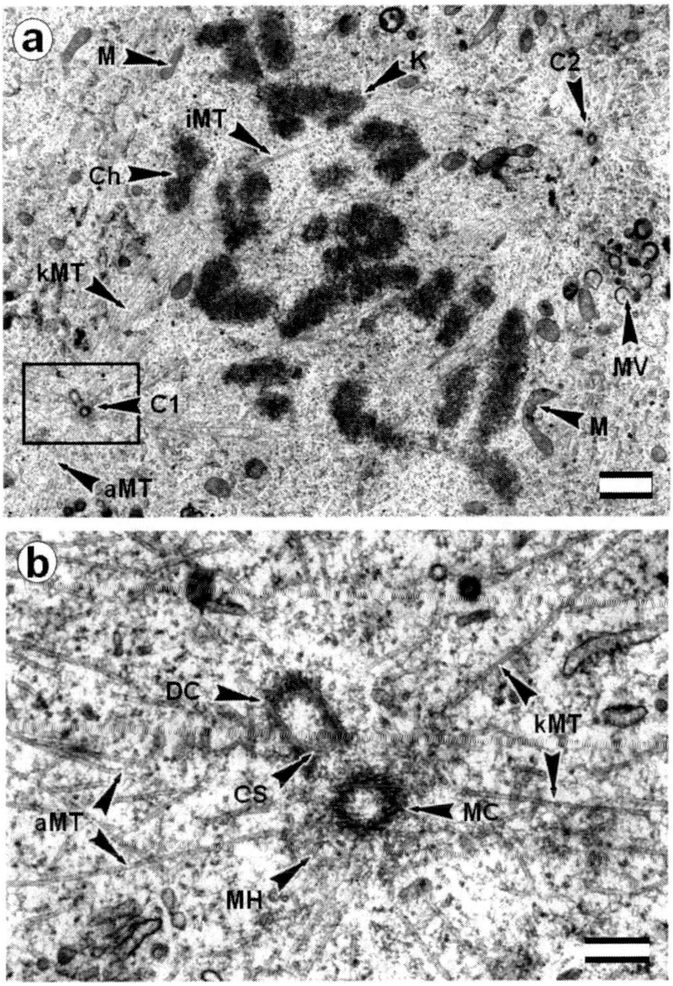

Figure 4.1. Ultrastructure of the cell and centrosome during early anaphase in a mitotic cell of the SVEV (pig kidney embryo) cell line. a – Cell at low magnification (the region of centrosome 1 is marked by a rectangle). b – Diplosome of the left spindle pole at high magnification. The mother centriole is fully enveloped by the mitotic halo, while the daughter centriole is partially immersed in the mitotic halo, with only its proximal end embedded. C1 – centrosome 1 at the left spindle pole; C2 – centrosome 2 at the right spindle pole; Ch – chromosomes; CS – cartwheel structure in the proximal lumen of the daughter centriole; DC – daughter centriole; K – kinetochore; M – mitochondrion; MH – mitotic halo; MC – mother centriole; MV – membrane vesicles; MT – microtubules (three types: aMT – astral MT; iMT – interzonal MT; kMT – kinetochore MT). Scale bars: 1 μm for a, 200 nm for b. Photos by R. Uzbekov, adapted from Uzbekov and Alieva (2013).

Proteins that are absent in the interphase centrosome emerge in the mitotic centrosome (Calarco-Gillam et al., 1983; Sager et al., 1986; Leslie and Wilson, 1987; Leslie, 1989, 1990; Alieva and Uzbekov, 2008). The spatial expansion and enhanced microtubule-polymerizing activity of the centrosome are facilitated by the recruitment of novel proteins from the cytoplasm (Leslie, 1990). The accumulation of at least a subset of these proteins is contingent upon the integrity of MTs: exposure to anti-MT agents inhibits their recruitment (Sager et al., 1986; Maekawa et al., 1989). Similar effects are observed with taxol treatment (Maekawa et al., 1989). Conversely, the accumulation of certain other proteins in mitotic centrosomes is independent of MTs; for instance, Aurora A kinase (Roghi et al., 1998).

Phosphorylation of centrosomal proteins also plays a pivotal role in spindle pole formation (Davis et al., 1983; Vandre et al., 1984a; Centouse et al., 1986). Further details regarding the proteins that appear in the centrosome during mitosis will be explored in subsequent chapters dedicated to the biochemical composition of the centrosome.

In prometaphase and metaphase cells, the centrosome nucleates MTs of several distinct types (Figure 4.1.). Some MTs are oriented along radii extending outward from the centrosome in directions opposite to the chromosomes; these are referred to as astral MTs. Other MTs connect the centrosome to kinetochores on chromosomes, hence designated as kinetochore MTs. Additionally, some MTs, which do not bind directly to kinetochores, can extend across the spindle apparatus and interact with similar MTs emanating from the opposite pole, reaching the opposite half-spindle. Electron microscopic studies of mitotic cells—previously observed *in vivo*—have demonstrated that for the initiation of chromosome movement during prometaphase, lateral interactions between a kinetochore and a single attached MT from the centrosome are sufficient (Rieder and Alexander, 1990). Mitotic spindle MTs exhibit dynamic instability and are more dynamic than interphase MTs, with growth rates averaging 14 μm/min and shortening rates averaging 16 μm/min (Hayden et al., 1990). Consequently, the rate of MT polymerization during prometaphase exceeds twice that observed in interphase cells (Schulze and Kirschner, 1987). The disassembly rate of these MTs is comparable to measurements obtained both *in vitro* (Mitchison and Kirschner, 1984b) and *in vivo* in interphase cells (Schulze and Kirschner, 1987).

The centrosome-associated MT array reaches its maximal density in metaphase and progressively diminishes during anaphase. Concurrently, the mitotic halo disappears, and by telophase, almost no MTs remain near the centrosome (Vorobjev and Chentsov, 1987). As previously noted, the daughter centriole typically elongates to match the mother centriole's size by late prometaphase or metaphase, maintaining its perpendicular orientation until mitosis completion (Robbins et al., 1968; Rattner and Phillips, 1973; Vorobjev and Chentsov, 1982).

Chapter 5

Pericentriolar Material and Intercentriolar Connections: Mechanisms of Centrosome Splitting and Separation

Rustem E. Uzbekov[1,*], PhD, HDR, IR
and Claude Prigent[2], PhD, HDR, DRCE
[1]Laboratory of Cell Biology and Electron Microscopy, Faculty of Medicine, University of Tours, Boulevard Tonnellé, France;
Faculty of Bioengineering and Bioinformatics, Lomonosov Moscow State University, Moscow, Russia
[2]L'équipe de recherche Cycle Cellulaire, Centre de Recherche en Biologie cellulaire de Montpellier, U'unité mixte de recherche 5237 CNRS;
University of Montpellier, Montpellier, France

5.1. Centriole Separation During the G1 Phase of the Cell Cycle

A diplosome comprises two interconnected, mutually perpendicular centrioles. This structure forms by the onset of mitosis and disassembles following mitotic exit. Although interphase cells occasionally exhibit divergent centrioles separated by several micrometers (μm), the two centrioles typically remain in close proximity. This observation raises critical questions: What maintains their proximity, and does a physical connection exist between them? Early ultrastructural studies provided clues. Amano (1957), in one of the first electron microscopy investigations of centrioles, described a filamentous ligament-like structure bridging the centrioles. Paradoxically, subsequent advances in fixation and staining techniques rendered these

* Corresponding Author's Email: rustem.uzbekov@univ-tours.fr

In: The Centrosome
Editor: Rustem E. Uzbekov
ISBN: 979-8-89530-815-8

connecting filaments less discernible in electron micrographs, complicating further characterization.

The physical connection between centrioles has been demonstrated in studies involving isolated centrosomes. During isolation procedures, mother and daughter centrioles frequently remain interconnected (Bornens et al., 1987; Tournier et al., 1991). These studies revealed that the intercentriolar linkage consists of a bundle of thin filaments, leading to its designation as a filamentous or fibrous ligament. Notably, the structural integrity of this filamentous bundle depends on the isolation methodology. For instance, in the presence of divalent cations, the filaments dissociate into a network of discrete aggregates composed of denser material. These aggregates, identified as satellites, are occasionally observed near centrioles *in situ* (Paintrand et al., 1992).

The dynamic nature of intercentriolar connections has been further investigated using cell lines expressing fluorescently tagged centrin, a protein intrinsically associated with centrioles (Piel et al., 2000). These experiments demonstrated that daughter centrioles in interphase cells can migrate significant distances away from their mother counterparts, suggesting that the intercentriolar ligament system is highly *labile*. Comparative analyses across cultured cell lines (Table 1) revealed that, in most cases, centrioles remain closely apposed at an average distance of 0.8–2 μm (Komarova et al., 1996; Jean et al., 1999; Uzbekov et al., 2002; Uzbekov, 2004).

Table 1. Prevalence of interphase cells with varying distances between mother and daughter centrioles (Uzbekov, 2004)

Cell culture	Part cells (%) with distance between centrioles			
	<1 μm	1–2 μm	2–5 μm	>5 μm
HeLa	96	4	–	–
XL2 (*Xenopus laevis*)	92	5	2	1
Grain-cells of rat cerebellum	91	7	1	1
L-cells	91	6	1	2
Vero (green monkey fibroblasts)	84	11	2	3
Chinese hamster cells	77	14	5	4
Primary rat fibroblasts	55	14	17	14
Embryonic pig kidney cells (SPEV)	28	40	24	8

However, exceptions exist: in embryonic pig kidney (SPEV) cells and primary rat fibroblasts, approximately *30% of interphase cells* exhibit centriole separations exceeding 2 μm (Komarova et al., 1996). This implies that *daughter centriole "wandering"* may be a cell type-specific phenomenon. In other cell types, such displacements occur infrequently and do not significantly alter the overall average intercentriolar distance.

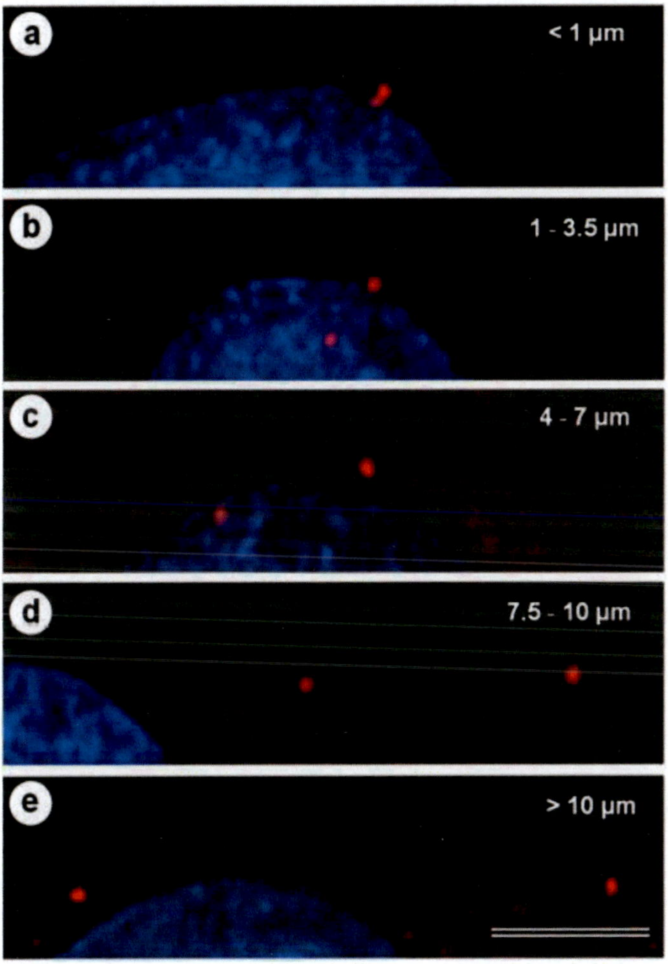

Figure 5.1. XL2 (*Xenopus laevis*) cell line. Centrosomes labeled with anti-pEg2 (*Aurora A kinase*) antibodies (red). Nuclei stained with DAPI (blue). Five panels illustrate distinct centrosome separation distance classes. Scale bar: 5 μm (Uzbekov et al., 2002).

Following mitotic exit, during early G1 phase of the cell cycle, centrioles remain closely apposed in most cell types. Subsequently, they lose their mutually perpendicular orientation and separate, typically maintaining a distance of <1 μm. This phase of the centriolar cycle, termed the "disorientation" stage (Robbins et al., 1968), precedes centriole duplication. In most cells, the separation of the two inherited centrioles occurs post-duplication, reflecting the divergence of newly formed centrosomes rather than individual centrioles. However, in rare instances where centriole separation exceeds the length of the mitotic spindle (an infrequent occurrence in most cell lines), centrosomes destined to form spindle poles undergo convergence rather than divergence prior to mitosis (Figure 5.1d, e).

The regulatory mechanisms coordinating these two processes - centriole separation in early G1 and centrosome divergence preceding mitosis - are not fully elucidated. However, emerging studies have begun to unravel aspects of their coordination, as discussed in subsequent sections.

It has been demonstrated that centriole disorientation depends on the activity of separase, a cysteine protease (Tsou and Stearns, 2006). Separase plays a critical role in sister chromatid separation during anaphase by cleaving cohesin complexes, which are responsible for chromatid cohesion (Nasmyth et al., 2000; Uhlmann et al., 2000; Waizenegger et al., 2000).

Notably, cohesin is not only localized along chromosome arms and at centromeres but has also been identified at the centrosome (Gregson et al., 2001). Tsou and Stearns (2006) proposed that separase-mediated cleavage of cohesin not only initiates sister chromatid separation but also concurrently triggers the disorientation of mother and daughter centrioles.

In human cells, cohesin is removed from chromosome arms during mitotic prophase via Plk1 (polo-like kinase 1) activity, while centromeric cohesin persists until anaphase onset. This delayed removal is mediated by shugoshin (Sgo1), a protein localized at kinetochores that protects centromeric cohesin from premature dissociation (Gimenez-Abian et al., 2004; Kitajima et al., 2004).

Wang et al., (2008) further revealed that sSgo1, a splice variant of Sgo1, localizes specifically to centrosomes. Depletion of Sgo1 or expression of a dominant-negative mutant induces premature sister chromatid separation and centriole disorientation, highlighting its dual role in checkpoint control and centrosome regulation.

Wong and Stearns (2003) hypothesized that the perpendicular orientation of centrioles restricts their duplication capacity. Thus, centriole disorientation may represent a licensing mechanism for S-phase duplication. Their work also demonstrated that while disorientation is independent of mitotic exit and Cdk2–cyclin E activity, it strictly requires separase, emphasizing the protease's central role in coordinating structural and regulatory events.

Astrin, initially characterized as a spindle-associated protein (Chang et al., 2001; Mack and Compton, 2001; Gruber et al., 2002), exhibits dynamic localization during mitosis, accumulating at spindle poles and kinetochores. Depletion of astrin leads to metaphase arrest, accompanied by premature centriole separation and loss of sister chromatid cohesion, suggesting its involvement in maintaining both chromosomal and centrosomal integrity (Thein et al., 2007).

Rootletin, identified as the primary structural component of the intercentriolar linker (Yang et al., 2002), forms extensive filaments upon overexpression in U2OS cells. Conversely, siRNA-mediated rootletin depletion induces centrosome separation, confirming its role in maintaining centriole cohesion (Bahe et al., 2005). Rootletin physically interacts with C-Nap1, a coiled-coil protein localized at the proximal ends of centrioles, which serves as an anchor for the fibrous linker (Fry et al., 1998a). During mitotic entry, Nek2 kinase phosphorylates C-Nap1, triggering its dissociation from centrioles and enabling centrosome separation (Mayor et al., 2000, 2002). This phosphorylation is counteracted by protein phosphatase 1 (PP1), and PP1 inhibition enhances Nek2-mediated C-Nap1 dissociation (Helps et al., 2000).

Intriguingly, Nek2 also phosphorylates β-catenin, a Wnt signaling effector (Clevers, 2006), which localizes to centriolar proximal/distal ends and intercentrosomal regions (Bahmanyar et al., 2008). β-catenin depletion results in monopolar spindles with duplicated but unseparated centrosomes (Kaplan et al., 2004), whereas expression of stabilized β-catenin mimics Nek2 hyperactivation, destabilizing centriole cohesion.

Triple immunofluorescence staining reveals partial colocalization of β-catenin and rootletin with C-Nap1, suggesting their assembly into a tripartite complex (Bahmanyar et al., 2008). C-Nap1 depletion reduces rootletin recruitment to the intercentriolar linker and shortens rootletin filaments between partially separated centrosomes (Bahe et al., 2005).

Bahmanyar et al., (2008) observed that β-catenin is absent in 34% of C-Nap1-depleted cells lacking intercentrosomal rootletin, whereas residual rootletin filaments retain β-catenin. Similarly, rootletin depletion abolishes β-catenin localization between centrosomes, indicating interdependency. During mitosis, β-catenin accumulates at centrosomes independently of the linker, coinciding with Nek2-mediated C-Nap1/rootletin dissociation. This aligns with studies showing that Nek2 overexpression induces centrosome separation, rootletin loss, and β-catenin enrichment at separated centrosomes. These findings support a model where C-Nap1 anchors rootletin filaments at centrioles, while β-catenin stabilizes intercentriolar connections. Nek2 activation during mitosis disrupts these interactions, enabling centrosome separation.

Ultrastructural analysis of interphase centrosomes reveals sparse electron-dense material between briefly separated centrioles (Figure 5.2.). However, serial sectioning identifies a transient fibrillogranular "cloud" bridging the gap, suggesting dynamic reorganization of linker components during the cell cycle.

The functional conservation of centriole linkage mechanisms is underscored by experiments demonstrating that centrosomes from diverse species (sea urchin, mouse, human) can induce spindle formation and parthenogenesis in amphibian oocytes (Maller et al., 1976; Karsenti et al., 1984b; Bornens et al., 1987). This remarkable universality highlights the evolutionary importance of centrosome-mediated microtubule organization.

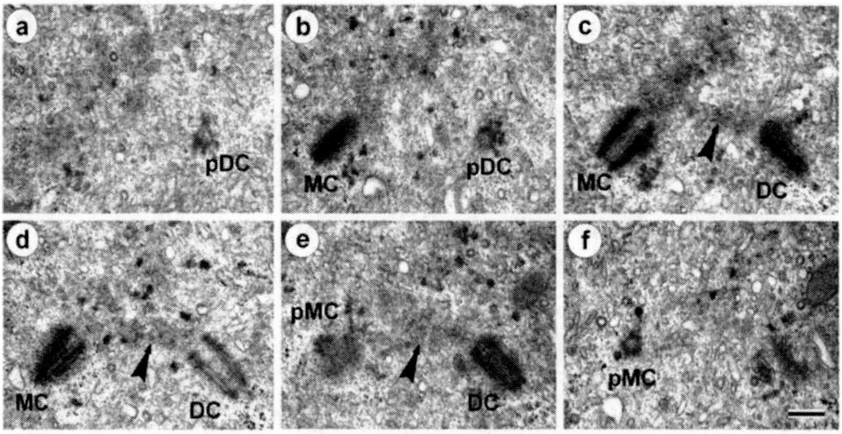

Figure 5.2. Centrosome ultrastructure during early S-phase. (a–f) Consecutive serial ultrathin sections through the centrosomal region. MC: mother centriole; DC: daughter centriole; pMC: procentriole associated with the mother centriole; pDC: procentriole associated with the daughter centriole. The proximal ends of the mother and daughter centrioles are connected by an electron-dense amorphous ligament (arrow). Scale bar: 200 nm.

No correlation was observed between centrosome-induced spindle formation efficacy and the cell cycle stage of the donor centrosomes (Tournier et al., 1989). However, centrosomal functionality exhibited variability: centrosomes from calf thymocytes, where mother and daughter centrioles adopt a linear configuration linked at their proximal ends (Komesli et al., 1989), failed to induce cleavage in Xenopus eggs (Tournier et al., 1991). Notably, microtubule (MT) nucleation kinetics in these centrosomes resembled those of KE37 human lymphoblastoma cells, which retain cleavage-inducing capacity. This functional deficit in calf thymocyte centrosomes appears unrelated to species or tissue specificity, as centrosomes from other bovine tissues and rodent/human thymocytes successfully triggered egg cleavage (Tournier et al., 1991). These findings suggest that the unique centriole linkage geometry in calf thymocytes impairs centrosomal activity.

During mitotic exit, centrioles lose their perpendicular orientation and separate, resulting in diplosome disengagement. While typically remaining closely apposed during interphase, centrioles occasionally exhibit asymmetric migration. Rattner and Phillips (1973) first documented centriole translocation to the cytokinetic furrow, later confirmed via GFP-tagged centriole tracking (Piel et al., 2000). The frequency of centriole separation varies markedly across cell lines (Komarova et al., 1996; Uzbekov, 2004) (Table 1).

Centriole separation is considered prerequisite for licensing new procentriole assembly (Tsou and Stearns, 2006). Premature separation disrupts mitotic fidelity, promoting multipolar spindle formation (Keryer et al., 1984; Alieva and Vorobjev, 1991).

5.2. Centrosome Splitting and Separation during the G2 Phase and Mitosis

It is essential to distinguish between centriole divergence in the G1 phase and centrosome separation in the G2 phase, where each centrosome comprises two centrioles.

In certain cell types with a high proportion of G1-phase cells exhibiting centriole divergence (Table 1), centrosomes are positioned at considerable distances from each other (Figure 5.1e). Remarkably, these centrosomes do not further separate but instead converge during spindle formation.

The mechanism of centrosome separation remains under active investigation, as it directly ensures bipolar spindle assembly and proper mitotic progression. Given the centrosome's role as the primary microtubule-organizing center (MTOC), MTs are likely integral to this process.

The canonical model posits centrosome separation as a result of antiparallel MT bundle sliding regulated by motor proteins. These motors translocate toward either MT plus- or minus-ends. Key players identified in centrosomes include:

1. Cytoplasmic dynein-dynactin-NuMA complexes (Gaglio et al., 1995–1997; Echeverri et al., 1996);
2. The minus-end-directed motor ncd/Kar3 (Matthies et al., 1996; Walczak et al., 1997);
3. Plus-end-directed kinesins of the bimC/Eg5 family (Blangy et al., 1995; Sawin et al., 1992; Wilson et al., 1997; Giet et al., 1999a).

These motors collectively ensure bipolar spindle formation, highlighting their critical role in mitotic fidelity.

During centrosome separation, motor proteins interact with and are regulated by auxiliary factors. The kinesin-like motor protein pEg5 localizes between centrosomes during G2 phase, coinciding with centrosome splitting (Figure 5.3.). Its activity depends on Aurora A kinase, which accumulates at centrosomes by S phase (Uzbekov et al., 2024b) and phosphorylates pEg5 both *in vivo* and *in vitro* (Giet et al., 1999). Aurora A depletion permits mitotic entry - evidenced by histone H3 phosphorylation, nuclear envelope breakdown, and chromosome condensation—but blocks centrosome separation, resulting in monopolar spindles surrounded by rosette-like chromosome arrangements (Cowley et al., 2009). These findings validate the proposed role of Aurora A in centrosome disjunction and spindle bipolarity (Giet et al., 1999a,b; Uzbekov et al., 2001; Uzbekov, 2004).

As centrosomes separate, they nucleate MT asters anchored via minus-ends, with plus-ends extending outward, often toward the opposing centrosome. Immunofluorescence data (Figure 5.3.) reveal XIEg5 motor enrichment at MT plus-ends. Since XIEg5 is a plus-end-directed kinesin (Sawin et al., 1992) capable of forming homodimers/tetramers with bidirectional motility (Cole et al., 1994), we propose that XIEg5 crosslinks antiparallel MTs from opposing centrosomes, generating outward forces to drive centrosome separation.

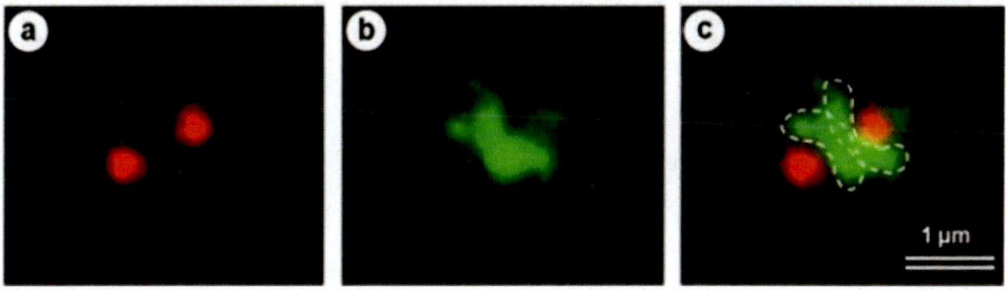

Figure 5.3. Double immunofluorescence of Aurora A kinase and Eg5 (KIF11) motor in late interphase XL2 cells. (a) Aurora A (pEg2) localization using monoclonal antibodies (red). (b) Eg5 (kinesin-like motor protein) detected by polyclonal antibodies (green). (c) Merged image showing colocalization of Aurora A and Eg5. Cells were analyzed during late S phase or early G2 phase. Scale bar: 1 µm. (Images: R. Uzbekov, adapted from Giet et al., 1999a).

The formation of the two spindle poles involves centrosome divergence, which occurs in two stages - splitting and separation (Uzbekov et al., 2002) - both of which depend, to varying degrees, on the integrity of MTs and actin filaments (Figures 5.4., 5.5.).

In interphase cells, MT depolymerization induced by nocodazole inhibits the splitting stage, whereas the separation of centrosomes that had already split prior to drug treatment accelerates (Uzbekov et al., 2002; Uzbekov, 2004). When MTs are completely disassembled by nocodazole and cells enter mitotic prophase, only 55% of cells exhibit centrosome divergence, compared to 94% in untreated controls.

Further centrosome movement during prophase often involves their curvilinear displacement around the nucleus. Since MTs themselves cannot adopt a curved trajectory unless their polymerizing plus-ends encounter a physical barrier or are influenced by motor proteins, it is likely that the centrosomes follow a curved path due to their physical association with the nucleus (Figure 5.6.). Additionally, the interaction of astral MT plus-ends with the cell cortex plays a significant role in spindle pole orientation and movement (Alieva and Vorobjev, 1991).

The observation that centrosomes co-fractionate with nuclei during cellular component separation by centrifugation supports the existence of a physical linkage between these structures (Dales et al., 1973; Nadezhdina et al., 1979; Maro and Bornens, 1980).

However, pretreatment of cells with cytochalasin - an actin filament-disrupting agent - prior to centrifugation destabilizes the nuclear-centrosomal complex (Maro and Bornens, 1980), enabling centrosome isolation as a distinct fraction (Mitchison and Kirchner, 1984a). Consequently, combined treatment with nocodazole and cytochalasin has become a standard procedure in modern centrosome isolation protocols (Mitchison and Kirchner, 1984a; Bornens et al., 1987; Chrétien et al., 1997).

These findings suggest that actin filaments tether centrosomes to the nuclear envelope, thereby channeling microtubule (MT)-generated repulsive forces along the nuclear membrane (Figure 5.6.).

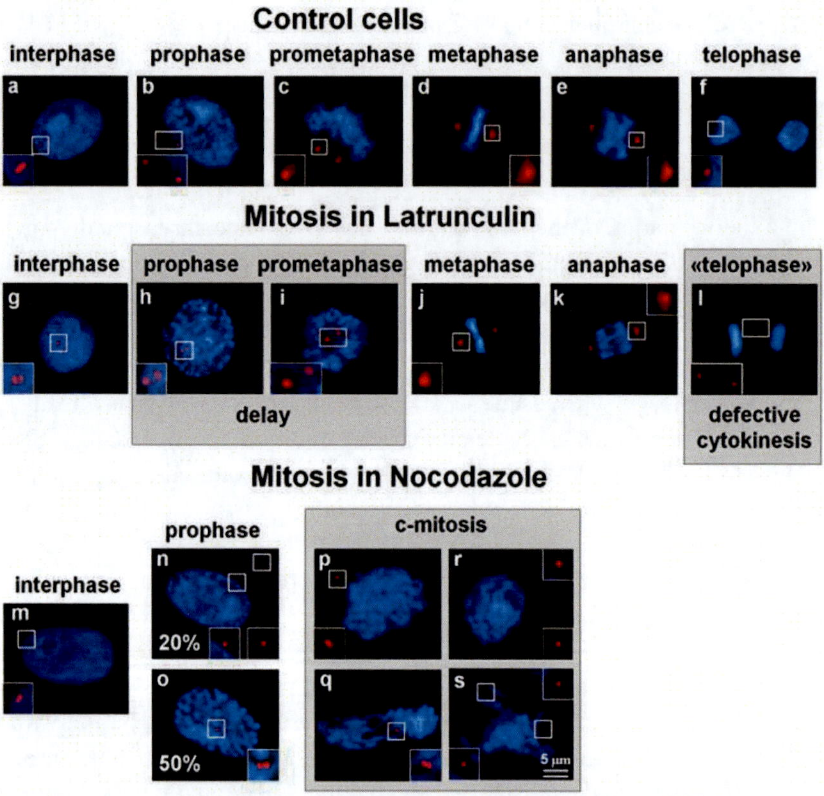

Figure 5.4. Centrosome positioning in XL2 cells under three conditions: normal mitosis (a–f), mitosis following actin filament depolymerization with latrunculin (g–l), and mitosis after microtubule (MT) depolymerization with nocodazole (m–s). Double immunofluorescence labeling for Aurora A kinase (pEg2; red) and nuclear DNA (DAPI; blue) (Uzbekov et al., 2002).

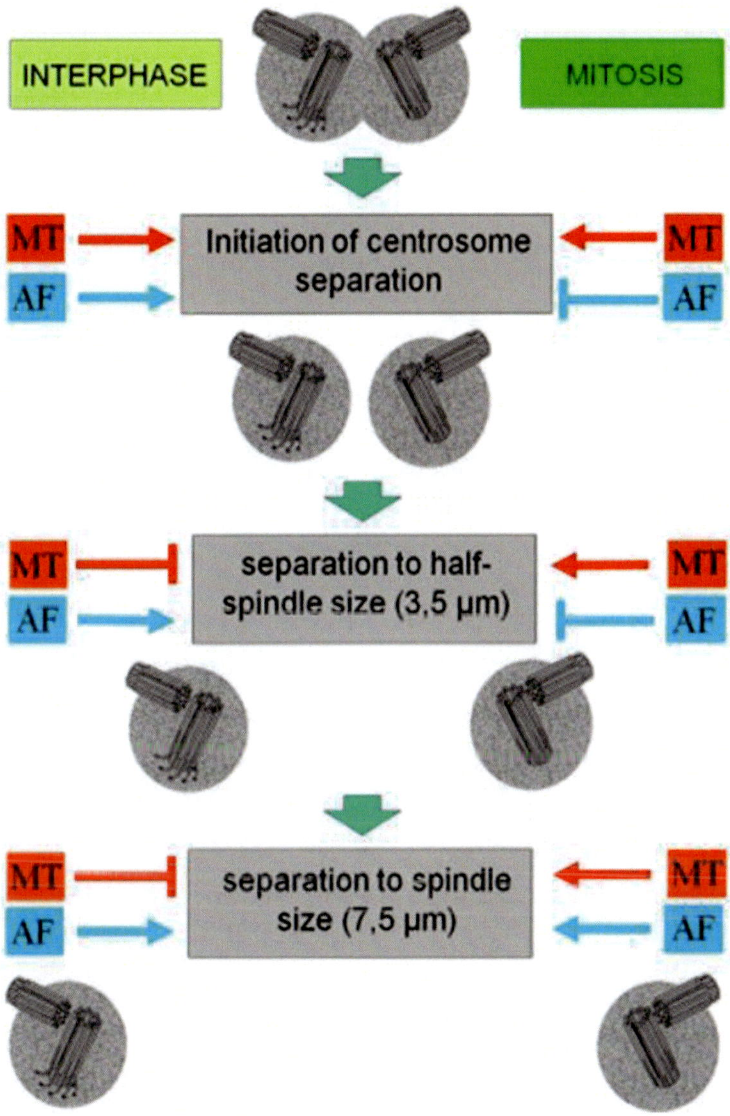

Figure 5.5. Roles of microtubules (MT) and actin filaments (AF) in centrosome disjunction during interphase and mitosis. Schematic representation of centrosome splitting (initial separation phase) and full separation. Left panel: interphase; right panel: mitosis. Adapted from Uzbekov et al., 2002.

This model is further supported by genetic evidence: mutation of the Drosophila twinstar gene, which encodes the actin-severing protein cofilin, delays centrosome movement and disrupts the interaction between nascent MT asters and the nuclear envelope (Gunsalus et al., 1995).

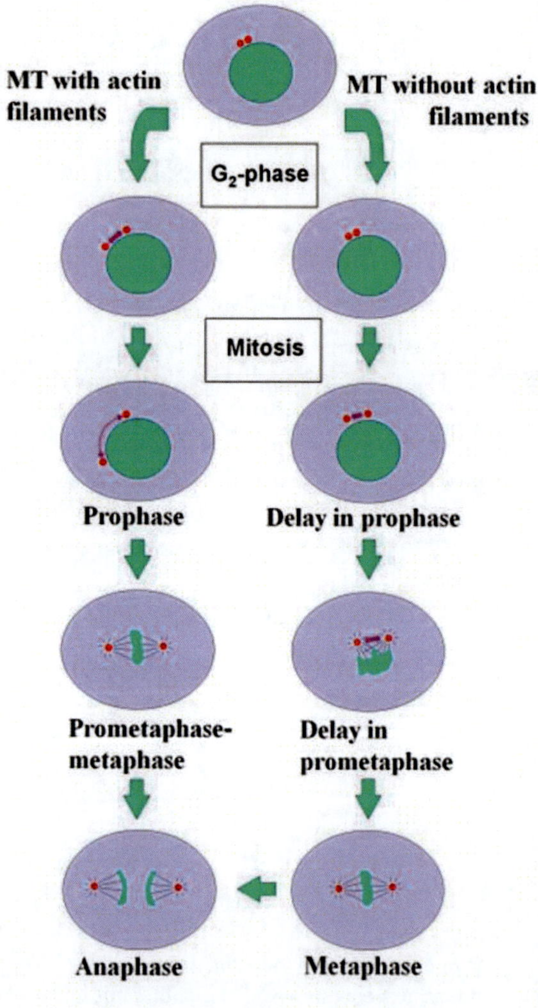

Figure 5.6. Centrosome dynamics and spindle formation under normal and actin-depleted conditions. Comparative analysis of centrosome movement and mitotic spindle assembly in G2 phase/normal mitosis (left) versus latrunculin-treated cells with depolymerized actin filaments (right). Chromosomes (green), centrosomes (red). Arrows indicate centrosome trajectories. Adapted from Uzbekov et al., 2002.

Direct experimental evidence comes from Euteneuer and Schliwa (1985), who demonstrated that in granulocytes, tetradecanoyl phorbol acetate (TPA)-induced centrosome splitting and separation requires intact actin filaments. Cytochalasin D-mediated actin depolymerization abolishes this effect, demonstrating that centrosome positioning depends not only on MTs but also on actin microfilaments. Consistent with this, actin disruption in yeast perturbs spindle pole body (SPB) motility and spindle orientation (Theesfeld et al., 1999). Multiple independent studies corroborate that centrosome separation involves actin filaments and associated motor proteins (Whitehead et al., 1996; Gavin, 1997; Forer and Pickett-Heaps, 1998; Mountain et al., 1999; Sharp et al., 1999; Silverman-Gavrila and Forer, 2000).

Our studies (Uzbekov et al., 2002) demonstrated that the percentage of cells with separated centrosomes significantly decreased following treatment with either nocodazole or latrunculin A. In most cases, cells lacking MTs or actin filaments failed to separate their duplicated centrosomes during interphase (Figure 5.4.).

Interestingly, MT depolymerization in cells where centrosomes had already initiated separation resulted in accelerated divergence. These findings indicate two distinct roles for interphase MTs: (1) they participate in the initial centrosome splitting, and (2) subsequently restrain the separation process.

The rise in p34cdc2 (CDK1) activity preceding G2/M transition triggers interphase MT depolymerization in untreated cells (Verde et al., 1990). We hypothesize that nocodazole-induced MT disassembly may partially mimic this physiological process, potentially serving as a regulatory signal for centrosome divergence.

In contrast to MTs, actin filaments appear to actively promote centrosome separation during interphase (Figure 5.4.), as evidenced by complete inhibition of centrosome divergence following latrunculin treatment (Uzbekov et al., 2002). As previously discussed, this process likely involves actin-associated motor proteins.

The mitotic phase presents a fundamentally different scenario. First, mitotic MT polymerization dynamics differ substantially from interphase patterns. Additionally, specific motor molecules become exclusively active during mitosis (Vernos and Karsenti, 1996). While MTs remain essential for proper centrosome separation throughout mitosis, actin filaments appear dispensable - though they may contribute during early mitosis prior to nuclear envelope breakdown (Figures 5.4., 5.5., 5.6.).

When actin filaments are absent, cells exhibit prolonged prophase and prometaphase durations (Uzbekov et al., 2002), suggesting impaired centrosome separation kinetics under these conditions. Remarkably, despite extended prophase duration, centrosome separation never exceeded 3.5 μm (approximately half the spindle length) before nuclear envelope breakdown (NEB) initiation. This observation raises a critical question: why does actin filament depolymerization delay cell cycle progression?

We propose that cells await a centrosome separation checkpoint signal before initiating NEB. In the absence of this signal - potentially caused by impaired actin-dependent centrosome separation - NEB is delayed to prevent formation of an aberrant spindle apparatus. The absence of complete mitotic arrest prior to cytokinesis in actin-depleted cells suggests the existence of redundant mechanisms for centrosome separation during mitosis. This hypothesis is supported by the observed centrosome separation occurring even during simultaneous depolymerization of both microtubules and actin microfilaments.

Our most intriguing finding (Uzbekov et al., 2002) revealed that cells compensate for prolonged prometaphase by significantly shortening metaphase duration. Consequently, the interval between NEB (prometaphase onset) and anaphase chromosome movement (metaphase completion) remains essentially unchanged.

These results demonstrate that latrunculin:

1. blocks centrosome separation during G2/M transition, and
2. partially inhibits mitotic centrosome separation through its action on actin microfilaments.

The centrosome may contain additional latrunculin-sensitive targets beyond actin microfilaments that could influence centrosome separation. Notably, the MT-dependent centrosomal protein centractin (Clark and Meyer, 1992), an actin homolog, represents one such potential target. Centractin serves as the core component of the dynactin complex - a multiprotein assembly comprising subunits of 160, 150, 62, 50, 45, 37, 32, 27, and 24 kDa (Gill et al., 1991; Paschal et al., 1993; Schafer et al., 1994). Electron microscopy reveals that centractin forms 37 nm minifilaments structurally resembling actin filaments (Schafer et al., 1994).

The dynactin complex serves as a critical regulatory hub for microtubule-based motor proteins through two primary mechanisms:

Interaction with minus-end directed motors: The complex directly binds cytoplasmic dynein, facilitating its motor activity and cargo attachment (Vallee, 1992; Vaisberg et al., 1993). This interaction is mediated through the p150Glued subunit and is essential for proper dynein function in intracellular transport and mitotic spindle formation (Vallee and Gee, 1998; Holleran et al., 1998).

Regulation of plus-end directed motors: The p150Glued subunit additionally interacts with kinesin-related proteins, as demonstrated by its binding to the bipolar kinesin Eg5 (Blangy et al., 1997). This interaction network is particularly significant given that p150Glued can simultaneously associate with both dynein (Karki and Holzbaur, 1995) and kinesins (Vaughan and Vallee, 1995), potentially coordinating their opposing activities.

While centractin's potential interaction with conventional actin filaments may contribute to centrosome separation, an alternative hypothesis suggests that latrunculin might directly target centrosomal centractin despite the current lack of direct experimental evidence. Previous investigations have established that classical actin-disrupting agents - cytochalasin D (which caps filament ends) and phalloidin (which stabilizes filaments) - fail to disrupt centractin filaments induced by overexpression (Holleran et al., 1996). However, latrunculin's distinct mechanism of actin sequestration through monomer binding suggests its effects on centractin might differ fundamentally from these other compounds.

Multiple lines of indirect evidence support the hypothesis of centractin involvement in centrosome separation:

1. Genetic studies in S. cerevisiae demonstrate that mutations in ACT3 (the yeast centractin homolog) lead to profound defects in spindle orientation and nuclear migration (Clark and Meyer, 1994)
2. Pharmacological inhibition with latrunculin produces remarkably similar phenotypes in yeast (Theesfeld et al., 1999), suggesting functional conservation
3. Structural modeling predicts that centractin impairment would likely destabilize dynactin complex integrity, thereby disrupting the coordinated activity of associated motor proteins that are essential for both centrosome splitting during interphase and proper spindle assembly in mitosis.

Chapter 6

Other Cell Structures Associated with the Centrosome

Rustem E. Uzbekov*, **PhD, HDR, IR**
Laboratory of Cell Biology and Electron Microscopy, Faculty of Medicine, University of Tours, Boulevard Tonnellé, France;
Faculty of Bioengineering and Bioinformatics, Lomonosov Moscow State University, Moscow, Russia

6.1. Cilia and Flagella – The Centriole as a Basal Body

Cilia and flagella, remarkable centriolar derivatives, represent essential components of the cellular motility apparatus. These structures were first identified by early microscopists in ciliated protozoa and spermatozoa - long before the discovery of centrioles themselves, whose small size made them difficult to resolve with light microscopy. Subsequent investigations revealed that each cilium or flagellum originates from a basal granule, now recognized as a centriole-derived basal body.

As discussed in Chapter 1, the seminal independent works of Henneguy (1898) and von Lenhossek (1898) first proposed the fundamental connection between mitotic spindle poles and ciliary basal bodies. Ultrastructural studies of sphagnum moss spermatozoa (Manton and Clarke, 1952; Manton, 1957) later revealed the characteristic axonemal structure with a "9×2+2" microtubule arrangement: nine outer doublets surrounding two central singlets.

Further research demonstrated that not all cilia are motile. Many cell types possess a single non-motile cilium associated with the mother centriole,

* Corresponding Author's Email: rustem.uzbekov@univ-tours.fr

In: The Centrosome
Editor: Rustem E. Uzbekov
ISBN: 979-8-89530-815-8

originally termed the "primary" or "abortive" cilium (Sotelo and Trujillo-Cenoz, 1958). The second half of this chapter will examine this important cilium type in detail.

Given their fundamental importance for both unicellular and multicellular organisms, cilia and flagella have been the subject of extensive research. These studies have been comprehensively summarized in numerous reviews and in the seminal monograph by Wheatley (1982). The current state of knowledge regarding their ultrastructure and biochemical composition has been thoroughly documented in the authoritative review by Fisch and Dupuis-Williams (2011).

Cilia and flagella perform three primary biological functions:

1. *Motility:* Many unicellular organisms rely on ciliary or flagellar beating for locomotion, while sperm motility is entirely dependent on flagellar movement.
2. *Fluid transport:* Coordinated beating of multiciliated epithelia generates directional fluid flow in various physiological contexts.
3. *Sensory transduction:* Primary cilia serve as specialized sensory organelles capable of detecting diverse extracellular signals.

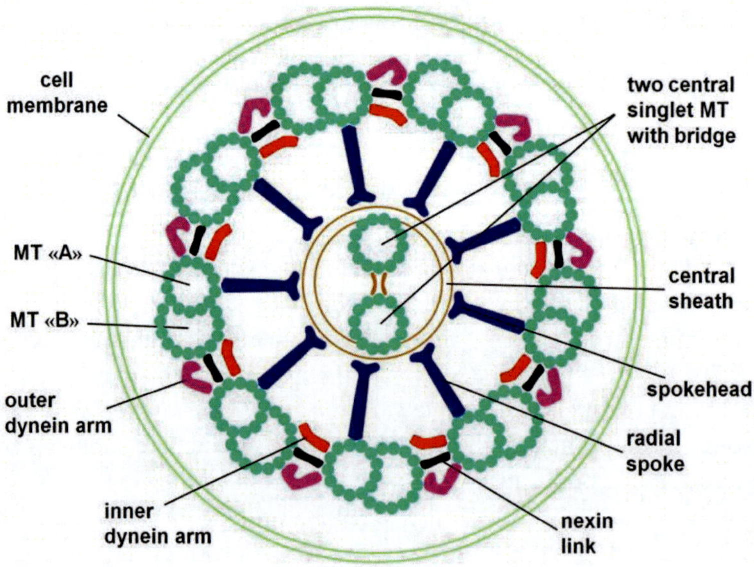

Figure 6.1. Schematic cross-section through the principal ("main") zone of a motile cilium, showing the characteristic 9x2+2 MT arrangement.

6.1.1. The Structural and Functional Diversity of Ciliary Organelles Specialized Derivatives of Primary Cilia

Primary cilia demonstrate extraordinary functional plasticity across different cell types, evolving into highly specialized sensory apparatuses. In the vertebrate retina, these organelles undergo profound structural modifications to become the light-capturing outer segments of photoreceptor cells, where they house the entire phototransduction machinery. Similarly, in the olfactory epithelium, primary cilia transform into elaborate chemosensory antennas, extending dendritic processes that contain odorant receptor proteins and downstream signaling components. Beyond their sensory roles, primary cilia function as exquisite mechanosensors in numerous tissues. The ciliated epithelia lining the respiratory tract, oviducts, and Eustachian tubes contain primary cilia that detect and respond to fluid flow through sophisticated calcium signaling mechanisms. Perhaps most remarkably, specialized ependymal cells in the brain ventricles and central canal of the spinal cord utilize their primary cilia as pressure sensors, contributing to cerebrospinal fluid homeostasis. These diverse functional adaptations of the primary cilium will be examined in greater depth in subsequent sections of this chapter.

The beating of motile cilia and flagella is provided by the displacement of MT doublets relative to each other, associated with the longitudinal movement of dynein handles along adjacent doublets in the presence of ATP (Figure 6.1.). It is believed that the outer dynein handles are responsible for the beat frequency, and the inner ones are responsible for force generation (Porter and Sale, 2000; Fisch and Dupuis-Williams, 2011).

Using cryo-electron tomography methods, new types of connections between axoneme components have been characterized (Nicastro et al., 2006; Heuser et al., 2009), which added new details to the understanding of how cilia beat. Cilia “beat” in one plane. The “rowing” occurs in the position of maximum extension, and the return to the initial position is by moving along the cell body, similar to the movement of the arms in breaststroke swimming.

6.1.2. The Molecular Machinery of Ciliary Motility

The elegant beating motion of motile cilia and flagella represents one of nature’s most sophisticated nanomachines. This movement arises from the precisely coordinated activity of dynein motor proteins, which generate sliding forces between adjacent MT doublets within the axoneme. ATP hydrolysis powers conformational changes in the dynein heavy chains, enabling them to "walk" along the B-tubule of neighboring MT doublets (Figure 6.1.). Contemporary studies demonstrate functional specialization among these dynein arms: outer dynein arms primarily govern metachronal rhythm and beat frequency, whereas inner dynein arms generate the principal power stroke (Porter and Sale, 2000; Fisch and Dupuis-Williams, 2011).

Advances in cryo-electron tomography (Nicastro et al., 2006; Heuser et al., 2009) have yielded unprecedented 3D reconstructions of axonemal ultrastructure, revealing previously uncharacterized linker proteins and radial spoke components that orchestrate the coordinated activity of thousands of dynein motors. These structural insights have clarified how stochastic molecular motions integrate into the planar beating patterns of intact cilia. The movement follows a biphasic mechanism, evoking competitive breaststroke swimming: a forceful effective stroke, during which the cilium fully extends to displace fluid, is succeeded by a recovery stroke, where the organelle bends to return to its initial position while minimizing backward fluid displacement.

6.1.3. Cilia in Human Health and Disease

The medical significance of cilia became apparent with the discovery that numerous genetic disorders, now collectively termed ciliopathies, result from defects in ciliary structure or function. These findings triggered an explosion of interest in cilia research, transforming these organelles from biological curiosities into major focus areas of biomedical investigation. Proteomic approaches (Arnaiz et al., 2009) have cataloged over 600 proteins comprising the ciliary proteome, with sophisticated biochemical studies elucidating the roles of individual components in ciliary assembly, maintenance, and signaling (King, 2010; Lindemann and Lesich, 2010). The intricate relationship between ciliary dysfunction and human disease pathogenesis will be explored in a special chapter.

6.1.4. Evolutionary Perspectives on Ciliary Terminology

The dual nomenclature of "cilia" and "flagella" represents a historical artifact from early electron microscopy, when researchers distinguished between:

- Singular or sparsely distributed organelles (termed flagella), typically exhibiting undulatory motion, as seen in sperm cells
- Numerous, closely packed organelles (called cilia) displaying coordinated beating patterns, characteristic of epithelial tissues.

In contemporary literature, these terms are used according to established conventions, despite their identical ultrastructural organization. For instance, we refer to the sperm's motile appendage as a flagellum, while describing similar structures in tracheal epithelium as cilia. This terminological distinction persists despite overwhelming structural evidence showing that both organelles share the canonical "9x2+2" microtubule arrangement (Figure 6.2.). The electron micrographs in Figure 6.2. demonstrate this remarkable conservation - panel a shows a human sperm flagellum, while panel b depicts a cilium from pig oviduct epithelium. Only subtle contextual clues (like surrounding structures) permit differentiation between these evolutionarily conserved organelles.

The preservation of this complex nanomachine across eukaryotic organisms - from unicellular protists to mammals - stands as one of the most striking examples of evolutionary constancy in cellular architecture. Yet beneath this structural conservation lies fascinating functional diversification, with variations in length, beating pattern, and regulatory mechanisms adapting these organelles to specific physiological contexts. The following sections will explore these specialized adaptations in greater detail, revealing how a fundamentally conserved structure has been exapted for diverse biological functions throughout evolution.

6.1.5. Structural Organization of Cilia and Flagella

The membrane of cilia is typically closely apposed to the axoneme, while in flagella, additional structures may lie between the membrane and axoneme, forming the fibrous layer of the flagellar main zone. In the middle piece of the sperm flagellum, between the axoneme and cell membrane, there are nine outer dense fibers and mitochondria (Figure 6.2a).

Early observations that motile cilia had two central MTs while non-motile primary cilia lacked them suggested motility required central singlets. However, exceptions exist in nature: non-motile cilia with central MTs (Flock and Duvall, 1965) and motile cilia without central MTs (Nonaka et al., 1998). Current understanding establishes that motility depends not on central MTs but on the presence of outer and inner dynein arms associated with peripheral MT doublets (Afzelius, 2004), as visible in Figure 6.1. and 6.2.

All cilia and flagella form from modified centrioles (called basal bodies), but the converse isn't true - not all centrioles form cilia. Certain cells like lymphocytes, hepatocytes, adipocytes and mononuclear leukocytes contain centrioles but never form cilia (Wheatley, 1982).

The wall of a cilium and flagellum, unlike the wall of a centriolar cylinder, consists not of 9 triplets but of 9 doublets of MTs. The MTs of each doublet represent continuations of the "A" and "B" MTs of centrioles. There are two principal types of axonemes distinguished by their MT composition and spatial organization, typically described using a formula that specifies the number and arrangement of MTs. Most motile cilia and flagella contain 9 peripheral MT doublets plus 2 central single MTs, designated by the formula ($9 \times 2 + 2$). This notation emphasizes that the central MTs, unlike the peripheral ones, exist as singlets rather than doublets (Figure 6.2.). Primary cilia, following the ($9 \times 2 + 0$) formula, lack central MTs. The abbreviated notation ($9 + 0$) or ($9 + 2$) commonly found in literature fails to accurately represent this structure, as the first number refers to doublets while the second indicates single MTs. Beyond these classical patterns, certain parasitic protozoa (gregarines) exhibit variant formulas including ($9 \times 2 + 5$), ($9 \times 2 + 0$), and ($6 \times 2 + 0$) in their gametes.

As in centrioles, MT "A" in cilia and flagella has a rounded shape in cross-section and 13 protofilaments. It is adjacent to MT "B," which contains 11 protofilaments of its own and 4 shared with MT "A" (Figure 6.1.). Since the total number of protofilaments in MT "B" is 15, its diameter slightly exceeds that of MT "A." The central MTs possess 13 protofilaments, like most cellular MTs, and a diameter of about 25 nm; both are surrounded by a common central sheath. Each MT "A" is connected to the neighboring MT "B" by a nexin bundle, as well as by the outer and inner dynein arms (Figure 6.1.). In addition, each MT "A" is linked to the central sheath (also referred to as the central capsule) via a radial spoke. The spokes consist of more than 20 highly conserved proteins (Piperno et al., 1981; Yang et al., 2006).

Unlike the MT triplets of the centriolar cylinder, the vector passing through the centers of two MT doublets is nearly perpendicular to the cilium's radius, meaning the doublets are arranged circularly without tilt. Although the diameter difference between MTs "A" and "B" may not be clearly visible in electron micrographs, MT "A" can always be identified by its protruding dynein arms (Figure 6.2.). This feature allows unambiguous determination of the cilium's orientation in cross-section. When dynein arms point clockwise (as in Figures 6.1. and 6.2.), the view shows the cilium section from the basal body side (cell side). Conversely, counterclockwise orientation indicates a view from the cilium's distal end. Thus, the dynein arm direction is opposite to the doublet twist direction (from MT "A" to MT "B") in the axonemal MTs.

It should be noted that Figures 6.1. and 6.2. display cross-sections through the cilium's central portion. However, analysis of longitudinal sections (Figures 6.3. and 6.4.) and cross-sections at different levels (Figure 6.5.) reveals that the cilium's structure varies along its length. The basal body-cilium complex can be divided into the following zones:

Zone I represents the *area of cilium attachment* to the cellular cytoplasm. Cilia and flagella connect to the cytoplasm through the basal body (centriole) via a system of rootlets and a specialized MT network. This system reaches remarkable complexity in certain flagellates and exhibits species-specific characteristics. In multicellular animals, the basal bodies of cilia possess electron-dense rootlets and are occasionally connected at their proximal end to striated rootlets (Figure 6.4.).

Zone II, the basal body zone, constitutes a modified centriolar cylinder similar to that found in the centrosome of dividing cells. However, unlike centrosomes, the basal body centriole in ciliated epithelial cells lacks a paired centriole. As centriole maturation requires more than one cell cycle, all centrioles forming cilia in ciliated epithelium can be considered "daughter centrioles." The formation process of such centrioles will be detailed in subsequent chapters.

We should note that in proliferating cells, cilia (primary cilia) formation occurs almost exclusively on mother centrioles. Another morphological distinction lies in the presence of rootlets, although striated rootlets occasionally appear in the centrosomes of dividing cells. The basal body foot represents a structure homologous to subdistal appendages (pericentriolar satellites), but unlike them, typically appears singly (or occasionally as a pair) on basal bodies.

Distal appendages (Sorokin, 1968) and alar sheets (wing-shaped sheets; Anderson, 1972) connect the basal body's centriolar cylinder to the cell membrane. These structures localize at the boundary between zones II and III, with one appendage per MT triplet (Figure 6.5.). The appendage heads (or knobs) mediate direct membrane attachment (O'Toole et al., 2007).

Several important centrosomal proteins localize to these appendages. Notably, CEP164 knockdown using specific siRNA prevents ciliogenesis (Graser et al., 2007). While ODF2 was initially implicated in the formation of both distal and subdistal appendages (Ishikawa et al., 2005), subsequent work by Thanos et al., (2013) challenged its role in distal appendage formation. The aberrant appendage morphology observed in Odf2-deficient cells (Ishikawa et al., 2005) may reflect broader defects in centrosome maturation rather than specifying ODF2's direct involvement in appendage composition.

Zone III - the transition zone - contains cilia with a (9×2+0) MT arrangement, featuring peripheral doublets but lacking central MTs. An electron-dense septum called the terminal plate (Gilula and Satir, 1972) separates the centriolar cylinder lumen from the ciliary lumen (Figures 6.3b, 6.4., 6.5m). Some specimens display an additional transverse septum apical to the terminal plate, termed the intermediate plate. Notably, the ciliary axoneme diameter in this zone (~165 nm) is smaller than both the centriole diameter (~200 nm), consistent with MT triplet-to-doublet conversion, and the subsequent main ciliary zone diameter (~200 nm). However, the ciliary membrane maintains a constant diameter of ~250 nm throughout (compare Figures 6.5o and 6.5m). This axonemal narrowing is clearly visible in Figure 6.4.

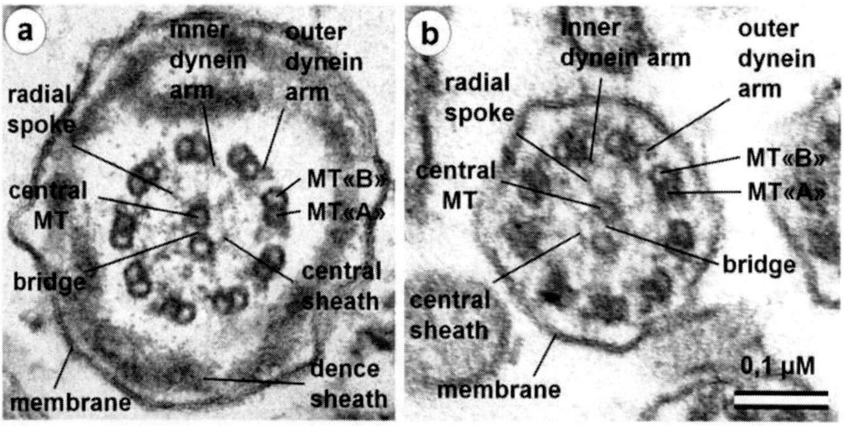

Figure 6.2. Comparative ultrastructure of (a) human sperm flagellum and (b) oviduct epithelial cilium from a suckling pig (Sus scrofa domesticus). Both cross-sections show: Plasma membrane surrounding the axoneme, nine peripheral microtubule doublets (microtubule A and microtubule B subfibers); Central pair of microtubules enclosed within a central sheath; Radial spokes connecting central and peripheral structures; Outer and inner dynein arms extending from microtubule A. View orientation: sections observed from basal body toward distal tip of cilium/flagellum (distal-to-proximal view would show mirror image). Scale bar: 200 nm. (From: Bragina et al., 2018; micrographs by R. Uzbekov).

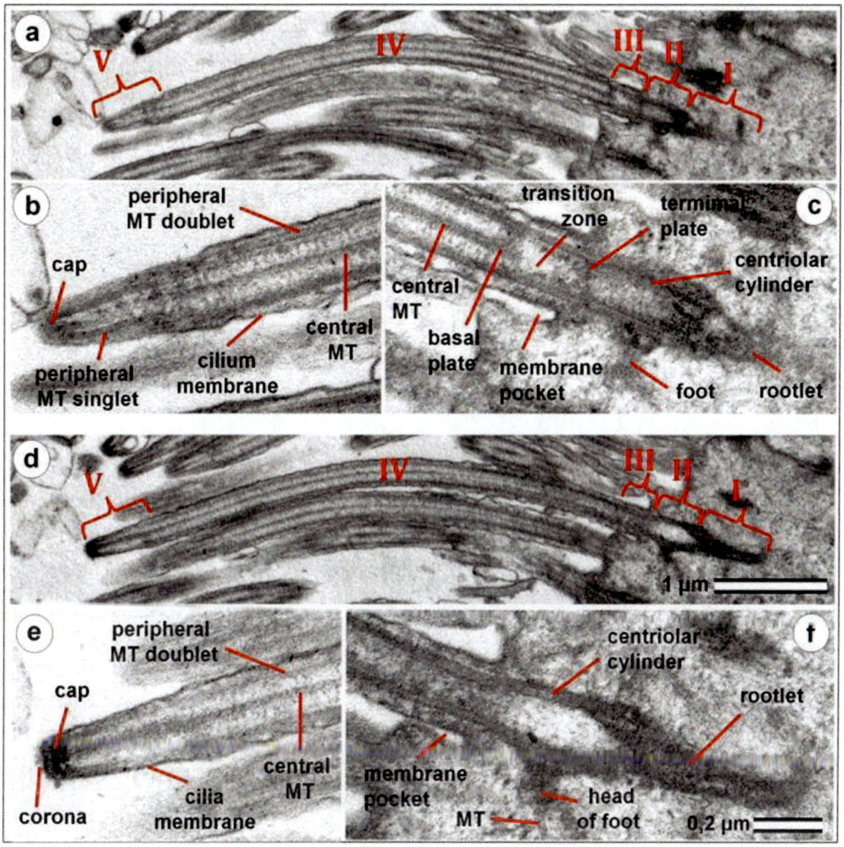

Figure 6.3. Serial longitudinal sections through a suckling pig oviduct epithelial cilium (a,g), with higher magnification views of basal (b,c) and apical (d,e) regions. Five structural domains are identified: I. Cytoplasmic anchoring zone (rootlets; c,f); II. Basal body (centriolar cylinder); III. Transition zone (lacking central MTs; c); IV. Principal ("main") zone; V. Apical zone (singlet MTs and ciliary cap). Key features: Membrane invagination forms a pocket (f) at transition zone; Single foot process (f; subdistal appendage homolog) associates with cytoplasmic MTs; Terminal plate demarcates basal body lumen; Central MTs originate at basal plate; Apical singlets and central MTs terminate at ciliary cap (e); Ciliary corona membrane projections (e); Scale bars: 1 µm (a,g); 200 nm (b-e). (From: Uzbekov and Alieva, 2013; modified).

During stress-induced ciliary loss, breakage occurs at the basal plate level, leaving the transition zone attached to the basal body (Quarmby, 2004). This process involves katanin, an MT-severing protein localized to the transition zone's distal portion (Lohret et al., 1999). In Chlamydomonas, flagellar resorption during mitosis involves breakage at the appendage level (Jonson and Porter, 1968).

Centriolar release from basal body function requires katanin activity, enabling centriolar migration for spindle formation (Rasi et al., 2009). The transition zone remnants become membrane-encapsulated and ejected, indicating its exceptional stability compared to other axonemal components (Parker et al., 2010).

The transition zone also contains distinctive MT doublet-membrane connections with various reported morphologies: bridges (Reese, 1965), champagne-glass structures (Gilula and Satir, 1972), radial linkers (Anderson, 1974), doublet outer projections (Cavalier-Smith, 1974), pegs (Dute and Kung, 1978), lateral projections (Breipohl et al., 1980), T-shaped fibres (Geimer and Melkonian, 2004), and Y-shaped connectors (O'Toole et al., 2007). We adopt the latter term (Figures 6.5m,o). These structures are particularly prominent in photoreceptor cilia, Leishmania flagella, and renal primary cilia (Horst et al., 1987; Gluenz et al., 2010).

Three specific proteins have been localized to the transition zone. A 210 kDa protein was found in the basal part of the transition zone, suggesting its involvement in zone formation (Lechtreck et al., 1999). Nephrocystin (NPHP1), an adaptor protein associated with signaling molecules involved in cell adhesion and actin cytoskeleton organization (Hildebrandt and Zhou, 2007), remains associated with the cilium after treatment with membrane-disrupting detergents (Fliegauf et al., 2006). Since NPHP1 loss leads to accumulation of ciliary transport cargo in the basal body region, this protein appears to regulate ciliary transport (Jauregui et al., 2008; Jiang et al., 2009). The centrosomal protein CEP290 localizes to Y-shaped connections between the axoneme and ciliary membrane (Craige et al., 2010). Its depletion eliminates not only Y-shaped connections but also wedge-shaped structures (termed by Ringo, 1967) - connections between the basal lamina, peripheral MT doublets and ciliary membrane (Craige et al., 2010). Both NPHP1 and CEP290 mutations are associated with human ciliopathies (Wolf and Hildebrandt, 2011).

At sites where Y-shaped connections contact the ciliary membrane in the transition zone (prior to its connection with the cell membrane), rows of globules called necklaces are observed. These were first described in freeze-fracture studies of ciliary structure (Andres, 1969; Gilula and Satir, 1972). Necklaces occur in both motile and non-motile cilia. While the number of globule rows varies widely (1-40), most studied specimens show 3-8 rows (reviewed in Fisch and Dupuis-Williams, 2011). This variability exists not only between species but also among different cell types within the same species and even within the same tissue, suggesting necklace dynamics reflect cellular functional states.

The necklace globules primarily function as a selective barrier, preventing large transmembrane protein complexes and lipids from migrating between the cell membrane and ciliary membrane (Boisvieux-Ulrich et al., 1977). Thus, necklaces maintain the ciliary membrane's unique composition. Interestingly, necklace formation precedes both ciliogenesis and transition zone development (Cordier and Haumont, 1979; Breipohl et al., 1980; Menco, 1980; Carson et al., 1981).

High-affinity calcium binding sites in the necklace region prompted the hypothesis that these globules might function as mechanical pressure-dependent calcium channels activated by ciliary beating (Gilula and Satir, 1972; Cordier and Haumont, 1979). However, since necklaces also occur in primary, non-motile cilia, this hypothesis was not further developed.

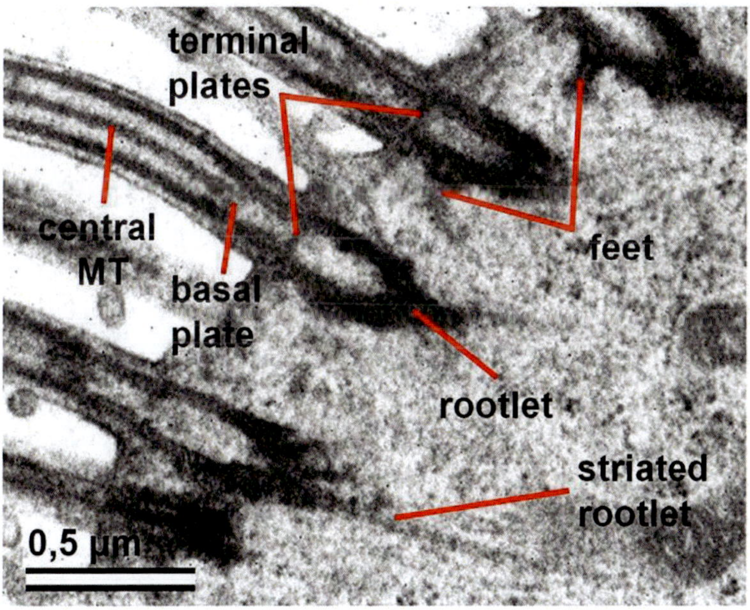

Figure 6.4. Apical cytoplasm of suckling pig (Sus scrofa domesticus) oviduct epithelium showing: basal bodies with uniformly oriented feet (subdistal appendage homologs); electron-dense rootlets (one exhibiting striated pattern); distinct terminal and basal plates. Scale bar: 500 nm. (From: Uzbekov and Alieva, 2013; modified).

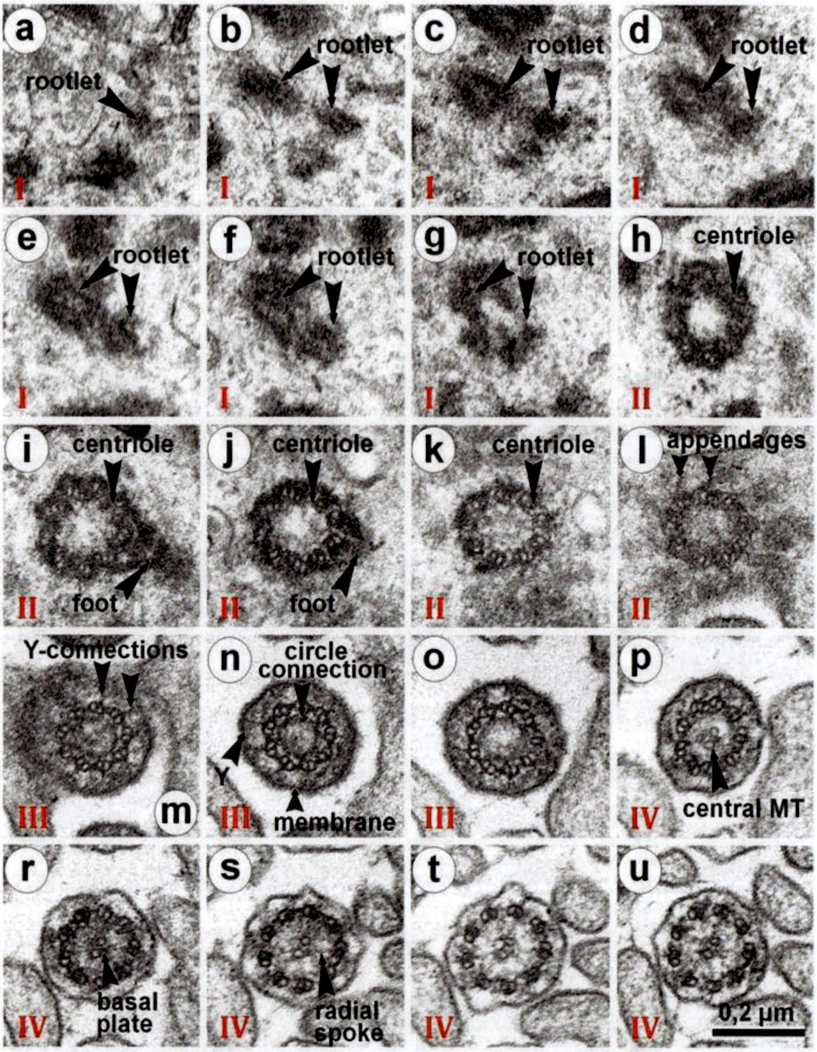

Figure 6.5. Serial cross-sections (20 consecutive) through a suckling pig oviduct cilium, showing progressive transitions between zones: I. Cytoplasmic anchoring (rootlets); II. Basal body; III. Transition zone; IV. Principal zone. Features: single foot per basal body, transitional appendages, axoneme-membrane Y-linkages (m), internal circumferential connections (n), central MT insertion in basal plate. Scale bar: 200 nm. (From: Uzbekov and Alieva, 2013; modified).

The GTP-binding cytoskeletal protein SEPT2, a septin family member (reviewed in Caudron and Barral, 2009) that interacts with actin, MTs and membranes, localizes to the transition zone. SEPT2 depletion causes ciliogenesis defects and disrupts the ciliary membrane's protein composition (Hu et al., 2010).

Zone IV - the "main zone" - represents the longest and most extensively studied region, where central MTs appear and the cilium adopts the (9×2+2) configuration. At the site of central MT emergence, a small amount of electron-dense material called the "axial granule," "axosome," or "basal plate" is observed (Gilula and Satir, 1972; *Figure 6.5p,c*). Unlike the terminal plate spanning the entire ciliary lumen, the basal plate occupies only its central portion.Structural connections between the basal plate and the inner axonemal compartment have been documented. In algae and lower plants, the basal plate exhibits greater complexity, comprising two layers of nine-rayed stars whose rays attach to MT "A" of doublets (Ringo, 1967). Longitudinal sections reveal an H-shaped morphology, with the crossbar representing the basal plate and the vertical (slightly inclined) lines corresponding to connecting filaments.

Immunostaining with γ-tubulin antibodies suggests the basal plate participates in central MT nucleation and stabilization (Silflow et al., 1999). Notably, Chlamydomonas vfl2 centrin mutants lacking the basal plate show central MTs protruding into the basal body (Jarvik and Suhan, 1991).

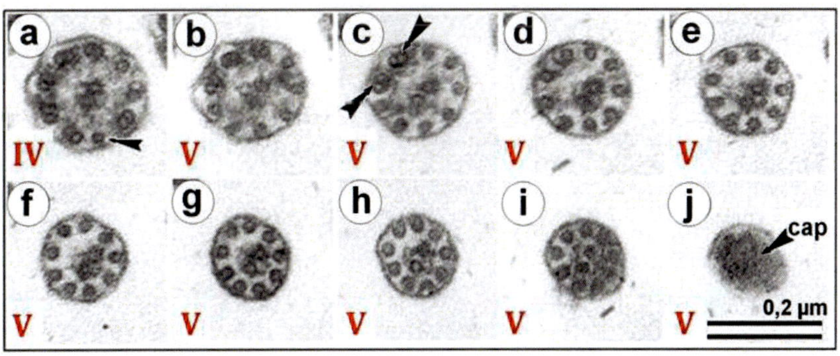

Figure 6.6. Serial cross-sections (10 consecutive) through ciliary tip, showing MT doublet-to-singlet transition: first MT singlet appears (panel a, small arrow); last MT doublets persist (panel c, large arrows). Zones: IV (principal); V (apical/singlet). Scale bar: 200 nm. (From: Uzbekov and Alieva, 2013; modified).

Zone V - the apical zone (MT singlet zone) - features disappearance of MT "B" from doublets, leaving only MT "A" (doublet-to-singlet conversion). While some reports indicate synchronous conversion across all doublets (Brightman and Palay, 1963; Reese, 1965), others demonstrate variable MT "B" lengths among doublets (Gibbons and Grimstone, 1960; Roth and Shigenaka, 1964). These discrepancies may reflect either interspecies variation or different ciliary functional states. In our model, doublets convert to singlets sequentially from the lower right (first emerging singlet indicated by arrow in Figure 6.6a) toward the upper left (last persisting doublets shown by arrows in Figure 6.6c). This progression vector lies perpendicular to the central MT axis, potentially explaining MT "B" length heterogeneity through ciliary shape modulation.

The most elaborate MT singlet zones occur in sensory cilia (Snow et al., 2004; Insinna and Besharse, 2008). Moreover, ciliary membranes at singlet zone levels show elevated signaling molecule concentrations (Menco et al., 1992; Matsuzaki et al., 1999).

These observations have led to hypotheses about this ciliary zone's specific role in cellular sensing (Fisch and Dupuis-Williams, 2011). Additionally, because MT "A" ends anchor to the ciliary membrane, the apical zone is believed to play a crucial role in motile cilia function. Ciliary MT assembly and disassembly occur at the distal tip structure called the ciliary cap or central cap (Figures 6.3d, 6.6i,j), which also regulates MT-dependent intraflagellar transport (reviewed in Pedersen and Rosenbaum, 2008).

Various organisms show distinct cap ultrastructural features: a granule, bilayer plate, wedge between central MTs, or filaments connecting the cap to MT singlet ends (Fisch and Dupuis-Williams, 2011). In developing ciliated epithelium, caps appear when cilia reach ~1.75 μm in length (Portman et al., 1987). Sperm flagella lack a unified central cap structure; instead, each singlet MT terminates at an individual membrane-associated granule (Woolley and Nickels, 1985).

The ciliary corona (Dirksen and Satir, 1972), alternatively termed ciliary claws (Morita et al., 1991), comprises cap-derived projections penetrating the distal ciliary membrane (Figure 6.3d). These detergent-resistant structures maintain attachment to the cap rather than the membrane itself (Kuhn and Engelman, 1978).

6.2. Primary Cilium

6.2.1. Structure of Primary Cilium

Many differentiated tissue cells contain single non-motile cilia, which grow exclusively from the more mature mother centriole (Sorokin, 1962). Sorokin (1968) proposed the term "primary cilium" for these rudimentary structures, though they are occasionally referred to as "abortive cilia" (Such, 1969; Murata et al., 1984).

As previously noted, primary cilia are absent in various leukocytes, hepatocytes, and adipocytes (Wheatley, 1969, 1982).

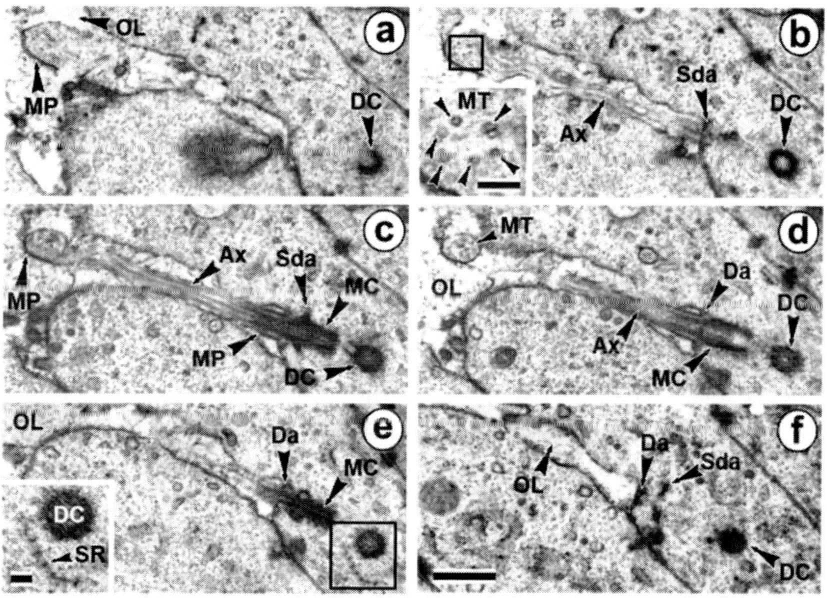

Figure 6.7. Six consecutive serial ultrathin longitudinal sections through a primary cilium in a one-day-old piglet oviduct cell. The curved cilium tip (b) demonstrates disorganized MT arrangement in the apical region. A striated rootlet appears near the daughter centriole (e, inset). Ax - primary cilium axoneme; Da - distal appendages; DC - daughter centriole; Sda - subdistal appendages; MC - mother centriole; MP - ciliary membrane pocket; MT - microtubules; OL - oviduct lumen; SR - striated rootlet. Scale bars: 0.5 μm (a-f); 0.1 μm (b and e insets).

Numerous *in vitro* cultured cell lines also exhibit primary cilia. This is particularly evident in various fibroblast lines, where literature reports indicate over 80% of cells may possess cilia (Wheatley, 1971, 1982). In epithelial cell cultures, the proportion is typically lower - approximately 30% in SPEV pig embryonic kidney cells (Alieva, Vorobjev, 1989a), primary endothelial cell cultures (Bystrevskaya et al., 1992), and PtK1 cells (Rieder et al., 1979).

While we previously described motile cilia morphology in pig oviduct epithelial cells, these same cells can form primary cilia during earlier postembryonic development. Ultrastructural comparison of both cilia types (Figures 6.3. to 6.6. versus Figures 6.7., 6.8.) reveals key differences: unlike motile cilia (which may number in the dozens or thousands per cell, as in ciliates), each tissue cell contains just one primary cilium. In primary cilia, the basal body (mother centriole) typically appears alongside its daughter centriole, whereas motile cilia basal bodies lack paired centrioles.

The primary cilium exhibits a conserved structure (Figure 6.7.). Its axoneme, extending from the centriole, consists of nine MT doublets - membrane-ensheathed continuations of MTs “A” and “B” from the centriolar cylinder. Most vertebrate cells *in vivo* develop primary cilia during various developmental stages, particularly in embryonic tissues and early postnatal development (Fonte et al., 1971; Jurand, 1974; Sorokin, 1962; Tucker et al., 1992). While most differentiated cells possess primary cilia (Latta et al., 1961; Dalen, 1981; Poole et al., 1985; Albrecht-Buehler, 1992; Bystrevskaya et al., 1992; Wheatley, 1994), some cell types rarely form them.

The proximal (cytoplasm-facing) end of centrioles associated with primary cilia (Figure 6.8d,e) typically lacks discernible associated structures. This contrasts markedly with basal bodies of motile cilia, which invariably connect to striated or non-striated rootlets at their proximal ends (Figures 6.3., 6.4.). In primary cilium-bearing cells, striated rootlets represent non-essential components. When present, they may either extend from daughter centrioles (Figure 6.8d,e) or reside within the pericentriolar cytoplasm (Figure 6.7.).

Spatial relationships between mother and daughter centrioles demonstrate significant structural flexibility, exhibiting configurations from parallel alignment to near-orthogonal arrangements. However, their proximal ends consistently remain in closer proximity than distal regions. The antiparallel organization shown in Figure 6.8. exemplifies this spatial relationship, displaying centrioles with opposing proximal orientations while maintaining parallel longitudinal axes. The distinctive triplet twist direction (Figure 6.8g) serves as a definitive ultrastructural indicator of centriole polarity, demonstrating mirror-image twisting patterns between paired centrioles.

Zone II of motile cilium basal bodies and mother centrioles associated with primary cilia show substantial structural conservation.

Both contain an electron-dense septal formation that clearly delineates the boundary between the centriolar lumen and developing ciliary compartment (Figures 6.3c, 6.4, 6.5m,n, 6.8k,m). Additionally, comparative analysis reveals that the transition zone of motile cilia (Figure 6.5n-p) and the proximal region of primary cilia share similar specialized structural features, most notably their characteristic Y-shaped connectors between the membrane and axoneme (Figure 6.8d,e).

There are several key ultrastructural differences between primary and motile cilia:

1. While centrioles of motile cilia typically possess one or two subdistal appendages (feet) (Figure 6.5i,j), those of primary cilia display multiple subdistal appendages (Figure 6.8j,l).
2. The mother centriole lumen contains a distinct lighter "channel" at its electron-dense distal end, a feature not observed in motile cilium basal bodies.
3. Although distal appendages show similar dimensions and organization in both cilium types, their heads appear more pronounced in primary cilia (Figure 6.8m)
4. The region where motile cilia form their central MT pair (Zone IV, Figure 6.5p) corresponds to where primary cilia exhibit disorganized doublet arrangement (Figure 6.8p), followed by gradual doublet loss (Figure 6.8c-x).
5. Unlike motile cilia which terminate in a specialized cap structure (Zone V, Figures 6.3d, 6.6i,j), primary cilia lack this terminal differentiation.

The primary cilium exhibits a conserved structural organization across different cell types. The well-characterized formation process (Sorokin, 1962) occurs through three distinct phases:

1. *Initiation phase:* A membrane vesicle (occasionally containing electron-dense inclusions) forms near the mother centriole's distal end. The emerging cilium bud penetrates this vesicle, creating a transient ciliary sheath (Figure 6.9.).
2. *Elongation phase:* Both the cilium shaft and its sheath extend, sometimes through fusion with additional membrane vesicles.

3. *Maturation phase:* The growing cilium contacts and fuses with the plasma membrane, achieving surface exposure (Figure 6.9.).

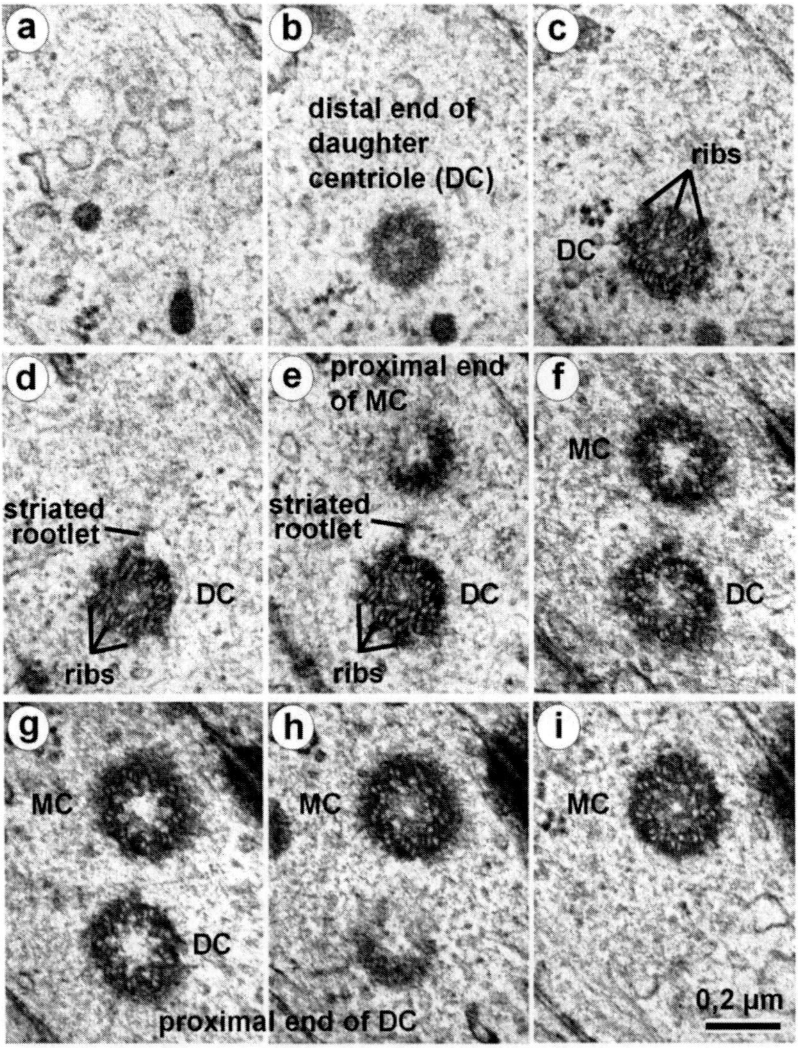

Figure 6.8. (Part 1). Twenty-one serial cross-sections through daughter (DC) and mother (MC) centrioles and primary cilium in piglet oviduct epithelium, progressing from cell interior toward cilium tip. The antiparallel centrioles show opposite triplet orientations (g). Rib structures (c) appear on the daughter centriole surface, while a thin striated rootlet connects their proximal ends (d,e). Image modified from Uzbekov and Alieva (2013).

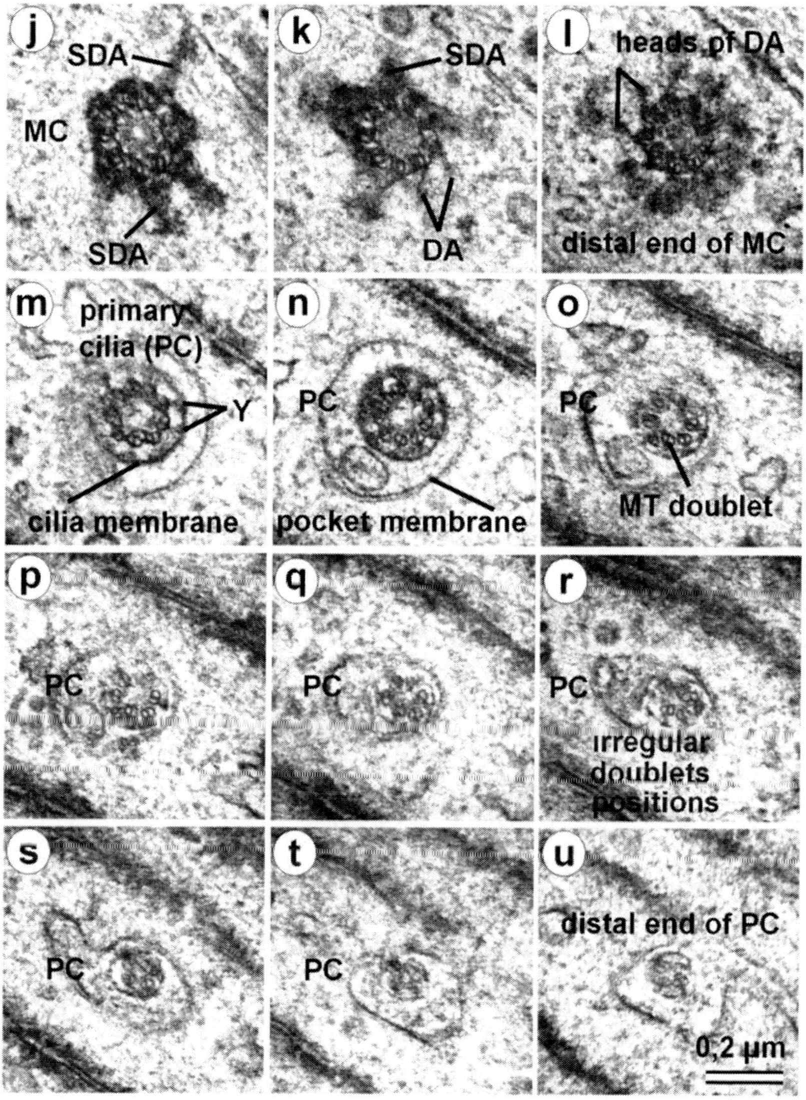

Figure 6.8. (Part 2). Seven subdistal appendages (SDA) at two levels attach to the mother centriole (MC). The transition zone (l) contains nine distal appendages (DA) with distinct heads. The primary cilium (PC) maintains regular structure for ~200 nm (m-o), followed by doublet misalignment, lumen narrowing, and progressive doublet loss (only four remain in final section). Y-shaped connectors link axoneme to membrane (m). Scale bar: 0.2 µm. Image modified from Uzbekov and Alieva (2013).

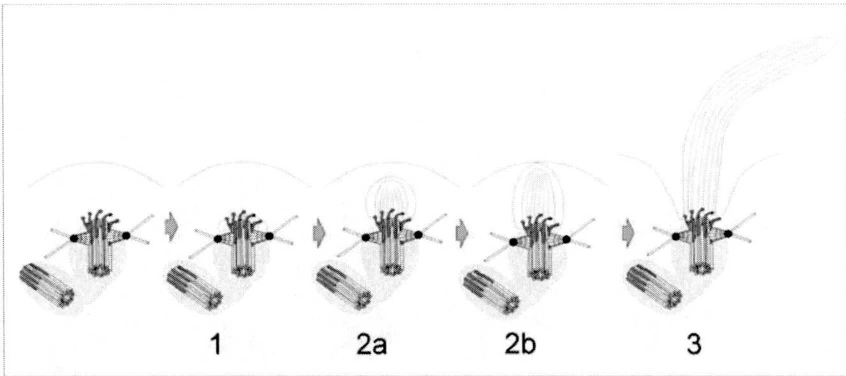

Figure 6.9. Primary cilium formation stages: (1) Vesicle formation at mother centriole distal end initiates cilium bud growth; (2a) Axoneme and sheath elongation; (2b) Sheath contacts plasma membrane; (3) Vesicle-plasma membrane fusion enables ciliary surface emergence.

Unlike motile cilia and flagella, primary cilia are non-motile. Some cell types terminate development after the second phase, maintaining entirely intracellular cilia. These structures assemble during early G1 phase (Archer and Wheatley, 1971) and disassemble by mitosis onset. In PtK1 cells, this disassembly progresses through prophase and prometaphase (Rieder et al., 1979).

Notably, Aurora A kinase - which localizes to centrosomes during S-phase (Uzbekov et al., 2024b) and mediates centrosome separation (Giet et al., 1999b) - also regulates primary cilium resorption (Pugacheva et al., 2007). Primary cilium MTs are enriched in acetylated α-tubulin, enabling specific detection with monoclonal antibodies (Cambray-Deakin and Burgoyne, 1987).

6.2.2. Functions of the Primary Cilium

Since the discovery of primary cilia in 1898 (Zimmerman, 1898), three principal hypotheses have been proposed regarding their function. The first suggests they serve as sensory organelles (Munger, 1958; Barnes, 1961; Poole et al., 1985; Wheatley et al., 1996). The second posits they are vestigial structures inherited from motile-cilium-bearing ancestors (Sorokin, 1962). The third proposes their involvement in cell cycle regulation (Tucker et al.,

1979a, 1979b). When these hypotheses were first formulated, experimental verification remained technically challenging.

Current evidence overwhelmingly supports the sensory function hypothesis. The essential role of primary cilia is now well-established for vertebrate visual, auditory and nervous systems (Pool et al., 1985; Roth et al., 1988; Schwartz et al., 1997; Handel et al., 1999; Brailov et al., 2000; Somlo and Ehrlich, 2001; Praetorius and Spring, 2001; Pazour et al., 2002a; Pazour and Witman, 2003; Nauli et al., 2011; Amador-Arjona et al., 2011; Blitzer et al., 2011). Over a dozen human diseases are now known to result from primary cilium defects, including polycystic kidney disease (Hughes et al., 1995; Mochizuki et al., 1996; Pazour et al., 2000; Somlo and Ehrlich, 2001; Park et al., 2011) and Bardet-Biedl syndrome (Pazour et al., 2000; Nachury et al., 2007).

The first experimental evidence establishing primary cilium function came from studies of Tg737 mutant mice that developed polycystic kidney disease (Moyer et al., 1994), where the mutation prevented normal primary cilium formation (Pazour et al., 2000). Subsequent research demonstrated that the Tg737 protein is homologous to IFT88, a flagellar transport protein essential for bidirectional particle movement along ciliary and flagellar MTs and crucial for their assembly (Pazour et al., 2000; Westlake et al., 2011). These findings confirmed that primary cilia serve sensory functions, operating as both sensory transducers (Roth et al., 1988; Schwartz et al., 1997; Seeger-Nukpezah and Golemis, 2012) and mechanoreceptors (Praetorius and Spring, 2001; Nauli et al., 2011; Abdul-Majeed and Nauli, 2011; Abdul-Majeed et al., 2012; Uzbekov et al., 2012).

Primary cilia exhibit cell-type-specific functional specialization. While early indirect evidence suggested their sensory role (Pool et al., 1985; Roth et al., 1988; Schwartz et al., 1997), conclusive demonstration came from studies revealing structural connections between primary cilia, centrosomes (Wheatley, 1982) and other organelles like the Golgi apparatus in chondrocytes and aortic smooth muscle cells (Poole et al., 1997). This led to the proposal that primary cilia serve as sensory components in a transport cycle, relaying extracellular information to centrosomes and Golgi complexes to optimize secretion of newly synthesized extracellular matrix components.

The current consensus recognizes primary cilia as specialized sensory antennae housing specific receptors (Seeger-Nukpezah and Golemis, 2012). For instance, kidney primary cilia contain polycystin-2, a Ca^{2+} channel component that initiates differentiation and proliferation signals (Pazour et al., 2002b). These cilia also function as mechanosensors, generating propagating

Ca^{2+} signals upon bending (Praetorius and Spring, 2001; Pazour and Witman, 2003). Other cell types show distinct receptor complements, such as somatostatin and serotonin receptors in neurons (Handel et al., 1999; Brailov et al., 2000). The non-random perpendicular orientation of primary cilium-forming centrioles relative to the long bone axis in osteocytes (Uzbekov et al., 2012) further supports their mechanotransduction role.

If we consider the primary cilium as an "antenna" integrated with the centrosome, it follows that the centrosome participates in intracellular signal transmission. This interpretation aligns well with the centrosome's central cellular position and its dynamic regulation of MT growth, suggesting it may function as both signal receiver and relay station, with the cilium serving as its sensory antenna. In this model, the functional sensory centrosome-cilium complex operates through the following sequence: (1) the ciliary axoneme membrane, through its specialized receptors (Handel et al., 1999; Brailov et al., 2000; Pazour et al., 2000, 2003; Abdul-Majeed and Nauli, 2011), detects extracellular signals; (2) these signals induce centrosomal morphological changes and enhance MT-organizing activity, manifested by increased numbers of subdistal appendages and greater numbers/lengths of radial MTs (Alieva and Vorobjev, 1994, 1995); and (3) the resulting elaborate MT network facilitates coordinated responses involving the nucleus and Golgi apparatus.

However, in cultured cells this integrated system often shows compromised functionality. The proportion of ciliated cells varies between passages, with peak occurrence in non-proliferating populations (Tucker et al., 1979a, 1979b; Wheatley et al., 1996). Concurrent with loss of tissue architecture, three-dimensional orientation and cell-cell contacts, cultured cells demonstrate: (1) reduced primary cilium incidence; (2) axoneme shortening; and (3) altered membrane positioning. These changes render the cilium physically unable to serve as an effective antenna, as retracted axonemes cannot contact the extracellular environment. Several factors contribute to this functional impairment: (1) partial cellular dedifferentiation; (2) artificial culture conditions maintaining constant extracellular parameters; and (3) consequent simplification of cellular responses compared to tissue contexts. As a result, cultured primary cilia may be non-functional or only partially functional, though they retain responsiveness to experimental manipulations (e.g., inhibitors or stimulators). The variable ciliation patterns following different treatments suggest that *in vitro* conditions may support only incomplete cilium formation (Alieva and Vorobjev, 2004). Therefore, while primary cilia function as essential sensory components of the centrosomal complex in tissues, their functionality is frequently compromised in isolated cell cultures.

6.3. Striated Rootlets

These structures, characterized by their distinctive transverse striations, are typically associated with basal bodies (Szollosi, 1964; Sakaguchi, 1965; Anderson, 1972; Lauweryns and Boussauw, 1973), though they occasionally appear near non-ciliated centrioles. The periodicity of these striations shows considerable variation across different cell types, ranging from 12 to 90 nm (Wolfe, 1972). Similarly, striated rootlets exhibit substantial diversity in both length and thickness (Figure 6.7e; Figure 6.10.).

High-resolution electron microscopy reveals that the transverse striations can sometimes be doubled (Szollosi, 1964). The most structurally complex striated rootlets occur in protozoans and algae, where their architecture and quantity are precisely regulated (Allen, 1971; Andersen, 1991; Mignot, 1996). Current evidence suggests these rootlets function primarily as mechanical anchors, maintaining stable connections between cilia/flagella and deeper cytoplasmic structures, particularly the nucleus (Mignot, 1996).

The thickness of striated rootlets typically reaches its maximum near centrioles or basal bodies and gradually decreases with increasing distance from these structures (Sakaguchi, 1965; Lauweryns and Boussauw, 1973). In certain algal species, these rootlets (alternatively termed rhizoplasts or parabasal fibers) may function as convergence sites for spindle MTs during mitosis, while the associated basal bodies remain positioned near the cell membrane at their wider ends (Slankis and Gibbs, 1972; Bouck and Brown, 1973).

In multicellular organisms lacking functionally active cilia, striated rootlets are nevertheless present, though they represent non-essential centrosomal components and never achieve the structural complexity seen in protozoan cells. Within tissue culture cells, these rootlets typically measure ≤0.7 μm in length and 0.05-0.1 μm in width, and unlike their protozoan counterparts, never establish regular connections with centriolar cylinders. However, they consistently localize near centrioles and, in contrast to primary cilia, may associate with either mother or daughter centrioles, or occasionally both simultaneously (Alieva and Vorobjev, 2004).

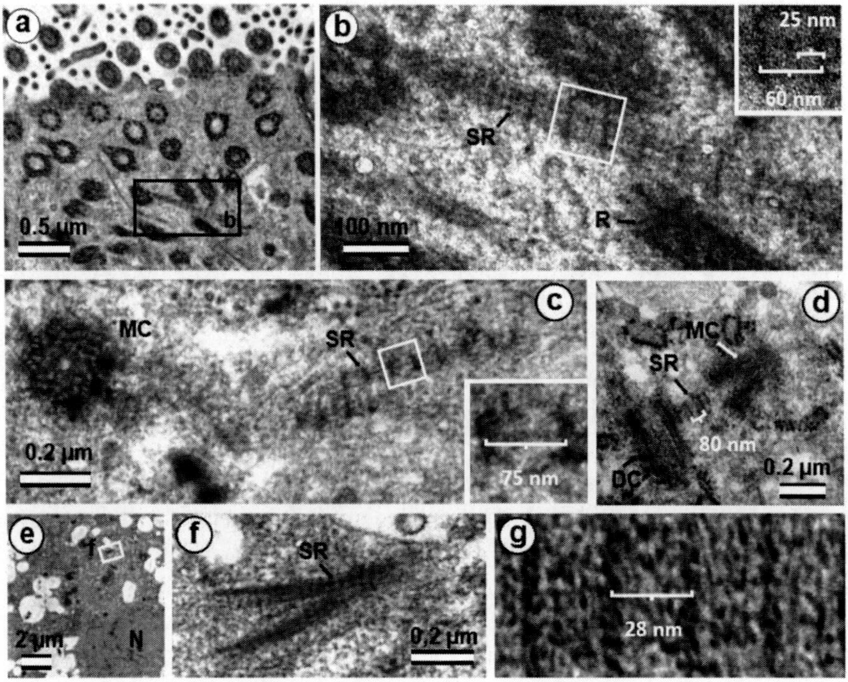

Figure 6.10. Striated rootlets in various cells: (a,b) Bovine oviduct ciliated cells (*in vitro* spheroids) show striated rootlets contacting non-striated basal body rootlets; (c) Mouse brain capillary endothelial cells (MBCEC) display striated rootlets near but not touching mother centrioles; (d) Rat osteocyte striated rootlets connect mother and daughter centrioles; (e-g) Parthenogenetic bovine embryo (4-cell stage) blastomeres contain striated rootlets despite centriole absence. Abbreviations: DC - daughter centriole; MC - mother centriole; N - nucleus; R - non-striated rootlet; SR - striated rootlet. Images by R. Uzbekov.

As previously established, the primary cilium represents an optional centrosomal component. Importantly, studies using SPEV cultured cells have demonstrated that the presence of striated rootlets and primary cilia constitute independent events - a cell may possess a primary cilium without striated rootlets, or conversely, contain striated rootlets without forming a primary cilium (Alieva and Vorobjev, 1989b). A key molecular component of striated rootlets is centrin (also known as centractin), a calcium-sensitive contractile filament protein (Salisbury and Floyd, 1978).

6.4. Other Transient Components of the Interphase Centrosome

In addition to the structures previously described, the centrosome contains several other components. Electron-dense spherical aggregates, comparable in size to pericentriolar satellite heads but lacking visible connections to centriolar cylinders, have been observed near centrosomes in certain cell types (de Thé, 1964; Vorobjev and Chentsov, 1977; Gosti-Testu et al., 1986). These structures frequently associate with MT ends and are hypothesized to participate in MT nucleation, leading to their designation as free MT nucleation centers (Vorobjev and Nadezhdina, 1986) or free MT convergence foci.

Paintrand and colleagues (Paintrand et al., 1992) propose that these free foci represent condensed pericentriolar material connected to centrioles via thin filaments, referring to them as *satellites* - a terminology also employed by other researchers (Rieder and Borisy, 1982; Baron and Salisbury, 1988).

Similar electron-dense formations with comparable morphology but lacking MT associations occasionally appear near centrosomes in various cultured cells. Direct evidence from one cell culture system demonstrates that such aggregates can result from retroviral infection (Lehmann-Che et al., 2007; Figure 6.11.).

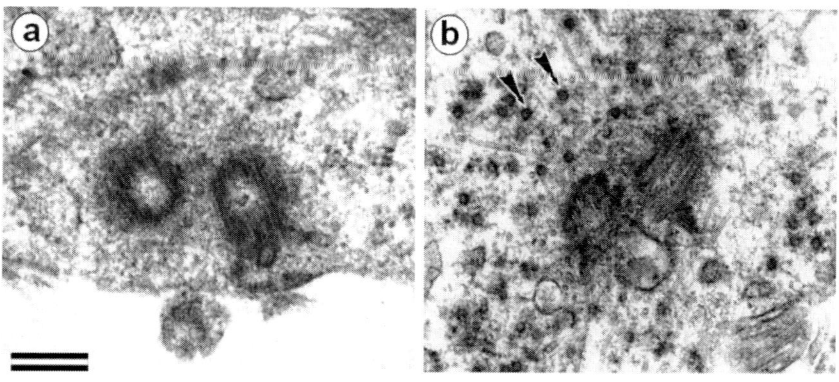

Figure 6.11. Centrosomal regions in MRC5 cells (human fetal lung fibroblast-like cell line): (a) Normal organization; (b) Fifteen days post-infection with prototype foamy virus (PFV), showing multiple viral particles (arrows) clustered around centrioles. Scale bar: 0.2 µm. Images courtesy of Professor Philippe Roingeard (University of Tours, France).

Several studies have reported the presence of a small granule approximately 0.1 μm in diameter located within the centriolar cylinder (Sorokin, 1962; Doolin and Birge, 1966; Stubblefield and Brinkley, 1967; Vorobjev and Chentsov, 1977). Furthermore, structural analysis of serial cross-sections revealed another intracentriolar component - a helical structure exhibiting approximately 10 turns along the proximal-distal axis, which the investigators termed a "nucleoid" (Stubblefield and Brinkley, 1967). While the existence of intracentriolar granules has been confirmed in multiple cell types, subsequent studies have failed to replicate the observation of this spiral structure, leaving its existence unverified.

Chapter 7

Centriole Formation Pathways and the Regulation of These Processes in Cells

Rustem E. Uzbekov[1,*], PhD, HDR, IR
and Claude Prigent[2], PhD, HDR, DRCE
[1]Laboratory of Cell Biology and Electron Microscopy, Faculty of Medicine, University of Tours, Boulevard Tonnellé, France;
Faculty of Bioengineering and Bioinformatics, Lomonosov Moscow State University, Moscow, Russia
[2]L'équipe de recherche Cycle Cellulaire, Centre de Recherche en Biologie cellulaire de Montpellier, U'unité mixte de recherche 5237 CNRS;
University of Montpellier, Montpellier, France

7.1. Variants of Centriole Formation

The process of centriole doubling, by analogy with DNA replication, is often termed "replication." This stems from the observation that a new organelle typically forms adjacent to an existing centriole. Similar to early DNA replication studies (Taylor et al., 1957; Meselson and Stahl, 1958), centriole doubling is semi-conservative - experiments using biotinylated tubulin showed that each daughter cell inherits one "old" and one "new" centriole (Kochanski and Borisy, 1990). However, closer examination reveals that DNA replication and centriole "replication" share little mechanistic overlap. The recognition of their fundamental differences emerged later, long after the term "centriole replication" was coined. Here, we use "duplication" as it more accurately reflects the process of centriole formation.

* Corresponding Author's Email: rustem.uzbekov@univ-tours.fr

In: The Centrosome
Editor: Rustem E. Uzbekov
ISBN: 979-8-89530-815-8

Extensive data on centriolar cylinder assembly in diverse cell types have now been accumulated. These mechanisms fall into two categories: (1) *de novo* formation, independent of pre-existing structures, and (2) assembly near parental centrioles or basal bodies.

Historically, the first 40 years of centriole research (late 1950s–late 1990s) focused on morphological aspects of duplication using electron microscopy. Only in the early 21st century did biochemical insights into its regulatory mechanisms emerge. This chapter follows a chronological framework, presenting modern regulatory pathways after a morphological overview.

7.2. *De Novo* Formation of Centriolar Cylinders of Centrosomes

De novo formation of centriolar cylinders in centrosomes was first described during parthenogenetic development in sea urchin eggs (Dirksen, 1961; Kallenbach, 1982), early embryogenesis in mice (Szöllösi et al., 1972), rats (Dvorak, 1978), rabbits (Szöllösi and Ozil, 1991), and the mollusc Lymnaea stagnalis (Kryuchkova et al., 1989). *De novo* formation of basal body-associated centriolar cylinders (which generate cilia) was also observed during differentiation of ciliated epithelial cells in the respiratory tract and oviduct (Stockinger and Cireli, 1965; Dirksen and Croker, 1966; Sorokin, 1968).

The *de novo* emergence of centrioles has been most extensively studied in mouse embryogenesis - a model system shared with rats. However, rodents (mice and rats) are “atypical” mammals regarding centriole inheritance and their role in early development. This stems from the complete disappearance of both centrioles during rodent spermatogenesis, whereas in most mammals, the proximal centriole persists in sperm and is transmitted to the oocyte during fertilization. The role of centrioles in early development of other mammals, including humans, will be addressed in a dedicated chapter. Below, we focus on *de novo* centriole formation in rodent early development.

In the mouse, the mitotic spindles of the first cleavage divisions are acentriolar (Szöllösi et al., 1972). These spindles become increasingly convergent with each division compared to the barrel-shaped meiotic spindle. Similar to meiosis, astral MTs are absent in acentriolar spindles (Schatten et al., 1985). In interphase blastomeres, MTs are homogeneously distributed, lacking visible convergence foci (Maro et al., 1985; Schatten et al., 1985; Houliston et al., 1987). The timing of centriole emergence remains inconsistent across studies: some authors identified centrioles at the 8–16 blastomere stage (Szöllösi et al., 1972), while others observed them only at the blastocyst stage (over 32 cells) (Schatten et al., 1988; Maro et al., 1991; Gueth-Hallonet et al., 1993; Abumuslimov et al., 1994, 1996). All studies, however, report asynchrony in centriole appearance across cells. In rats (Dvorak et al., 1978) and mice (Maro et al., 1991), centrioles were first noted in trophectoderm cells. Later, Abumuslimov et al., (1994) demonstrated simultaneous centriole emergence in some trophectoderm and inner cell mass cells of mouse embryos. In both cases, the earliest centriolar cylinders exhibited immature or procentriole-like structures: shortened length, a lumenal "axis with spokes" ("cartwheel structure"), and incomplete triplet MTs in the cylinder walls (Abumuslimov et al., 1994). Notably, the proximal end of some centrioles, containing the "axis with spokes," was attached to electron-dense globules. Early centrioles do not appear to participate in spindle pole formation or interphase MT organization (Abumuslimov et al., 1994). Only later do non-centriolar, yet functional centrosomes integrate with centrioles, forming the unified structure typical of adult cells (Abumuslimov et al., 1994, 1996).

Normally, proliferating somatic cells do not form centrioles *de novo*. However, using an original experimental model, Khodjakov et al., bypassed cellular regulatory mechanisms, inducing *de novo* centriole formation in cultured cell lines (Khodjakov et al., 2002). As previously demonstrated, blocking DNA synthesis for multiple cell cycles allows centrioles to duplicate at a reduced rate (Balczon et al., 1995). In contrast, Khodjakov et al., laser-microirradiated centrioles in cells pre-synchronized with hydroxyurea for 1.5 cell cycles. This synchronization resulted in approximately half of the cells arresting at the G1/S boundary, with the remainder distributed across early, mid-, and late S phases.

Since all cells passed the restriction point (located in late G1, when centriole duplication initiates), factors required for duplication were already present in the cytoplasm, and duplication mechanisms were activated. However, with centrioles destroyed, templated duplication was impossible. Under these conditions, new centrioles formed similarly to those in mouse blastomeres or basal bodies (see below): electron-dense, gamma-tubulin-containing aggregates appeared first, followed by centriolar cylinders, which occasionally exhibited morphological defects (Khodjakov et al., 2002). Thus, this study directly demonstrated that cells can reconstruct centriole architecture without a pre-existing morphological template.

7.3. *De Novo* Formation of Centriolar Cylinders of Basal Bodies

Unlike centrosomal centriolar cylinders, basal body-associated centriolar cylinders can form simultaneously in large numbers within cells (Stockinger and Cireli, 1965; Dirksen and Croker, 1966; Sorokin, 1968). This enabled detailed characterization of structures preceding centriolar cylinder assembly (Dirksen and Croker, 1966; Sorokin, 1968; Dirksen, 1971). Initially, the centrosome contains two centrioles (Figure 7.1., stage 1). Subsequently, fibrogranular aggregates (~1 μm or larger in diameter) emerge near centrioles in the apical cytoplasm (Figure 7.1., stage 2). Dirksen and Croker (1966) termed these structures "proliferative elements." Each proliferative element comprises numerous electron-dense granules (40–80 nm diameter) embedded in a less electron-dense fibrillar matrix. Granules often align in chains along the periphery of the fibrillar material. Proliferative elements localize near Golgi apparatus cisternae (Sorokin, 1968) and regions of the rough endoplasmic reticulum (Dirksen, 1971).

Proliferative elements in developing rat lung epithelium (Stockinger and Cireli, 1965; Dirksen and Croker, 1966; Sorokin, 1968) and newborn mouse oviduct epithelium (Dirksen, 1971) exhibit highly similar morphology. Subsequent studies of ciliated epithelium formation across organisms confirmed the universality of these centriolar cylinder precursors (Dirksen, 1991). Dirksen (1971) observed direct associations between proliferative elements and mature centrioles (Figure 7.1., stage 2).

Although the author does not specify the centriolar end contacting the proliferative element, the provided micrograph suggests association with the proximal end, as the ciliary sheath - a distal-end marker - is visible at the opposite pole.

During maturation, proliferative elements segment into individual globules, forming electron-dense spherical structures (~400 nm diameter). Dirksen and Croker (1966) termed these "condensation forms," while Sorokin (1968) proposed "deuterosomes." A deuterosome comprises a dense core (100-150 nm diameter) surrounded by a looser corona, from which 1-14 procentrioles emerge simultaneously (Sorokin, 1968) (Figure 7.1., stage 3). The assembly initiates with an "axis with spokes," followed by growth of internal MT A-subfibers at spoke tips. These MTs subsequently form doublets and ultimately triplet MTs.

In short procentrioles, MT doublets and triplets are initially arranged radially, gradually twisting as they elongate, resembling mature centrioles. In cross-sections, procentrioles are often symmetrically arranged, with their central axes oriented toward the deuterosome center. As procentrioles grow, their distal ends diverge radially. Upon reaching half their mature length, they detach from the deuterosome and continue elongating (Figure 7.1., stage 4). Sorokin (1968) reported uniform procentriole lengths within a single deuterosome, but noted variations across different deuterosomes in the same cell. Conversely, Kalnins and Porter (1969) observed consistent centriolar cylinder lengths within a cell but discrepancies between neighboring cells. Unlike "proliferative elements," deuterosomes (termed "condensation forms" by Dirksen and Croker, 1966) never associate with mature centrioles (Dirksen and Croker, 1966; Dirksen, 1971, 1991). Dirksen and Croker (1966) designated the "condensation form" with associated growing centrioles as a "generative complex."

Dirksen (1971) proposed that condensation form material is consumed during centriole assembly, as procentriole growth correlates with cavity expansion within the generative pole. The same study identified another generative complex type, where procentrioles originate directly from mature centrioles, coexisting with deuterosome-derived complexes in the same cell. This highlights the absence of strict determination in centriolar cylinder reproduction mechanisms, even within a single cell.

In ciliated epithelial cells, the simultaneous formation of up to eight new centriolar cylinders on a single maternal centriole has been documented (Sorokin, 1968). This observation indicates that duplication is not governed by a singular template on the maternal centriole but is instead regulated by intracellular mechanisms.

The newly formed centriolar cylinders migrate to the cell surface, where they generate cilia. These cilia acquire motility once they reach approximately one-sixth of their final length (Sorokin, 1968) (Figure 7.1., stage 5).

Each mature basal body possesses a rootlet - a fibrillar structure extending from the proximal end of the centriolar cylinder.

The rootlet tapers distally from the basal body and features one or two "feet" homologous to the subdistal appendages (pericentriolar satellites) of centrosomal centrioles. A schematic outlining the sequential stages of basal body centriole formation is provided in Figure 7.1.

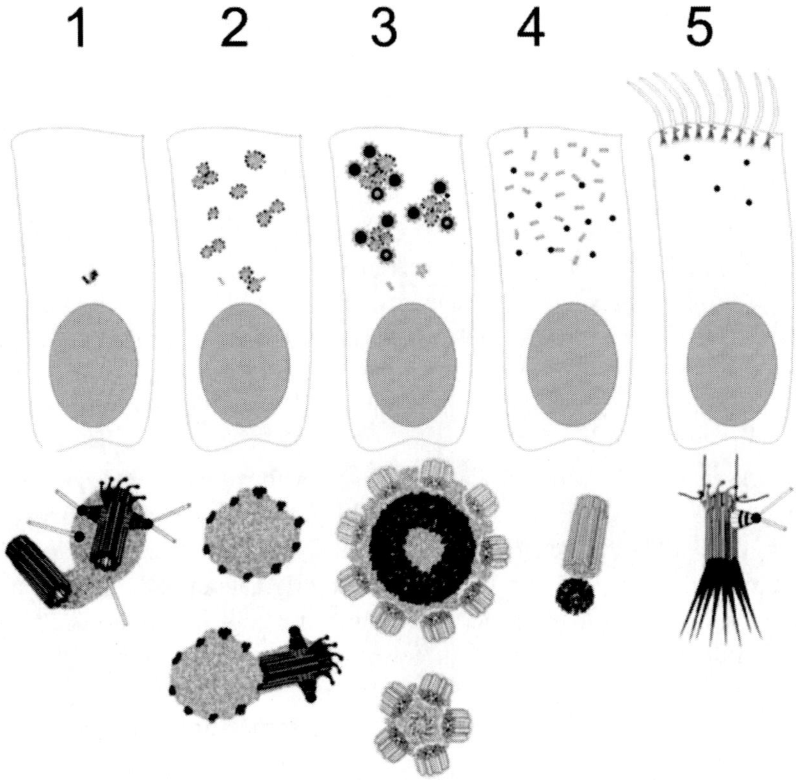

Figure 7.1. Formation of basal bodies in ciliated epithelial cells (Sorokin, 1968; Dirksen, 1971). The upper section depicts the overall cellular morphology, while the lower section provides magnified views of centrioles and their "predecessors" at corresponding stages. At stage "1," the centrosome exhibits a structure characteristic of interphase cells. At stage "2," numerous fibro-granular "proliferative elements" emerge near the proximal end of the mature centriole and within the apical cytoplasm. Stage "3" involves the formation of generative complexes, surrounded by developing procentrioles, adjacent to these elements. Multiple procentrioles may also assemble around the mature centriole. Subsequently, the centriolar cylinders detach from the generative complex, elongate to full length (stage "4"), migrate to the cell membrane, and initiate cilia formation (stage "5").

In summary, the reproductive capacity of centrioles is independent of their triplet microtubule (MT) structure. Information encoding the triplet MT arrangement can be transmitted via deuterosomes (during ciliogenesis) or through non-centriolar precursors (as observed in oogenesis and embryogenesis).

7.4. Duplication of Centrioles near Pre-existing Centriolar Cylinders

In most cells, new centriolar cylinders form in proximity to pre-existing ones. The mechanism of centriolar duplication is unique: a new centriole arises perpendicularly to the maternal structure rather than via replication of its template. In certain fungi and mosses, centriole doubling occurs through procentriole formation near the maternal centriole's distal end, resulting in a mirror-symmetric daughter centriole with antiparallel triplet MTs (for details on daughter centriole positioning, see Vorobjev and Nadezhdina, 1986). Notably, maternal and daughter centrioles remain separated by a fibrillar matrix that connects both cylinders (Gall, 1961).

Centriole duplication was first described in Mixine spermatocytes, where large centriolar cylinders are visible via light microscopy (Schreiner and Schreiner, 1905). These authors demonstrated that during meiosis, daughter centrioles form perpendicularly to maternal ones at pachytene and elongate to half their length by late diplotene (Figure 1.3. Schreiner and Schreiner, 1905).

In somatic cells, duplication occurs during interphase. A single "procentriole" (Gall, 1961) typically forms per maternal centriole. This ensures a strict correlation between centriole number and cell ploidy: diploid G1-phase cells possess two centrioles, while G2-phase cells contain four (Robbins et al., 1968). However, recent studies reveal exceptions. For instance, diploid female and haploid male wasps (with haplo-diploid sex determination) exhibit identical centriole numbers (Uzbekov et al., 2024a). Deviations also occur during germ cell development (see The Centrosome in Germ Cells) and in terminally differentiated cells.

In muscle fiber syncytia, centrioles are lost (Przybylski, 1971), likely due to their exit from the mitotic cycle (Lash et al., 1957). Myoblasts, which retain proliferative capacity, retain centrioles (Przybylski, 1971). Notably, polyoma virus-induced DNA synthesis in muscle cell nuclei fails to restore cell cycle progression beyond metaphase, potentially due to centriole absence (Yaffe and Gershon, 1967).

In contrast, cardiomyocytes preserve centrioles (Przybylski, 1971), consistent with their retained mitotic potential (Overy and Priest, 1966; Rumyantsev and Snigerevskaya, 1968).

The principal stages of procentriole formation resemble those observed during basal body assembly in ciliogenesis. During the initial stage, an electron-dense disk, reportedly measuring approximately 70 nm in thickness (Dippell, 1968) or less than 50 nm (Uzbekov, 2007), is detectable near the proximal pole of each maternal centriole (Figure 7.2a). Subsequently, this disk translocates distally from the maternal centriole.

In cross-sections aligned parallel to the procentriole's longitudinal axis, a central discontinuity becomes apparent within the disk (Figure 7.2b), likely corresponding to the central hub of the cartwheel structure, which serves as a scaffold for subsequent MT polymerization (Uzbekov, 2007).

During the G1 phase of the cell cycle, Plk4 kinase localizes around existing centrioles as discrete foci. A fluorescent labeling reagent, SiR650-TetH, was developed based on centrinone's high affinity for Plk4. High-resolution microscopy revealed that this reagent enables precise localization of Plk4 to centrioles, demonstrating a 9-fold symmetry in its spatial arrangement (Salim et al., 2023). Upon interaction with the STIL protein, a single STIL-Plk4 focus forms (Ohta et al., 2014; Kratz et al., 2015; Moyer et al., 2015), initiating the assembly of one procentriole per maternal centriole. The 9-fold symmetry of the procentriole is established by the attachment of nine SAS-6 protein dimers, bound via their lateral regions to the STIL-Plk4 complex (Kitagawa et al., 2011; van Breugel et al., 2011). Thus, the Plk4-STIL-HsSAS-6 complex acts as a nucleation site for procentriole assembly.

During the subsequent stage (Figure 7.2c), the procentriole elongates to 100 nm. Initially, nine MT singlets form at the disk periphery, progressing to nine doublets, and finally to nine MT triplets (Dippell, 1968; Wolfe, 1972). According to our studies (Uzbekov, 2007; Uzbekov and Alieva, 2010), the first two stages of procentriole formation occur in G1 phase, with the structure corresponding to Figure 7.2c emerging at the G1/S transition.

Notably, the procentriole diameter remains approximately 150 nm even at early stages, suggesting that in SPEV (pig embryo kidney) cells, the initial formation of a long central hub does not precede peripheral triplet MT growth.

MT singlet, doublet, and triplet elongation likely occurs concurrently with cartwheel structure maturation (discussed further in this chapter). Significant procentriole elongation (>100 nm) commences only after cells enter S phase (Uzbekov, 2007).

During the subsequent stage (Figure 7.2c), the procentriole elongates to 100 nm. Initially, nine MT singlets form at the disk periphery, progressing to nine doublets, and finally to nine MT triplets (Dippell, 1968; Wolfe, 1972). According to our studies (Uzbekov, 2007; Uzbekov and Alieva, 2010), the first two stages of procentriole formation occur in G1 phase, with the structure corresponding to Figure 7.2c emerging at the G1/S transition.

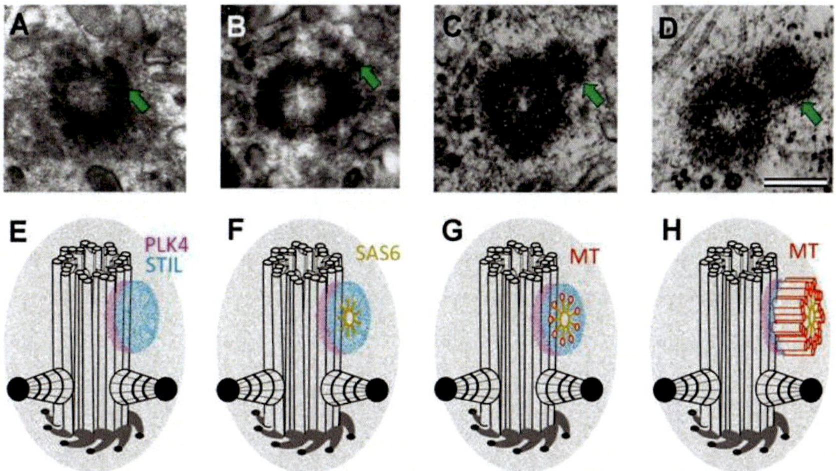

Figure 7.2. Initial stages of procentriole formation in SPEV culture cells (Uzbekov, 2007; Prigent and Uzbekov, 2022). Panels A–D show electron micrographs; panels E–H depict corresponding schematic diagrams illustrating stage-specific localization of centriolar proteins. A–C and E–F: G1-phase cells. Image C corresponds to a cell fixed 5 h 48 min post-mitosis (approximately 2 h prior to S-phase onset; G1 duration in this cell line is 7–10 h, Sakharov and Voronkova, 1993). D and H: early S-phase cell (10 h 45 min post-mitosis). Four stages of procentriole assembly: A – dense disk formation; B – cartwheel structure assembly; C – MT nucleation; D – MT elongation. Arrows indicate developing procentrioles. Scale bar: 0.2 µm.

Notably, the procentriole diameter remains approximately 150 nm even at early stages, suggesting that in SPEV (pig embryo kidney) cells, the initial formation of a long central hub does not precede peripheral triplet MT growth.

MT singlet, doublet, and triplet elongation likely occurs concurrently with cartwheel structure maturation (discussed further in this chapter). Significant procentriole elongation (>100 nm) commences only after cells enter S phase (Uzbekov, 2007).

The triplet MTs continue to elongate, typically reaching maternal centriole length by mitotic onset. Exceptions exist; for instance, in KE37 cells, daughter centrioles remain moderately shorter than maternal counterparts during mitosis, achieving full length only in the subsequent G1 phase (Chretien et al., 1997). In epithelial cells, centriole replication occurs apically, whereas fibroblasts initiate duplication near the nucleus (Sorokin, 1968). Remarkably, even when centrioles are separated by substantial distances (>10 µm), procentriole formation proceeds synchronously, with both nascent structures exhibiting near-identical elongation rates (Uzbekov, unpublished data). Polyploid megakaryocytes and hepatocytes, which maintain ploidy-correlated centriole numbers, also demonstrate synchronous duplication across all centriolar cylinders (Moskvin-Tarkhanov and Onishchenko, 1978; Onishchenko and Chentsov, 1986).

Studies employing centrin fused to green fluorescent protein (GFP) suggest that older centrioles may initiate procentriole assembly marginally earlier (White et al., 2000). However, these findings do not fundamentally alter the paradigm of a unified intracellular trigger for duplication.

Thus, centriole duplication exhibits no spatial correlation with intracellular positioning.

7.5. Coordination Between Centriolar and Cell Cycles

The centriolar and cell cycles are typically synchronized; their uncoupling leads to division abnormalities, notably multipolar mitoses (Mazia, 1963). Multipolar mitoses arise during prolonged nuclear cycle arrest (Mazia, 1961; Onishchenko et al., 1979; Khodjakov et al., 1986) or transient mitotic inhibition (Keryer et al., 1984; Sluder and Begg, 1985; Alieva and Vorobjev, 1989c, 1991). Potential causes include spindle disorganization - disrupted MT-centriole/kinetochore linkages, centrosome separation, loss of astral MTs, or premature maturation of daughter centrioles (Mazia, 1961; Keryer et al., 1984; Khodjakov et al., 1986; Alieva and Vorobjev, 1989c, 1991).

A key question in understanding centriole duplication regulation is identifying the cell cycle stage when duplication initiates. Early light microscopy studies in Mixine spermatocytes revealed procentriole formation during or immediately before meiosis (Schreiner and Schreiner, 1905) (Figure 1.3.). In Xenopus laevis embryos, centrioles duplicate rapidly during the first 12 cleavage divisions, with new centriolar cylinders assembling within 30 minutes between divisions (Graham and Morgan, 1966). Similarly, Spisula solidissima oocyte extracts form centrioles within 4 minutes post-activation, doubling by 15 minutes as the meiotic spindle completes assembly (Palazzo et al., 1992). These findings suggest that centriole growth is limited not by time but by the availability of molecular components and activation signals. While DNA replication accelerates via additional replicons during cleavage (reducing S-phase duration 10-20-fold), centriole maturation in somatic cells spans nearly two cycles. The mechanisms enabling such accelerated centrosome maturation remain unresolved.

In somatic cells, DNA replication is confined to interphase (Howard and Pelc, 1953), creating a "complete" cell cycle with four phases: mitosis, S-phase (DNA replication), G1 (pre-S), and G2 (post-S) (Figure 7.3.). G1 is further subdivided into postmitotic (G1PM) and presynthetic (G1PS) phases. The transition from G1PM to G1PS occurs at the restriction point (R), determined experimentally by serum dependence in cultured cells (Zetterberg and Larsson, 1995).

Initially, it was postulated that centriole duplication in somatic cells occurs immediately before mitosis, akin to the process in blastomeres; however, subsequent studies disproved this hypothesis (Robbins et al., 1968; Stubblefield, 1968).

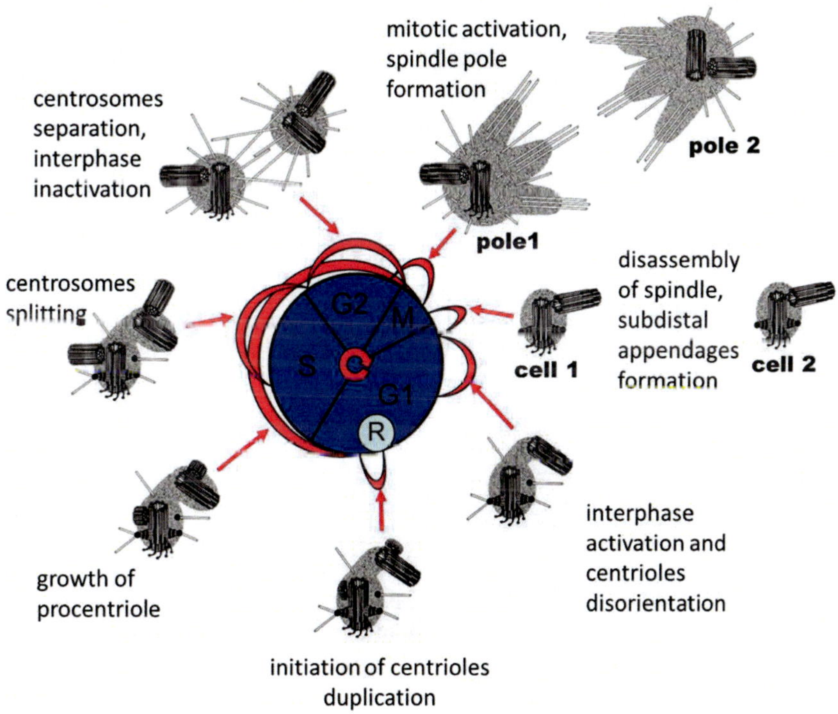

Figure 7.3. Correlation between the centriolar and cell cycles (nuclear phases) in an animal cell. G1, S, G2, and M (mitosis) denote the phases of the nuclear cycle; R indicates the restriction point (adapted from Prigent and Uzbekov, 2022).

Unlike DNA replication, which can be tracked via labeled precursor incorporation, centriole duplication onset (excluding giant centrioles) requires ultrastructural analysis. Electron microscopy of synchronized cells revealed correlations between cellular and centriolar cycles. Stubblefield (1968) proposed subdividing the centriolar cycle - a term introduced by Van Beneden and Boveri - into phases: maturation, reproduction, and divergence, analogous to the nuclear cycle.

Robbins et al., (1968) first comprehensively described the centriolar cycle in somatic cells.

They observed that the earliest post-mitotic centrosomal reorganization in G1 phase involves loss of orthogonal centriole orientation, adding a 'disorientation' phase to the established classification (Robbins et al., 1968).

Unlike sequential nuclear phases, centriolar phases often overlap. For instance, in some cell types, centriole separation initiates in G1 phase (Komarova et al., 1996), while in others, the centrosome disjunction phase coincides with the maturation phase (Uzbekov et al., 2002). A unified model of the centrosome cycle and its coordination with the cell cycle is presented in Figure 7.3.

A central debate in correlating nuclear and centriolar events concerns temporal precedence: does centriole duplication precede DNA replication? Robbins et al., (1968), analyzing partially synchronized cells (20% in S-phase), found procentrioles adjacent to maternal centrioles in 65–70% of S-phase cells. Although this suggested centriole duplication precedes DNA replication, the nuclear-centric paradigm of the time precluded definitive conclusions. Earlier, Mazia (1961) inferred from indirect evidence that mitotic center reproduction may initiate before DNA replication.

The temporal precedence of centriole duplication over nuclear DNA replication raised questions about their interdependence, prompting investigations into their regulatory relationship.

Independence of centriole duplication from DNA synthesis was first demonstrated in L929 cells, where arabinosyl cytosine-induced DNA replication inhibition did not impair procentriole assembly (Rattner and Phillips, 1973). The authors further observed procentriole initiation in a subset of G1-phase cells. Subsequent studies in enucleated sea urchin embryos or aphidicolin-treated cells revealed near-normal centriole duplication, leading Sluder (1989) to propose exclusive cytoplasmic control of this process.

Similarly, prolonged hydroxyurea/aphidicolin-mediated DNA synthesis blockade in CHO cells reduced centriole replication rate but did not abolish it (Balczon et al., 1995). Direct evidence emerged from SPEV cells tracked *in vivo*: autoradiography-electron microscopy integration revealed centriole duplication onset 2–3 hours prior to DNA replication (Uzbekov, 2007) (Figure 7.4.). These findings establish that initial centriole duplication stages occur during G1 phase. The observed autonomy of centriole duplication initially fueled hypotheses about centrosomal nucleic acids, given their role in replicative processes.

As previously established, centrioles can form independently of pre-existing centriolar cylinders. Consequently, the nine-triplet MT structure of centriolar cylinders is not inherently required for their formation. What then serves as the template for centriole self-duplication?

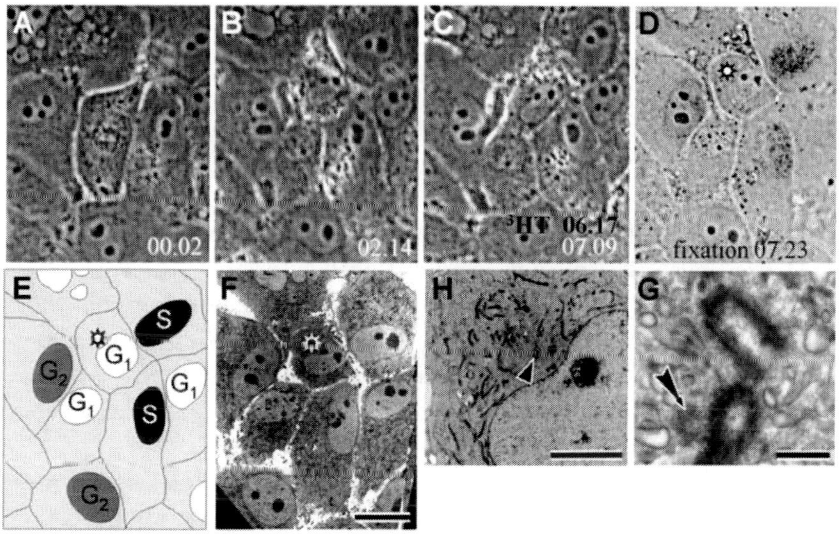

Figure 7.4. Centriole duplication in a SPEV cell during the G1 phase. The cell was tracked *in vivo* (A–C, Phase contrast) starting from mitotic anaphase. [3H]thymidine was added to the culture medium 1 h 6 min before fixation. Cells were fixed, dehydrated, and embedded in EPON resin. Following glass removal, the resin surface was coated with photographic emulsion. After 10 days of exposure, the emulsion was developed to visualize silver grains. Cells with precisely timed cell cycle stages were identified through live-cell imaging (D, E) and analyzed by electron microscopy (F–G). The asterisk-marked cell (D–F), fixed 7 h 23 min post-anaphase, showed no nuclear labeling, confirming its G1-phase status. Procentrioles were detected in this cell (H, arrowhead; G, arrow) (adapted from Uzbekov, 2007). Scale bars: A–F, 20 μm; H, 4 μm; G, 0.2 μm.

The hypothesis of centriole-associated DNA, whose replication might initiate centriole duplication, emerged logically. Early studies reported trace DNA in basal body fractions or via histochemical methods. Without delving into conflicting reports, it is now widely accepted that centrosomes and basal bodies lack DNA (Dirksen, 1991). In contrast, RNA has been consistently detected in centrosomes and basal bodies using diverse methodologies (Heidemann and Sander, 1977; Rieder, 1979; Berns et al., 1981), though its functional relevance remains debated.

The discovery of five centrosome-specific RNAs (cnRNAs) in Spisula solidissima oocytes (Alliegro et al., 2006) revitalized Went's (1977) hypothesis of a semi-autonomous centrosomal "genome," akin to mitochondrial or plastid systems. Notably, cnRNA11 harbors a conserved reverse transcriptase domain (Alliegro et al., 2006). Comparative PCR revealed cnRNA enrichment in centrosomal fractions, alongside ribosomal 18S RNA (Alliegro et al., 2006). *In situ* hybridization confirmed cnRNA colocalization with γ-tubulin-labeled centrosomes.

Peptides derived from cnRNA sequences were used to generate monoclonal antibodies; however, no corresponding proteins were detected, suggesting cnRNAs are non-translated (Alliegro and Alliegro, 2008). Sensitivity limitations of immunoblotting, however, cannot exclude low-abundance protein synthesis. Subsequent work identified nuclear DNA homologs of cnRNAs, challenging the notion of centrosomal RNA autonomy. The authors propose retrotransposition from cnRNAs into the nuclear genome (Alliegro and Alliegro, 2008).

Centrosomal RNAs may fulfill structural or catalytic roles analogous to ribosomal RNAs. Intriguingly, the primordial ribosome core, originating in a pre-protein era, was composed entirely of RNA (Bokov and Steinberg, 2009).

However, as demonstrated by the mechanism of duplication regulation discovered in 1999 (Hinchcliffe et al., 1999; Lacey et al., 1999; Meraldi et al., 1999), RNA within the centriole appears nonessential for duplication, and the matrix is likely protein-based (see below). Three independent studies established cytoplasmic regulation of duplication initiation (Hinchcliffe et al., 1999; Lacey et al., 1999; Meraldi et al., 1999). Key stages include:

1. Late G1 phase (restriction point): A biochemical cascade is triggered, dependent on cell size, growth factors, and intercellular interactions.
2. Phosphorylation of pRB by Cyclin D/CDK4/6: This releases EF2 transcription factors, activating Cyclin E/A synthesis and initiating centriole duplication.

Thus, DNA replication and centriole duplication share a common regulatory pathway.

Cyclin E localizes to the centrosome via its 20-amino acid CLS (centrosomal localization sequence) domain, essential for both centrosomal anchoring and DNA synthesis initiation (Matsumoto and Maller, 2004). Wild-type (but not mutant) CLS peptides disrupt endogenous Cyclin E/A centrosomal binding and inhibit DNA replication. Overexpression of Cyclin E-CDK2 accelerates S-phase entry, causing polyploidy.

Notably, Cyclin E retains centrosomal localization and S-phase promotion even when CDK2 binding is disrupted, but not when the CLS domain is mutated. This indicates a CDK2-independent, CLS-dependent mechanism for S-phase regulation.

Cyclin E colocalizes with the DNA replication factor MCM5 at centrosomes, interacting via its CLS domain. MCM5 overexpression inhibits centrosome reduplication in S-phase-arrested CHO cells (Matsumoto and Maller, 2004). Intriguingly, procentriole assembly can occur in CDK1-deficient cells (Vidwans et al., 2003), suggesting cyclin-dependent kinases (CDKs) prevent centriole reduplication within a single cycle. These findings implicate cell cycle regulators in centrosome duplication control.

At the beginning of this century, studies identified centriole duplication primer proteins across organisms: BLD-10 in Chlamydomonas (Matsuura et al., 2004); SPD-2, ZYG-1, DSAS-6, and DSAS-4 in Drosophila embryos (Leidel et al., 2005; Basto et al., 2006; Rodrigues-Martins et al., 2007). Subsequent work revealed that overexpression of their homologs - Plk4, SAS-4, and SAS-6 - in CHO cells induces centriolar precursor complexes resembling those in ciliogenesis (Kuriyama, 2009). These findings suggest conserved priming proteins across taxa, though incomplete sequence homology implies divergent duplication mechanisms (Azimzadeh and Bornens, 2007; Strnad and Gönczy, 2008; Song et al., 2008).

Research has focused on identifying duplication-specific proteins via RNA interference (RNAi), mutagenesis, and antibody blocking. In Caenorhabditis elegans, functional genetic screens identified five essential proteins: ZYG-1 kinase (O'Connell et al., 2001; O'Connell, 2002), coiled-coil SPD-2 (Kemp et al., 2004; Pelletier et al., 2004), SAS-4 (Kirkham et al., 2003; Leidel and Gönczy, 2003), SAS-5 (Delattre et al., 2004), and SAS-6 (Dammermann et al., 2004; Leidel et al., 2005). SPD-2 and ZYG-1 initiate procentriole assembly, recruiting SAS-5 and SAS-6 to form the central hub (Delattre et al., 2006; Pelletier et al., 2006). SAS-4, the final component, binds

SAS-5/SAS-6 and mediates MT assembly around the central structure (Kirkham et al., 2003; Salisbury, 2003; Pelletier et al., 2006).

In human cells, the ZYG-1 kinase homolog is PLK4 (Habedanck et al., 2005), while in Drosophila, this role is fulfilled by SAK/PLK4 (Bettencourt-Dias et al., 2005). Over 30 proteins have been implicated in centriole duplication across eukaryotic systems (Strnad and Gönczy, 2008).

The formation of the "spoke axis" hub was initially hypothesized to depend on SAS-6 or its homologs (Strnad and Gönczy, 2008). SAS-6 levels correlate with procentriole elongation, consistent with a structural role (Dammermann et al., 2008). This hypothesis is further supported by its proximal localization near maternal centrioles and absence during early G1 phase, preceding procentriole assembly (Strnad et al., 2007; Song et al., 2008).

In bld12 mutants (lacking the SAS-6 homolog), basal bodies exhibit disrupted central hubs, with centriolar walls containing 2–3 MT triplets. Rarely observed closed centriolar cylinders display 7–11 MT triplets, highlighting SAS-6's essential role in maintaining structural integrity (Nakazawa et al., 2007). Notably, giant centrioles in Acerentomon microphinus possess up to 14 MT doublets and enlarged hubs (~30 nm diameter), while spoke dimensions remain unaltered (Riparbelli et al., 2009).

Early models posited that SAS-9, organized as nine 6-nm subunits, forms a 15-nm-diameter ring (O'Toole et al., 2003; Nakazawa et al., 2007). However, electron microscopy (EM) studies demonstrated that the cartwheel hub measures ~30 nm in diameter (Uzbekov and Prigent, 2007). This discrepancy was resolved by the discovery that the cartwheel comprises nine radially arranged SAS-6 dimers rather than SAS-9 monomers.

In 2011, two independent studies (Kitagawa et al., 2011; van Breugel et al., 2011) demonstrated through crystallography and electron microscopy (EM) that SAS-6 forms stable dimers. These dimers exhibit lateral interactions mediated by their N-terminal domains *in vitro*, with an average angle of ~40° between their tail regions. This geometry allows nine dimers to assemble into a closed ring structure (Figure 7.5.), measuring approximately 24 nm in diameter. However, this dimension is slightly smaller than the hub diameter observed *in vivo* by EM. The discrepancy may arise from conformational adjustments in SAS-6 globular domains induced by binding partners *in vivo*, which could expand the central hub. Furthermore, interspecies variability in SAS-6 primary structure may account for subtle differences in hub diameter across eukaryotes.

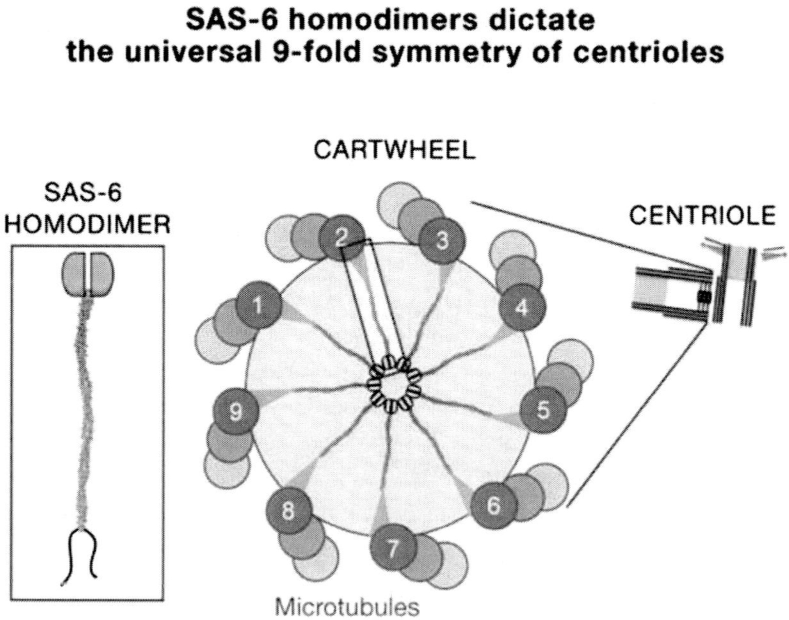

Figure 7.5. Model of SAS-6 dimer interactions during axial structure formation in the procentriole (Kitagawa et al., 2011).

7.6. Why Is Only One Procentriole Formed?

Despite advances in understanding the structural basis of centriole formation, the mechanisms restricting procentriole assembly to a single site near the proximal end of the maternal centriole remain unresolved.

The maternal centriole likely serves as a platform for recruiting gene products expressed during mid-G1 phase. These components presumably interact near the centriolar surface to initiate procentriole assembly. Two regulatory models, analogous to other biological systems, have been proposed:

1. Unidirectional inhibition: A single initiation site actively suppresses the formation of additional sites, akin to the block to polyspermy during fertilization.
2. Dominant primer selection: Multiple initiation sites form, but one becomes dominant and suppresses others, resembling oocyte maturation prior to ovulation.

Early hypotheses posited a unique structural matrix limiting procentriole formation. However, experimental evidence contradicts this model. During basal body formation, up to eight new procentrioles can assemble around a single centriolar cylinder. Furthermore, no ultrastructural markers predicting the procentriole assembly site are observed prior to its emergence (Dirksen, 1971; Kuriyama, 2009).

Thus, spatial constraints do not limit procentriole number; instead, precise regulatory mechanisms enforce singularity. This is exemplified by ciliogenesis in respiratory epithelial cells, where multiple centriolar cylinders form, contrasting with single procentriole assembly in somatic cells (Sorokin, 1968). The overexpression of key regulators (Plk4, SAS-6, SAS-4) in CHO cells induces ectopic procentriole formation (Kuriyama, 2009), with morphologically detectable precursors (fibrogranular aggregates) resembling those in ciliated epithelia (Dirksen, 1971). These findings suggest that negative feedback regulators, such as the ubiquitin ligase Slimb (Rogers et al., 2009), inhibit the activity or recruitment of duplication factors at the maternal centriole.

Consequently, the dominant primer selection model is more plausible. Additional initiation sites may fail to reach detectable sizes or undergo rapid disassembly, evading morphological observation.

7.7. How Does a Procentriole Grow?

Experimental studies have proposed two distinct models of procentriole growth (Figure 7.6.).

The first model (Strnad and Gonczy, 2008) posits a two-step mechanism: Central tube formation: Dependent on SPD-2, ZYG-1, SAS-5, and SAS-6. SPD-2 recruits ZYG-1 to the centrosome, enabling subsequent localization of SAS-5 and SAS-6 (Delattre et al., 2006; Pelletier et al., 2006). SPD-2 and ZYG-1 localize to centrioles prior to tube assembly, suggesting regulatory rather than structural roles. SAS-5 and SAS-6 appear concomitantly with tube initiation.

MT triplet assembly: Requires SAS-4. In Caenorhabditis elegans, procentrioles contain nine MT singlets rather than triplets (Pelletier et al., 2006).

The second model (Figure 7.6.) proposes that procentrioles initiate as planar disks, not tubes. These disks grow radially through sequential layering, with simultaneous elongation of MT triplets and the central hub (Uzbekov, 2007; Uzbekov and Alieva, 2010).

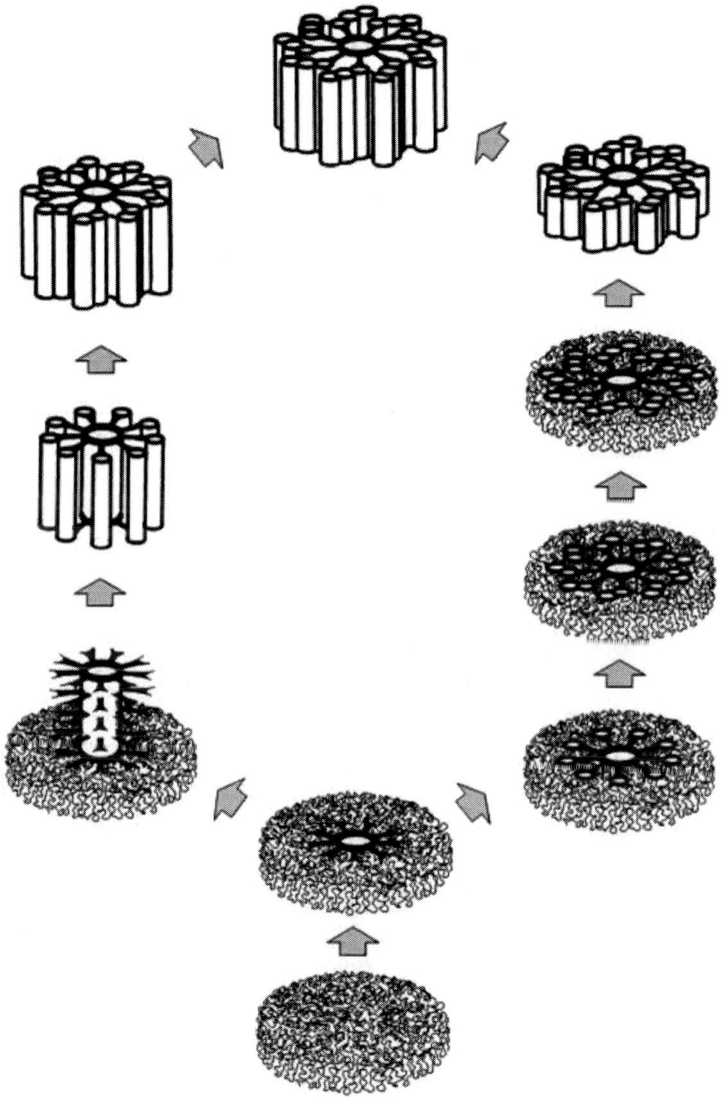

Figure 7.6. Two models of procentriole assembly. *Model 1* (left): Formation of an elongated central tube, followed by lateral addition of MT singlets, doublets, and triplets. *Model 2* (right): Coordinated elongation of the procentriole and MT triplets through sequential layer assembly. Distal view of procentriole.

Ultrastructural data support this model: early procentrioles exhibit disk-like morphology with a central cavity (Figure 7.2b).

While interspecies variation in early procentriole assembly may exist, the disk-layering model aligns more closely with ultrastructural and experimental observations. Recent insights into SAS-6's role in establishing ninefold symmetry (Kitagawa et al., 2011; van Breugel et al., 2011) raise critical unresolved questions: How is the initial SAS-6 disk anchored to the maternal centriole? What regulates SAS-6 disk stacking during elongation? These questions likely involve known regulatory proteins (e.g., SPD-2, ZYG-1, SAS-4) and await resolution through future studies.

CPAP (centrosomal P4.1-associated protein), a key regulator of centriolar cylinder elongation, binds tubulin dimers to facilitate procentriole elongation and ensure centrosome duplication (Kohlmaier et al., 2009; Schmidt et al., 2009; Tang et al., 2009). CPAP expression is cell cycle-dependent, peaking during S/G2 phases and undergoing degradation in late mitosis. Depletion of CPAP blocks centrosome duplication (Tang et al., 2009), while its overexpression induces procentriole-like structures (PLSs) containing stabilized MTs and centriolar proteins (Kohlmaier et al., 2009; Schmidt et al., 2009). A tubulin-binding-deficient CPAP mutant (CPAP-377EE) fails to promote PLS elongation, confirming the necessity of tubulin interaction. CPAP activity is antagonized by CP110, a distal centriole capping protein: CP110 depletion mimics CPAP overexpression, triggering procentriole overelongation (Schmidt et al., 2009). Similar defects, including abnormal daughter centriole elongation, arise from nocodazole-induced MT depolymerization (Alieva and Vorobjev, 1989c, 1991) or CP110 displacement. Excessive CPAP also induces supernumerary procentrioles near elongated centrioles, leading to multipolar spindles and cytokinesis failures (Kohlmaier et al., 2009). Thus, CPAP governs both procentriole elongation (Tang et al., 2009) and numerical control (Kohlmaier et al., 2009).

Chapter 8

The Role of the Centrosome in the Organization of the Microtubule System in the Cell

Rustem E. Uzbekov*, **PhD, HDR, IR**
Laboratory of Cell Biology and Electron Microscopy, Faculty of Medicine, University of Tours, Boulevard Tonnellé, France;
Faculty of Bioengineering and Bioinformatics, Lomonosov Moscow State University, Moscow, Russia

Historically, cell biology has established that the centrosome serves as the dominant MT-organizing center (MTOC) in most cells. During mitosis, the centrosome nucleates MTs to form the mitotic spindle, while in interphase, it organizes a radial MT network (Figure 8.1., 8.2.).

The canonical radial MT organization model, widely accepted since Porter's electron microscopy studies (1966), posits that MTs emanate from the centrosome, positioned near the cell's geometric center, toward the periphery (Karsenti et al., 1984a; Doxey, 2001; Desai and Mitchison, 1997; Keating and Borisy, 1999). MTs, as polar polymers, exhibit distinct plus and minus ends. Their minus ends remain anchored to the centrosome, which stabilizes them against disassembly (McIntosh and Euteneuer, 1984; Keating and Borisy, 1999; Moritz and Agard, 2001; Howard and Hyman, 2003). This "authoritarian" centrosome-centric model (Burakov and Nadezhdina, 2006) applies to cells such as CHO and NRK, where nearly all MTs are centrosome-bound (Komarova et al., 2002a). Free MT minus ends in these cells arise infrequently, either through MT detachment from the centrosome (Keating et al., 1997; Waterman-Storer and Salmon, 1997) or MT breakage (Dye et al.,

* Corresponding Author's Email: rustem.uzbekov@univ-tours.fr

In: The Centrosome
Editor: Rustem E. Uzbekov
ISBN: 979-8-89530-815-8

1992; McNally and Vale, 1993). However, both processes are rare (Keating et al., 1997; McNally et al., 2000; Vorobjev et al., 1997; Waterman-Storer and Salmon, 1997), making plus-end dynamics the primary determinant of MT network architecture. For example, in CHO cells (~20 μm radius), 68% of MTs extend to the cell edge without pauses or depolymerization within 1–2 minutes (Komarova et al., 2002).

Centrosome-independent MT nucleation occurs during specific differentiation processes. In syncytial myotubes, MT nucleation shifts to nuclear surfaces as centrosomes disassemble (Tassin et al., 1985; Connolly et al., 1986; Bugnard et al., 2005). Similarly, polarized epithelial cells (Reilein and Nelson, 2005), pigment cell melanosomes (Malikov et al., 2004), and dipteran salivary gland cells exhibit MT convergence at nuclear membrane foci and cytoplasmic sites (Abumuslimov et al., 1991). This parallels plant cells, where centrosome-absent MTs originate from nuclear membranes (Lambert, 1993). Recent studies reveal additional MT-organizing centers, including the Golgi apparatus, even in centrosome-containing cells (Chabin-Brion et al., 2001; Efimov et al., 2007).

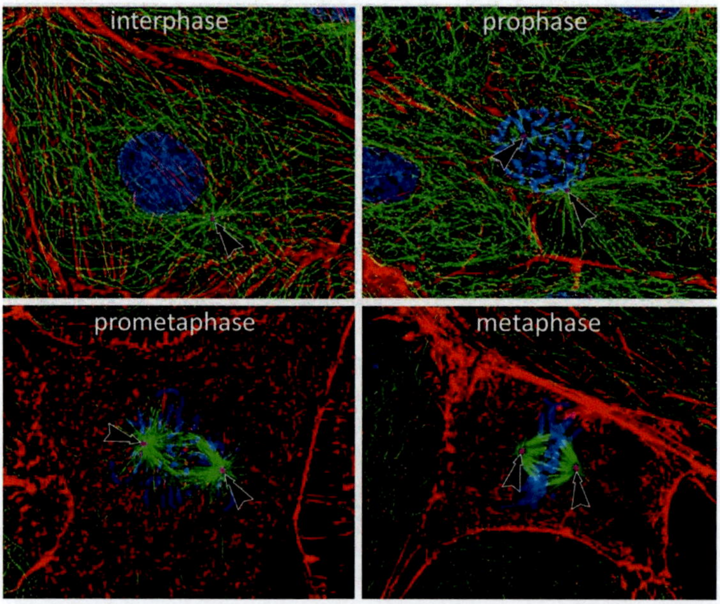

Figure 8.1. The MT system in the PtK1 cell line at the end of interphase and at the initial stages of mitosis. MTs are shown in green, actin in red, DNA staining (DAPI) in blue, and gamma-tubulin staining revealing the centrosome (shown with arrows) in purple. Photographs were kindly provided by A. L. Khodjakov.

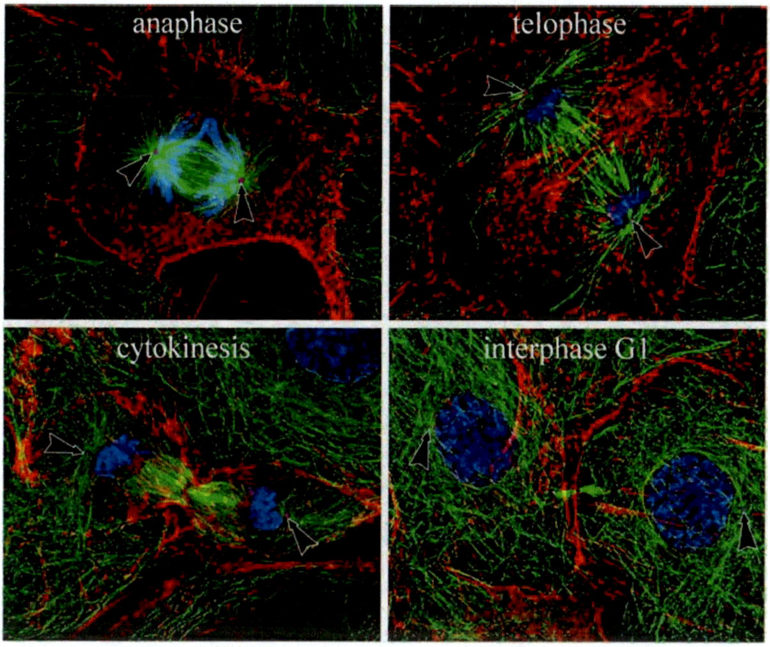

Figure 8.2. The MT system in a PtK1 cell at the end of mitosis and at the beginning of interphase. MTs are shown in green, actin in red, DNA staining (DAPI) in blue, and gamma tubulin staining in purple, revealing centrosomes (arrows). Photographs were kindly provided by A. L. Khodjakov.

Free MTs, uncoupled from centrosomes, dominate many specialized cells (Bre et al., 1987; Rodionov et al., 1999; Mogensen et al., 2000; Bartolini and Gundersen, 2006), often exceeding centrosome-bound MTs in number (Vorobjev et al., 1997). These populations exhibit distinct organizational patterns within single cells (Mogensen, 2004; Bartolini and Gundersen, 2006).

The centrosome anchors MT minus ends while permitting plus-end elongation via tubulin subunit addition (McIntosh and Euteneuer, 1984; Moritz and Agard, 2001; Howard and Hyman, 2003). This architecture characterizes keratocytes (Vorobjev et al., 2003), melanophores (Rodionov et al., 1994, 1998), and cultured CHO/NRK cells with predominantly centrosome-anchored MTs (Komarova et al., 2002). The γ-tubulin ring complex (γ-TuRC) mediates both MT nucleation and anchoring (Moritz et al., 1998, 2000; Oegema et al., 1999), while the ninein complex solely anchors MTs (Mogensen, 2004). Additional centrosomal proteins contribute to these functions (Dammermann and Merdes, 2002; Uetake et al., 2004; Delgehyr et al., 2005; Yan et al., 2006). In differentiated cells, cytoplasmic localization of these complexes dictates MT system organization. For instance, apical-basal MT orientation in polarized epithelia requires apical anchoring complex redistribution (Mogensen et al., 2002; Mogensen, 2004; Bartolini and Gundersen, 2006).

Thus, the centrosome's role in interphase MT organization is context-dependent, contrasting its mitotic dominance and challenging mid-20th-century assumptions.

Chapter 9

In Quest of the Nature of the Centrosome Centering

Anton Burakov[*], PhD
A. N. Belozersky Institute of Physico-Chemical Biology, Lomonosov Moscow State University, Moscow, Russia

This chapter describes the fascinating history of biologists' search for the reasons why one particular organelle got its name. This history has been as intriguing as the history of cell biology as a whole and had its ups and downs and unexpected twists in the plot. Let's start by asking the first question:

9.1. Is the Centrosome Deservedly Named?

In his short communication published at the end of the 19th century (Boveri T., 1887b) Theodor Boveri wrote that «the centrosome represents the dynamic centre of the cell; its division creates the centres of the forming daughter cells, around which all other cellular components arrange themselves symmetrically" (Scheer U., 2014). These words were written long before the discovery of microtubules, since the term "microtubules" itself was introduced in 1963 by Myron Ledbetter and Keith Porter (Hepler et al., 2013), and therefore long before the discovery of the function of the centrosome as the center of their organization.

It was later discovered that the centrosome is a membraneless organelle that functions as the main organizer of the microtubule cytoskeleton in all

[*] Corresponding Author's Email: antburakov@libro.genebee.msu.ru

In: The Centrosome
Editor: Rustem E. Uzbekov
ISBN: 979-8-89530-815-8

eukaryotes (Azimzadeh and Bornens, 2007; Bornens, 2021). The point at the center of microtubule convergence in interphase cultured cells became a familiar sight, and the name of the organelle, literally meaning "central body," turned out to be completely natural, although there was ample evidence of its movement in various cells (see below). Rarely does anyone bother to confirm the truisms. Therefore, there are few experimental studies aimed at statistically establishing how true the statement is that the "central body" is in the center. An example of a thorough analysis of the centrosome location is a work in which the authors analyzed mouse embryonic fibroblasts growing in a sparse or confluent monolayer, as well as the either disc-shaped or triangular cells produced by the micropatterning (Hale et al., 2011). The authors studied the position of the centrosome in the cells relative to the centroid, that is, the center of mass, or, in other words, a point inside a geometric figure whose coordinates correspond to the arithmetic mean of the coordinates of all points of the figure. It turned out that the statement about the central position of the centrosomes is most true for cells in a confluent monolayer, as well as for the round-shaped cells, where the centrosome was located closer to the centroid than in triangular cells or cells of a different shape in a sparse culture. The centrosome was displaced from the centroid by less than 10% of the average cell radius in approximately 75% of round cells and only 40% of triangular cells (Hale et al., 2011). Overall, these data illustrate well the thesis that the centrosome is indeed usually located in the central region of the cell, but also that there are forces that move it as needed.

Here it is necessary to stipulate that usually, speaking about the positioning of the centrosome and its central position, it means its location on the cell area in a plane parallel to the substrate. However, even if we take as an example a single layer of cubic or cylindrical epithelium, we can distinguish the basal-apical and orthogonal planar planes of cell projection. The central position of the centrosome in the cell on the planar plane does not at all mean its symmetrical location on the orthogonal plane (Figure 9.1a). Indeed, the centrosome is usually localized in the apical part of the epithelial cells, i.e., on the side of the cells facing the organ cavity and remote from the intercellular matrix. In poorly differentiated epithelial cells, for example, in the intestinal crypt, the centrosome is located closer to the center of the cells, and as differentiation proceeds, for example, in the intestinal villi, it moves to the apical part (Komarova and Vorob'ev, 1993; Tang and Marshall, 2012; Burute et al., 2017; Goldspink et al., 2017a; 2017b).

Another example of a non-central centrosome location when viewed from the side rather than from above is its location at the base of the primary cilium, i.e., directly under the plasma membrane, rather than in the depths of the cell. For example, when cilia are induced in cultured cells by serum starvation, the centrosome also shifts to the part of the cell that corresponds to the apical side of the epithelium, i.e., furthest from the substrate (Pitaval et al., 2017). Conversely, in actively proliferating cultured cells, the centrosome is usually located in the part of the cell that is closest to the substrate. If we look beyond cultured cells, in the tissues of the body the centrosome forms the primary cilium, projecting above the surface of the epithelial or endothelial layer or into the nephron duct (Dionne et al., 2018).

Despite this, in general, when observing non-polarized immotile cells with an upright light microscope, we come to the conclusion that the centrosome indeed has deservedly received its name. And the next question immediately arises.

9.2. What is the Reason for a Non-random Centrosome Position?

So, if the centrosome is indeed often located in the central region of the cell, what makes it there? One possible answer suggests itself at the first glance at an interphase cell, when the presence of a massive organelle essential for all eukaryotes and often located in the central region immediately becomes obvious, namely, the cell nucleus. Indeed, the centrosome is usually located near the nucleus, so it is easy to suggest that such a small organelle as the centrosome can simply stick to the nuclear envelope and thus occupy a position in the central region as a passive passenger. But if the connection between the centrosome and nucleus does not provide exhaustive answers to all questions about the centrosome position, then it is necessary to recognize the presence of some forces that could move the centrosome independently of the nucleus.

Let us first consider the first answer, that is, the connection of the centrosome with the nucleus. Indeed, the thesis of the connection of the centrosome with the nucleus has all the grounds. The centrosome is indeed specifically connected to the nucleus by two coexisting mechanisms: one involves microtubules and dynein, and another is mediated by different types of molecular scaffolds, including actin microfilaments (Maro and Bornens, 1980; Burakov and Nadezhdina, 2013), and is usually located at a distance of about 0.2 μm from the nuclear envelope. In addition to the fact that the connection of the nucleus and centrosome is carried out by microtubules and actin, the location in the cell of both the nucleus and the centrosome is determined by the coordinated action of cytoskeletal structures (Kiss et al., 2014). When the nucleus has a multilobulated morphology, as in neutrophils, the centrosome is located inside the nucleus courtyard (Uzbekov et al., 1989; Chiplonkar et al., 1992). The relative position of the nucleus and centrosome was also studied in the above-mentioned model of round-shape and triangular cells grown on a micropatterned substrate (Hale et al., 2011). It was shown that the connection between the centrosome and the nucleus weakens and the distance between these organelles increases upon disruption of microtubules by nocodazole, actin microfilaments by latrunculin B, and upon depletion of lamins A of the nuclear membrane - that is, in all cases when cytoskeletal structures could not attach to the nucleus. In triangular cells, the distance between the centrosome and the nucleus increased, particularly when the protein kinase GSK-3, which in turn regulates dynein activity, was inhibited. Knockout of the tubulin chaperone protein TBCCD1 was also shown to increase the distance between the centrosome and the nucleus (Gonçalves et al., 2010). In addition, there is evidence that centrosome attachment to the nucleus can moderate changes in the forces acting on it (Salpingidou et al., 2007; Lombardi et al., 2011). Thus, the nucleus and the centrosome are indeed linked to each other, and in some cases the centrosome does indeed act as a passive satellite of the nucleus, attached to its envelope. But another thing is also true - the presence of an undoubted connection between the centrosome and the nucleus does not in any way cancel the presence of forces that move the centrosome around the cell independently. And the existence of such forces has been clearly illustrated by numerous and well-documented studies.

It is well known that cell nuclei are capable of moving through the cytoplasm. In general, the movement of the nucleus through the cell depends on both actomyosin and microtubules and dynein, depending on the direction of movement (Zhu et al., 2017). At the same time, independent movement of the centrosome and nucleus has been demonstrated during neuronal migration, when the centrosome moves uniformly, remaining in the center of the cell, while the nucleus can perform saltatory movements, either ahead of the centrosome or lagging behind it (Distel et al., 2010). Similar oscillatory movements are performed by the nucleus and centrosome when glioma cells move along thin lines coated with fibronectin (Kiss et al., 2014). A good illustration of the movement of the centrosome relative to the nucleus is its displacement to the leading edge of the motile fibroblasts or astrocytes (Luxton and Gundersen, 2011; Barker et al., 2016). This location of the centrosome is usually shown in schematic drawings of a moving fibroblast. At the same time, in some cell types or when cells move along certain substrates (for example, when fibroblasts move along thin strips of fibronectin substrate surrounded by a non-adhesive surface), the centrosome can even be localized at the back of the cell behind the nucleus (Luxton and Gundersen, 2011; Pouthas et al., 2008; Doyle et al., 2009; Zhang and Wang, 2017). All these results indicate that centrosomes and nuclei in a wide variety of cells often move in the cytoplasm asynchronously.

Displacement of the centrosomes relative to nuclei can also be observed in several other processes, such as planar cell polarity (PCP), i.e., uniform polarization of cells in the plane of the cell sheet (Jones and Chen., 2007; Goodrich and Strutt, 2011; Carvajal-Gonzalez et al., 2016; Roman et al., 2019). In PCP, centrosomes are shifted to one edge of the cells, which is observed, for example, in Drosophila wing cells, where centrosomes are shifted to the distal edges of the cells. In vascular endothelium, centrosomes are usually shifted to the edge of the cells facing the heart (Rogers et al., 1985). Some researchers noted that the centrosomal displacement, which is necessary for their subsequent proximal-distal polarization during PCP development, is actin-dependent (Garrido-Jimenez et al., 2018), and also that actin filaments are involved in the centrosome movement to the apical region of the cell (Haag et al., 2018).

Another example of centrosome displacement relative to nuclei is the epithelial-mesenchymal transition (EMT), which accompanies embryogenesis and carcinogenesis, in which the centrosome in an epithelial cell moves from the apical to the basal part, which becomes the anterior part during further migration through the extracellular matrix (Burute et al., 2017).

The third example of the displacement of the centrosome relative to the nucleus is the formation of an immune synapse, when the centrosome is displaced to the caudal region of the lymphocyte, which is accompanied by the destruction of the connection of the centrosome with the nucleus (Kupfer et al., 1986; Sancho et al., 2002; Martín-Cófreces et al., 2008; Martín-Cófreces et al., 2011; Kloc et al., 2014). After the establishment of the T-cell immune synapse, the centrosome undergoes extreme relocalization from above the nucleus to the membrane. This centrosome positioning allows for minus-end-directed transport of cytotoxic vesicles along centrosomal microtubules, facilitating secretion into the synapse (Geiger et al., 1982; Stinchcombe et al., 2006). This centrosome displacement toward the immune synapse is usually compared to the centrosome displacement toward the apical edge of the cell during cilia formation (Cassioli and Baldari, 2019) rather than to fibroblast movement, where the centrosome is often displaced toward the leading edge (Barker et al., 2016).

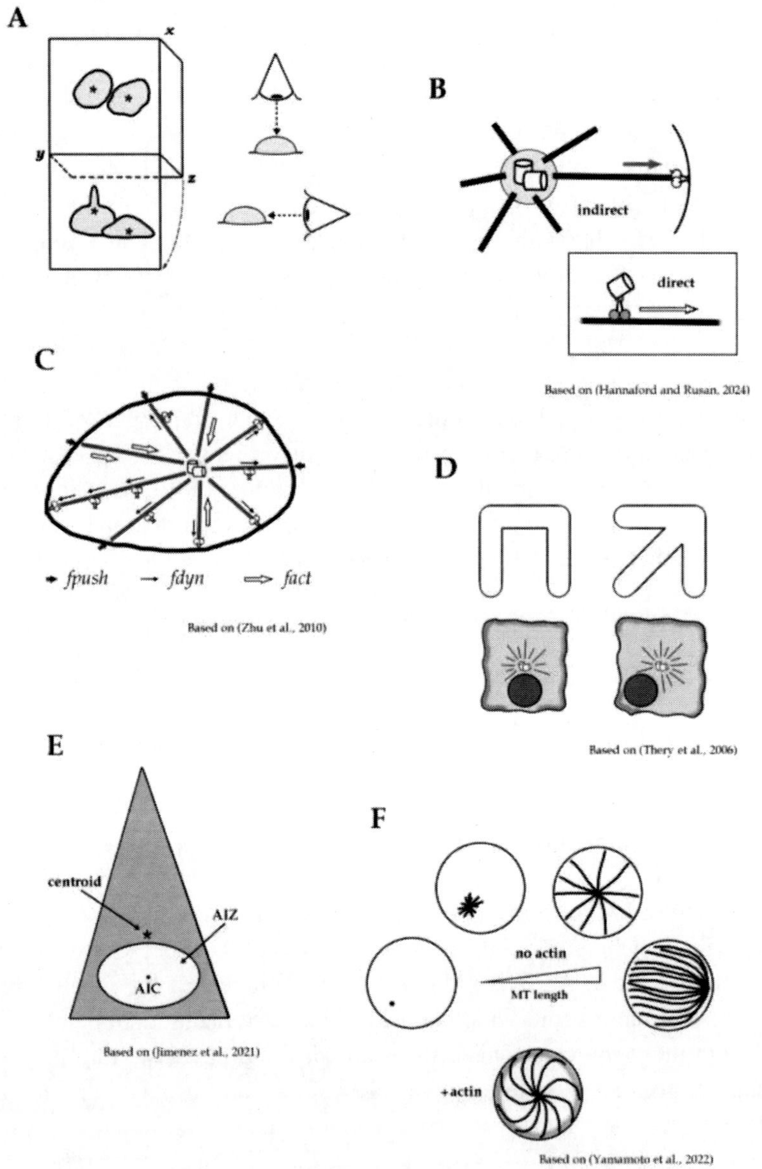

Figure 9.1. Key concepts in discussing the nature of centrosome positioning. *A*. The central position of the centrosome can be relative, i.e., differ in different planar planes, and depend on the position of the observer. Thus, in a centrosome shifting

from the basal to the apical part of the fibroblast during ciliogenesis, its position remains central in XY coordinates while becoming asymmetric along the Z axis.
B. Indirect and direct transport of the centrosome. The first one occurs when a force is applied to a microtubule attached to the centrosome - for example, when dynein, anchored in the cell cortex, pulls the centrosome towards itself. The second variant is observed when the centrosome acts as a cargo moved by motor proteins along microtubules - as happens when kinesin is moving the centriole toward the microtubule plus end.
C. Distribution of three principal types of forces applied to centrosomal microtubules: length-dependent force fdyn by outward-pulling dyneins, actin–myosin drag force fact, and decentering inward-pushing force generating by microtubule dynamics fpush.
D. Nucleus positioning depends on actin polarization. Cells were seeded on adhesive micropatterns of different forms (shown on the top). The nucleus-centrosome axis is directed toward more adhesions (dark gray border) with the centrosome located in the centroid and the nucleus shifted toward free edges (edge without border).
E. Centrosome with aster of microtubules is located not in the cell centroid but in the center of the inner region defined by the acto-myosin network, which is devoid of actomyosin bundles and enriched with dynein. In non-round cells, the actin inner zone (AIZ) may be located asymmetrically, and the actin inner center (AIC) may be noticeably shifted relative to the centroid.
F. The success of artificial centrosome positioning in microwell by pushing forces generated by microtubule plus-ends critically depends on the length of microtubules, which should roughly correspond to the radius of microwell. At the same time, the symmetric cortical branched actin meshwork favors microtubule aster centering and makes this process much less sensitive to microtubule length.

The final example of a natural biological process that demonstrates the presence of forces capable of moving the centrosomes, which is absolutely not dependent on the presence of a nucleus (simply because of its absence), is the movement of the spindle poles in dividing cells. The movement of centrosomes during determining the position of the spindle poles in stem cell mitosis was studied in the works of Yukiko Yamashita with co-authors (Yamashita et al., 2003; Chen et al., 2018). These works showed that a Drosophila male germ cell adheres tightly to its niche (epithelium) and divides strictly perpendicular to it, with the proximal daughter cell remaining in the niche as a stem cell, while the distal one begins to differentiate - and such cells do not begin to divide until the spindle poles take their correct (proximal and distal) position (Venkei and Yamashita, 2015), and the positioning of the centrosome in this system occurs with the participation of external signals emanating from the niche cells (Chen et al., 2018). At the end of the conversation, one cannot help but mention another well-known example of active movement of centrioles independent of the nucleus: the post-telophase migration of the mother centriole along midbody microtubules toward the site of abscission and then back to the cell center as cytokinesis is completed (Piel et al., 2000). We will not debate the centrosome movements any further here, which were discussed in detail in a very informative review of Hannaford and Rusan (Hannaford and Rusan, 2024).

All the above examples serve as evidence that although the centrosome is indeed linked to the nucleus and can sometimes follow it as a passive passenger, there are also mechanisms in cells that cause its movement independently. In this case, the connection with the nucleus will be a hindrance and hide the presence of these forces, thus complicating the movement of the centrosome. This problem can be solved by studying the centrosomes in cells lacking a nucleus, i.e., cytoplasts. Such an analysis was carried out several years ago in the work of Jimenez with co-authors (Jimenez et al., 2021), who studied the location of the centrosomes in cytoplasts, which in addition were spread out on a micropatterned substrate, predetermined its shape. We will return to the results of this work below, and having answered positively to the question of the existence of forces that localize the centrosomes independently of the nucleus, we will now ask about.

9.3. Physical Nature of the Forces that Center the Centrosome

As we have seen, it has long been obvious that the centrosome not only holds on to the nucleus and thus passively accompanies its movements, but is also capable of moving independently of the nucleus and at the same time occupying a central position in the cells. This suggested that such a central position is ensured by the work of some special mechanisms that coordinate the forces that ensure the movement of the centrosomes. As the methods used in cell biology developed, researchers received more and more new data on this issue, and scientists' ideas about the mechanisms of centrosome positioning thus developed progressively.

The major role of the centrosome in interphase cells, aside from the positioning external structures such as cilia and flagella, is to act as a microtubule organizing center, as was captured in early drawings long before microtubules as such were described (Wilson, 1925). Since the interphase centrosome is often located at the center of a symmetrical microtubule aster, it was logical to assume that the location of this organelle, from which numerous dynamic structures emanate, somehow depends on them. Therefore, the attention of researchers was primarily focused on the positioning of the centrosome as the center of the microtubule aster, rather than as a separate discrete organelle. Indeed, it was shown that it is the microtubule aster that holds the centrosome in place in cells when the cell is enucleated by centrifugation in the presence of cytochalasin (Mitchison and Kirschner, 1984). Centrosome positioning has been shown later to require a polarized radial array of cytoplasmic microtubules (Euteneuer and Schliwa, 1992; Koonce et al., 1999). Following the discovery of dynamic microtubule instability and the existence of various motor proteins, it was proposed that the forces acting on cytoplasmic MTs might include pushing by microtubule growth at the plus ends (Inoue and Salmon, 1995), since microtubule growth can generate significant mechanical force (Dogterom and Yurke, 1997). Then it was shown that microtubule polymerization forces are sufficient to center centrosome asters *in vitro* (Faivre-Moskalenko and Dogterom, 2002). Two another forces, also applied to the centrosomes along the microtubules, were considered: the second force was the pulling applied to the microtubules by motor proteins anchored in the cortex (Dujardin and Vallee, 2002; Gundersen, 2002) or in the cytoplasm (Hamaguchi and Hiramoto. 1986), and the third force was actin dependent movement (Mikhailov and Gundersen, 1995.; Waterman-Storer and Salmon, 1997; Yvon and Wadsworth, 2000; Salmon et al., 2002), in which the centripetal actin flow create a significant mechanical force leading to centripetal movement of microtubules in contact with actin filaments (Salmon et al., 2002).

Numerous studies have been published in support of dynein's involvement in centrosome positioning. Early works showed that dynein was required for centrosome reorientation during cell polarization, for the asymmetric positioning of mitotic spindles in S. cerevisiae and *C. elegans* (Eshel et al., 1993; Li et al., 1993; Skop and White, 1998; Gonczy et al., 1999), and for the positioning of astral microtubules and mitotic spindles in *Dictyostelium* amoeba and mammalian cells (Koonce et al., 1999; O'Connell and Wang, 2000). Numerous studies have shown that inhibition of dynein resulted in imbalance leading to microtubule bending and incorrect centrosome positioning (Holey et al., 1997; Koonce et al., 1999; Wu et al., 2011). An additional argument in favor of considering dynein as one of the main participants in centrosome positioning was its proven involvement in centrosome movement during the formation of the immune synapse. It is now known that the displacement of the centrosome during the formation of the immune synapse or primary cilium is accompanied by the accumulation of dynein in the plasma membrane region where the centrosome is displaced. This accumulation of dynein is due to diacylglycerol in the plasma membrane and, reciprocally, the distribution of myosin (Liu et al., 2013). Cortical dynein accumulates at the center of the immune synapse, where it interacts with centrosomal microtubules, leading to centrosome migration (Combs et al., 2006; Yi et al., 2013; Sanchez et al., 2019; Gros et al., 2021). The microtubule tip protein complex (+TIP) is also activated and promotes microtubule capture by dynein via a search-and-capture mechanism (Lomakin et al., 2009) and the development of traction forces (Yi et al., 2013; Sarkar et al., 2019). It should be noted that the reorganization of the actin cytoskeleton also plays a significant role in the formation of the immune synapse. When lymphocytes are activated, the Arp2/3 complex moves from the centrosome to the immune synapse, which leads to the detachment of the centrosome from the nucleus and facilitates its movement to the synapse area (Obino et al., 2016).

But let us return to the positioning of the centrosome in the center of the cell. Before continuing to analyze the above three forces and their influence on the position of the microtubule aster, we make an important reservation—we are talking here about the centrosome as active organelles that act as microtubule organizing centers. However, no one prevents an inactive centrosome from acting as a regular cargo that can be transported by motor proteins in much the same way as vesicles are moved during vesicular transport. In the review of the mechanisms of centrosome positioning (Hannaford and Rusan, 2024), the authors use the concepts of indirect and direct centrosome transport. In the first case, various factors, such as microtubule dynamics against some obstacles or dynein anchored somewhere, move centrosomes by acting on centrosomal microtubules. In the second case, centrioles lacking PCM or centrosomes lacking centrosomal microtubules move along the cytoskeleton as cargo - and authors call this centrosome transport "direct motor transport" (Figure 9.1B). This is observed in *C. elegans* sensory neurons and *D. melanogaster* neural stem cells, where centrioles move along the microtubule network via direct motor transport, and this requires direct interactions between centriole components and the motor proteins dynein or kinesin-1 (Li et al., 2017; Hannaford et al., 2022). In any case, in the discussion below we consider only indirect transport, since we are interested in the positioning of the functionally active centrosome, which, through the nucleation and anchoring, organizes a radial aster of dynamic microtubules.

Since all three mechanical forces listed above were presumably applied to the radial microtubules extending from the centrosome and oriented along them, it seemed logical to introduce an imbalance of forces into a symmetrical system with a central location of the microtubules aster in order to clarify the positioning mechanisms. For this purpose, microtubules were locally disrupted at one edge of the cells by bringing a microcapillary with a nocodazole to them, which led to a rapid local depolymerization of tubulin. These experiments were also accompanied by inhibition of the activity of dynein and actin centripetal flow, and as a result, it was concluded that all three forces act simultaneously on the centrosome: pushing forces generated by the growing microtubules plus ends and the actin flow created by myosin contraction, and pulling forces created by dynein (Burakov et al., 2003). Subsequently, in silico modeling was performed, where the distribution of dynein not only in the cell cortex but also along microtubules was taken into account (Zhu et al., 2010). A role of dynein in centrosome centering was supported by several experimental studies, in which dynein activity was inhibited by small-molecule inhibitors or by introducing inhibitory proteins into the cell (e.g., the CC1 fragment of the p150Glued protein), and a sharp shift of the centrosome from the center was observed (Palazzo et al., 2001; Burakov et al., 2003; Brito et al., 2005; Wu et al., 2011). It has been confirmed that the cell cortex in cultured cells contains many dynein molecules coupled with its cofactor dynactin (Mazel et al., 2014) and, in addition, many dynein molecules are located on intracellular membrane structures. When *Dictyostelium* has two centrosomes in a binucleate cell, they form two independent stars, each occupying its own territory in the cytoplasm. Laser ablation of one of the stars leads to a rapid centering of the second, which was well explained by the action of pulling forces from dynein (Odell et al., 2019).

As a result, a servomechanical model was proposed where the self-regulating mechanism of centrosome positioning in the cell centroid was based on three forces applied along the radial microtubules attached to the centrosome. The first force is the pulling force developed by cytoplasmic dynein anchored both in cell cortex and in the deep regions of the cytoplasm. The second is dragging of the microtubules toward the cell center by an actomyosin-driven actin flow. And the pushing forces generating by growing plus ends of dynamic microtubules which can encounter obstacles in the cytoplasm or in the cell cortex and thus push the centrosome toward the center; modeling showed that this factor rather introduces a destabilizing effect into the static position of the centrosome. The general equation of these three forces can be written as follows:

$$Ftot = Fdyn + Fact + Fpush$$

where *Fdyn*, *Fact* and *Fpush* are the sum of the forces from dynein, actin centripetal flow, and microtubule plus-end pushing forces, respectively.

Obviously, *Fpush* depends on the *fpush*, developed by a single microtubule, on the number of dynamic microtubules and the depth of dynamic instability; *Fact* depends on *fact* (the effect of actin flow per unit length of microtubules) and the length of microtubules to the second degree; *Fdyn* depends on *fdyn* (the strength of one dynein molecule) and the length of microtubules (Figure 9.1C).

To simplify, if the centrosome is shifted from the center of the cell, for example, to the right, then it has longer microtubules on the left, and then, both the outward (pulling to the left) *Fdyn* and inward (pushing to the right)

Fact are larger on the left; shorter microtubules on the right develop increased *Fpush*, pushing to the left. The solution of the derived equations with parameters satisfying the experimental data was carried out. In the best way, they corresponded per one microtubule to *fdyn* proportional to 3 *fpush*, and fact proportional to 8 *fpush*. In a computer simulation, the centrosome behavior was fully reproduced. The centering mechanism was predicted to be robust: all that was needed for centrosome centering was for the total dynein force to be greater than 1 motor traction force per microtubule. Centrosome centering according to this mechanism was reproduced in silico in round, elliptical, square, and fan-shaped cell models. An interesting suggestion was made, which was subsequently reproduced *in vivo:* when dynein was inhibited, the centrosome shifted to the long edge of an elongated cell. This was a consequence of the strict dependence of the Fact value on the distance to the cell edge in this model (Zhu et al., 2010).

Fact are larger on the left; shorter microtubules on the right develop increased *Fpush*, pushing to the left. The solution of the derived equations with parameters satisfying the experimental data was carried out. In the best way, they corresponded per one microtubule to *fdyn* proportional to 3 *fpush*, and fact proportional to 8 *fpush*. In a computer simulation, the centrosome behavior was fully reproduced. The centering mechanism was predicted to be robust: all that was needed for centrosome centering was for the total dynein force to be greater than 1 motor traction force per microtubule. Centrosome centering according to this mechanism was reproduced in silico in round, elliptical, square, and fan-shaped cell models. An interesting suggestion was made, which was subsequently reproduced *in vivo*: when dynein was inhibited, the centrosome shifted to the long edge of an elongated cell. This was a consequence of the strict dependence of the Fact value on the distance to the cell edge in this model (Zhu et al., 2010).

This model had no obvious contradictions, but there were some questions that required further clarification. Namely, should astral microtubules terminate strictly on the plasma membrane, or can they slip along it? What is the role of other microtubule motor proteins that can develop pushing forces? Can/should microtubules be twisted near the centrosome? Can/should they bend under the action of a pushing force? And can dynein really develop such significant forces?

9.4. Further Development and Refinement of the Model

The authors also examined cells seeded in aAs mentioned above, to disturb the central position of centrosomes in cells, it is necessary to disrupt the symmetry of the dynein or its activating factors arrangement and/or the actomyosin system. Supporting data were obtained using cells growing on a micropatterned substrate (Théry et al., 2006). The cells were seeded on crossbow-shaped adhesive patterns, which forced them to take on the shape of a fan, with the sides being unattached to the substrate. This resulted in an asymmetric distribution of actin, actin-binding proteins, and microtubule plus ends. The nucleus in these cells was displaced from the centroid to the narrow part of the cell, and the nucleus-centrosome axis was formed in such a way that it was directed from the narrow to the wide part of the fan. The authors also examined cells seeded in adhesive patterns in the form of the letters X, Π, and K, and the form of short arrows. In all cases, the cells had a square shape, but their non-attached to the substrate edges were located differently (Figure 9.1D). It turned out that the nucleus-centrosome axis was in all cases directed toward the greater number of adhesions, with the centrosome always located at the cell centroid and the nucleus shifted toward the free edges. The authors suggested that actin- and microtubule-binding proteins, such as APC, are concentrated at the adhesive edges of the cells and capture the plus ends of the microtubules, arresting their growth and subjecting them to tensile forces. Along the non-adherent edges of the cell, microtubules continue to grow without binding to the cortex until they reach the next adhesive zone (Théry et al., 2006).

A whole series of subsequent studies were focused on dynein and microtubule dynamics and relatively little on the role of the actomyosin system in centrosome behavior, although, as we already have seen, this is a significant factor. Many aspects of the forces applied to the microtubules have been studied by in silico modeling or *in vitro* experiments using microchambers loaded with isolated centrosomes and microtubule proteins.

In one of the works, the authors focused on the bending of microtubules after their disruption by laser ablation and showed that this bending is determined by tensile and buckling dynein forces applied along the entire length of the microtubules (Wu et al., 2011). In silico modeling demonstrated that in a small cell, the centrosome can be placed into the center just by microtubule dynamics (i.e., pushing forces) alone, although in larger cells and a simulated viscous medium such as the cytoplasm, the pulling by dynein must be applied (Wu et al., 2011). In an *in vitro* system, centrosomes and tubulin were placed into microfabricated chambers or liposomes, where a microtubule star was formed, and the walls of the chambers or liposomes were coated with dynein on the inside surface (Laan et al., 2012a, b). Dynein has been shown to activate microtubule catastrophes growing toward a barrier coated with it, and it significantly improves the centering of the aster in a small volume. Asymmetry in the distribution of dynein leads to the decentering of such an aster. On the other hand, a special problem remained to explain the process of centrosome centering in very large cells - for example, the movement of the spermal aster associated with the male pronucleus and formed by the centrosome of the spermatozoon to the female pronucleus located in the center of the oocyte of the sea urchin, *Xenopus* or *C. elegans* (Wühr et al., 2009). Microtubules of the spermal aster may not reach the oocyte cortex with their ends, but the aster together with the pronucleus moves towards the oocyte center at a fairly high speed (Shinar et al., 2011; Longoria and Shubeita, 2013; Tanimoto et al., 2016; Tanimoto et al., 2018). It was reported that this movement depends on microtubules and dynein but does not depend on actomyosin (Tanimoto et al., 2016; Tanimoto et al., 2018). So, it was concluded that such movement of the microtubule aster may be due to dynein-dependent movement of small vesicles along microtubules to the aster center; vesicles move inward, and reactive forces pull microtubules outward (Shinar et al., 2011; Longoria and Shubeita, 2013; Kimura and Kimura, 2011a,b). Mutations and knockdown of genes encoding several proteins involved in minus-end transport (dynein light chain DYRB-1, rilp-1, rab-7, and rab-5) in *C. elegans* resulted in the absence of centrosomal centering in the zygote (Kimura and Kimura, 2011a, 2011b). It should be noted that the spermal aster moves in the cytoplasm of the zygote along with the male pronucleus, i.e., the entire structure has a significant mass. Under this condition, microtubule pushing forces should be clearly insufficient for the movement, and calculations indicate that the centrosome alignment process cannot proceed without dynein action, with a dozen dynein molecules being sufficient to generate the required forces (De Simone et al., 2018). A recent study used targeted segregation of specific microtubule populations and cytoplasmic flow analysis to determine the precise contributions of dynein-mediated cytoplasmic pulling forces, cortical pulling forces, and microtubule pulling forces to centrosome alignment. This study suggests that dynein-mediated cortical pulling forces are the primary mechanism of centrosome alignment in the single-cell *Caenorhabditis elegans* embryo (Wu et al., 2023).

In another study, the *in silico* modeling of centrosome centering in cells of various shapes was performed without taking into account actomyosin forces but taking into account the possibility of microtubule pivoting around the centrosome and their gliding along the cell cortex (Letort et al., 2016). In all cases, the modeled microtubules were longer than the cell radius. Modeling showed that the microtubule aster moves toward the center of a round-shaped cell due to the pulling force of dynein, generated by both cortically located dynein and dynein distributed throughout the cytoplasm. Using models of ellipsoidal, rectangular, and triangular cells, it was calculated that with a uniform distribution of dynein in the cytoplasm, it is possible to achieve a close-to-central position of the centrosome in all cases. However, the greatest distance from the center of the aster to the cell centroid was obtained in triangles, especially in a triangle with a wide base. With the cortical arrangement of dynein in triangles, the simulated center of the microtubule aster was always shifted from the centroid. It also turned out that centering due to pushing forces is possible only in the absence of pivoting and sliding of microtubules, i.e., with rigid fixation of both ends of the microtubules. In general, the centering of the aster was facilitated by decreasing the rigidity of the microtubules, increasing their dynamics, and decreasing their number. The introduction of asymmetric pulling forces created by dynein located on one side of the round cell led to a sharp decentering of the aster (Laan et al., 2012a, b). Thus, a consensus has been reached that a homogeneous distribution of minus-end microtubule motors in the cytoplasm or in the cell cortex can ensure centrosome centering (Kunse et al., 1999; Kimura and Kimura, 2011; Wu et al., 2011; Laan et al., 2012), while a heterogeneous distribution of motors due to local clusters can locally increase the pulling forces and ensure microtubule-organizing center decentering up to contact with the plasma membrane (Dujardin et al., 2003; Yi et al., 2013). The regulation of centrosome positioning by such pulling forces looks robust: *in silico* simulation demonstrate that the processes of centering, in the case of homogeneous motor distribution, or peripheral positioning, in the case of heterogeneous motor distribution are both weakly sensitive to the changes in microtubule length (Letort et al., 2016). In contrast, the centering mechanism based on microtubules pushing forces appears to be much less robust, as it is much more sensitive to the length of the centrosomal aster microtubules. Applying pushing forces to microtubules will be ineffective if they are too short to reach the cell boundary. Longer microtubules can provide centering if their length exactly matches the length of their confining region. However, microtubules that extend this critical length cause an abrupt transition from centering to decentering (Holy et al., 1997; Faivre-Moskalenko and Dogterom, 2002; Pinot et al., 2009; Laan et al., 2012).

9.5. Actin Network Turns out to be More Important than Initially Thought

It is obvious that even the most careful computer modeling cannot replace the setting up of experiments that yield new results. The role of the actin cytoskeleton was taken into account in the classical model of centrosome positioning, but gradually data began to accumulate that this role is much more significant than it seemed at first glance.

Quite a lot of data has been collected on the role of actin in the location of centrosomes in a cell. In particular, the separation of centrosomes before mitosis depends not only on the integrity of the microtubule network but also on the integrity of actin filaments (Uzbekov et al., 2002). Microtubules extending from the centrosome can interact with actin through either specific crosslinkers or non-specific steric interactions (Hebert et al., 2012). Thus, dense-growing actin networks can apply pushing forces to microtubules (St Johnston and Sanson, 2011; Nunes de Almeida et al., 2019). Over the past decade, many examples of physical interactions between MTs and actin filaments have been collected (Jiang et al., 2012; Lopez et al., 2014; Colin et al., 2018; Farhadi et al., 2018; Dogterom and Koenderink, 2019; Inoue et al., 2019). The nice evidence for the existence of direct microtubule-actin crosstalk has emerged recently: it has been shown that actin crosslinker protein anillin can directly crosslink microtubules and actin filaments, can cause sliding of actin filaments on the microtubule lattice, and can transport actin filaments by the growing microtubule tip (Bareja et al., 2025).

A major advance in the study of the actin cytoskeleton's influence on processes involving microtubules has been the development of an adhesive micropatterning technique, which has allowed researchers to control cell shape, adhesion, and spindle orientation (Théry, 2010; Fink et al., 2011). In its simplest form, this technique allows for the standardization of these parameters to prevent artifacts from influencing experimental results. However, micropatterning has been effectively used to study how modulation of these parameters affects centrosome positioning within the cell (Jimenez et al., 2021; Pitaval et al., 2010, 2017; Lima et al., 2024).

In the study mentioned above (Jimenez et al., 2021), authors studied the location of centrosomes in cytoplasts where the nucleus could not distort the location of the centrosome; and which, in addition, were spread out on a microstructured substrate that strictly predetermined their shape. It was shown that in an asymmetric cytoplast there is a small shift of the centrosome from the centroid, which is proportional to the asymmetry of the actin cytoskeleton. To describe the architecture of the actin cytoskeleton, the authors introduced new concepts of the actin inner zone (AIZ), i.e., the central zone of the cytoplast free of actin bundles, and the geometric center of the AIZ, the actin inner center (AIC) (Figure 9.1E). In a symmetric cell, the AIC is the same as the centroid. It turned out that the position of the centrosome corresponds to the position of the AIC in both symmetric and asymmetric cells. When the cell is treated with a ROCK kinase inhibitor (Y27632), which inhibits F-actin assembly and retrograde actin flow and removes actin bundles from the edges bordering the AIZ, actin is redistributed uniformly throughout the cell, while the AIC shifts and begins to coincide with the centroid—and the centrosome moves there too. Since, according to the current model, centrosome centering is due to forces applied to astral microtubules, the authors studied their architecture depending on actin localization and found that microtubules are straight and radial in the AIZ (indicating their tension and the force applied to them), but are tangled outside this zone. Thus, even if MTs are tangled outside the AIZ, their radiality in the central zone will ensure successful centering. Experiments with cytoplasts of different sizes, but with approximately the same AIZ, showed that the accuracy of centrosome positioning depends on the size of the AIZ and not on the overall cell size. When the microtubules disassemble and the pulling forces exerted by the peripherally located dynein/dynactin disappear, the centrosome leaves the AIC but not the AIZ and moves chaotically within it. Thus, the conclusion from this work was that the role of the actomyosin system is to determine the size and shape of the AIZ, which is the "legitimate field of action" of the radial microtubules that pull the centrosome due to dynein forces and thus place it at the center of the AIZ. In addition, a decade ago it was shown that the centrosome initiates the polymerization of not only microtubules but also actin filaments (Farina et al., 2016). Soon after, it was shown that actin filaments nucleating at or near the centrosome can influence the nucleation and growth of centrosomal microtubules (Inoue et al., 2019). They physically block the formation of centrosomal microtubules at the earliest stages of their polymerization. Thus, centrosomal actin can influence the magnitude of the microtubule pushing forces generated by newly formed microtubules. So, not only peripheral actin limits the AIZ and is involved in centrosome centering, but also centrosomal actin, which limits the nucleation of microtubules on the centrosome and thus influences the magnitude of the forces applied to it.

A further development of the research on the role of actin in centering processes was the work where authors reconstructed the interaction of dynamic microtubule asters with actin networks, conducting this analysis *in vitro* in cell-sized wells using purified proteins and microfabrication techniques (Yamamoto et al., 2022). Since the cell boundaries are formed by lipids, authors used a method of coating microwells with a lipid bilayer (Zieske and Schwille, 2014) and then covering the upper surface of the well with a mineral oil. In order to standardize the number of microtubules in the centrosome aster, the study was performed using artificial microtubule organizing centers (so-called aMTOCs) created from polystyrene beads with short stabilized microtubule stumps, which efficiently generated 15–20 dynamic MTs on each aMTOC. It was decided not to use a preparation of purified centrosomes, since purified centrosomes generate quite variable numbers of MTs (Inoue et al., 2019). Yamamoto with co-authors analyzed the efficiency of aMTOC centering depending on the ratio of microtubule length to the well radius, increasing the concentration of tubulin and thus increasing the length of microtubules extending from the aMTOCs. While the length of the aster microtubules was twice shorter than the well radius, aMTOC took a random position. At a higher tubulin concentration, when the radial microtubules reached the edges of the well, most of the aMTOCs reached the center of the well and remained there. However, as the concentration was further increased and the microtubule lengths began to significantly exceed the well radius, microtubule elongation and sliding along the well edges broke the symmetry of the network, and aMTOC moved away from the center (Figure 9.1F). These results are consistent with those previously reported in water-in-oil droplets (Pinot et al., 2009), but previous numerical simulations (see above) suggested that friction along the cell boundaries could prevent symmetry breaking by enhancing the vortex-like structure in the network that maintains MTOC centering (Letort et al., 2016). It is known that actin form dense networks that can restrict microtubules lateral translocation and aster displacement in the cells (Field and Lenart, 2011; Blanchoin et al., 2014). The authors of this study tested how actin networks affect aMTOC centering by growing actin filaments simply by spontaneous actin polymerization or by nucleation-promoting factor attached to the lipid in the presence of the Arp2/3 complex - to restrict the cytoplasmic pool and promote the assembly of a dense cortical actin layer in the vicinity of the lipid layer. When microtubules are not added to the system, unbranched actin filaments in bulk had no visible effect on polystyrene bead positioning, with bead positioning being slightly centered in the presence of cortical actin, as the thickness of the cortex limited the available space for beads, consistent with previous data (Jimenez et al., 2021).

The most interesting result of this work is that it was shown that the actin meshwork in the volume worsens the centering process and actually limits the movement of aMTOC at the optimal length of microtubules, while the unbranched actin meshwork did not affect the elongation of microtubules at all. At the same time, the cortical branched actin meshwork favors aster centering and stabilizes the position of aMTOC by limiting the slippage of long (longer than the well radius) microtubules, while asters from microtubules with a length equal to the cell radius still demonstrate the correct centering mechanism in the presence of cortical actin. These results suggest that cortical actin can provide reliable centering, which is much less sensitive to microtubule length, while an asymmetric cortical actin meshwork causes aster decentering. The main conclusions from this work were that actin networks make aster positioning both more robust and versatile, because in the actin-free system microtubule asters displayed transitions from centering to decentering depending on microtubules length. At the same time, aMTOC can hold its position even without the microtubules thanks to the bulk actin filaments, which resisted aster displacements. In turn, cortical actin networks favored aster centering over a wide range of aster microtubule lengths.

9.6. Flip Flop: Data Clarification Leads to a New Theory

Anyone who has studied physics knows that classical mechanics is a self-consistent theory, meaning that no statements within it contradict each other. However, there came a time when the sum total of accumulated observations led to the theory of relativity, which fundamentally changed the way physicists perceived the world and seriously limited the scope of application of Newton's laws. It seems that a similar event is currently taking place in the development of cell biologists' ideas about the mechanisms of centrosome centering.

In the generally accepted model of centrosome centering, which depends on the microtubule radial aster placed in the actin inner zone of the cell and subjected to forces from growing plus ends, anchored dynein, and the actomyosin network, the balance of real physical forces, however, has not been directly assessed, and the parameters involved in the centering have been logically deduced from the cell's response to biochemical changes or geometric constraints. Furthermore, in contrast to the classical radial arrangement of straight microtubules depicted in the drawings of Theodore Boveri (Wilson, 1925), usually quite knotty morphology of microtubules in the cells seems incompatible with most models based on pushing or pulling forces along them (Zhu et al., 2010; Letort et al., 2016). In their fresh work, Schaeffer and co-authors have studied in detail the origin and spatial distribution of mechanical forces acting on the centrosome in cultured cells, which led them to very interesting and surprising conclusions (Schaeffer et al., 2025).

The surprises began immediately when the authors used laser nanoablation to cut microtubules near the centrosome. Suddenly, it turned out that ablation of a few microtubules in both cells and cytoplasts had no immediate significant effect on the position of the centrosome. The authors then removed by laser nanoablation about half of the microtubules reaching the centrosome and continued to remove all microtubules reappearing in this region over the next few minutes, but this resulted in only a subtle shift of the centrosome less than half a micron over 3–5 minutes. Since these data challenged all previous models of the balance of forces at the centrosome, the experiments were repeated in cells with different architectures of the tubulin cytoskeleton, i.e., PtK2, RPE1, and MEF.

To jointly investigate the interacting tubulin and actin cytoskeleton, authors then microinjected small doses of labeled tubulin and actin, which allowed them to distinguish polymerization, depolymerization, or pauses of microtubules from the translocation events, while actin speckles revealed translocation of filaments along stress fibers and fiber displacements. Cells were seeded on specially shaped adhesive micropatterns to induce the controlled formation of contractile actin bundles. Local ablations in those regions led to either microtubule disassembly with no detectable effect on actin bundles or to the recoil of both microtubules and actin bundles. These experiments showed that microtubules were not free in their behavior in the cytoplasm but rather were attached to actin bundles along their entire length. Furthermore, using a special pharmacological cocktail blocking both actin filament dynamics and translocation (Peng et al., 2011) showed that, unlike control cells, the centrosome did not exhibit relaxation after laser ablation in cases where the actin network was "frozen." Since this further cast doubt on the possibility that any forces could be generated along microtubules and propagate to the centrosome, the authors' next step was to re-examine the role of dynein in microtubule aster centering.

For this purpose, two potent and specific dynein inhibitors were used, one of which blocks the activity of dynein ATPase (Steinman et al., 2017), and the second blocks dynein binding to microtubules without affecting ATPase activity (Höing et al., 2018). Treatment with either inhibitor resulted in neither a change in the distance between the centrosome and the centroid nor in a complete dispersion of the Golgi, as would occur with complete inactivation of dynein. At the same time, microinjection of blocking antibodies against dynein resulted in a clear dispersion of the Golgi but also did not affect the position of the centrosome, although a disruption of the pericentrosomal microtubule network was observed with a loss of its astral organization in PtK2 and MEF, but not in RPE1 cells. In all three cell lines, the centrosome remained close to the cell center. This suggests that centrosome centering in adherent cells in culture does not require dynein binding or activity, further casting doubt on the actual role of microtubules in centrosome positioning.

Microtubules are known to be involved in multiple signaling pathways, including the regulation of actomyosin contractility (Krendel et al., 2002), and therefore microtubule disruption by nocodazole can result in increased actomyosin contractility (Rafiq et al., 2019; Seetharaman et al., 2022). Authors of this work simultaneously treated cytoplasts with nocodazole and the Rho kinase inhibitor Y27632, resulting in most microtubules being depolymerized and only a few short microtubules remaining on the centrosome, unable to interact with either the cell periphery or most of the intracellular space. Despite this (and in addition, in the absence of a nucleus to which the centrosome could attach), the centrosome was still correctly positioned in the centroid. At the same time, after treatment of cells with cytochalasin-D, the position of the centrosome was randomized, to the point that several centrosomes were even observed at the cell periphery. Overall, these results suggest that the actin network, rather than the microtubule network, appears to be required to maintain the centrosome at the center of the cells.

These unexpected results seemed to contradict the previous studies that suggest that microtubules play a main role in generating and transmitting forces that regulate microtubule aster positioning in adherent cells (Zhu et al., 2010; Wu et al., 2011; Hale et al., 2011), including the previous findings of the authors about the centrosome displacement following dynein inactivation (Jimenez et al., 2021). Authors speculated that the difference in results between the previous and new sets of experiments might be due to some difference in the mechanisms by which the centrosome is maintained and initially established. The main difficulty is to distinguish these mechanisms from each other, because in the latter case, the study needs to start from a clearly non-centered position, which is rarely observed under normal conditions. However, they discover a way out and find the cells with the centrosomes placed close to the periphery: an intermediate stage of the enucleation protocol, that is, immediately after the centrifugation and before washing out the inhibitors affecting the actin and tubulin cytoskeleton. Observations showed that washing out cytochalasin D, Y27632, and nocodazole leads to the centrosome movement from the cell periphery to the centroid within a few hours. Surprisingly, washing out the anti-actin drugs only demonstrates that the centrosome centering was possible, although incomplete, in the absence of microtubules. And vice versa, replacing the enucleation buffer with an actin "freezing" buffer followed by washing out nocodazole demonstrated that despite normal microtubule recovery, the microtubule aster did not center. Although microtubules penetrated the entire cytoplasm, the forces generated by their polymerization or by molecular motors acting on them were insufficient to move asters among the actin network with frozen dynamics. In order to reveal the true role of the actin in centrosome centering, authors observed the change in shape and rearrangements of the actin network of cytoplasts during washing out the anti-cytoskeleton drugs after enucleation. It turned out that when the cytoplasts changed shape, their centroid shifted towards the centrosome. Obviously, when the actin network was "frozen" and no rearrangements leading to a change in shape could occur, the centroid did not move anywhere. Calculations showed that, on average, the movement of the centroid was responsible for almost 50% of the centering process. It means that centrosome centering depended more on the reorganization of the cell around the centrosome than on its own displacement, to the point that centrosome centering can sometimes be observed in the absence of any significant centrosome displacement.

The final stunning conclusion from all these experiments, according to the authors, was that the position of the centrosome appeared to be determined by the architecture and flow of the actin network rather than by a balance of pushing and pulling forces in the microtubule network. Despite its unexpectedness, it is consistent with the numerous reports showing that microtubules are anchored to actin filaments. The authors of this study believe that in adherent cells, microtubules appear embedded inside a dense actin network, which most likely generates a mechanical coupling that prevents autonomous displacements of the microtubule aster. In contrast, in mitotic cells, the actin network is more concentrated along the periphery than in the cytoplasm so microtubules are less constrained. In addition, the new concept explains well the possible mechanics of the processes that occur when it is necessary to quickly reposition the centrosome - these are changes in the actin network in the cortex and in the depth of the cell and modulation of the activity of the centrosome, which, if necessary, can lose the microtubule aster that interferes with relocation and quickly move where it is needed.

9.7. So What Next?

As we have seen, all the history of centrosome positioning research is interesting and exciting. Starting from the observation that this unique organelle is located in the center of the cell, as cell biology techniques developed and knowledge about intracellular structures expanded, cell biologists gradually advanced step by step toward understanding why this happens.

At all times, the tools and methods used by scientists determine the accuracy of the results they obtain and their further interpretation. When examining fixed and stained by iron hematoxylin cells with a bright-field upright microscope, it was initially possible only to document the interesting fact of the presence of some bodies at the spindle poles in dividing cells or in the central region of an interphase cell and note that certain threads radiate from them. Much later, these threads, which will be called microtubules, will be examined in living cells using an inverted fluorescent microscope, and biologists will study how the central position of the microtubule aster will be affected by various inhibitors, microinjections of antibodies, or local application of nocodazole. A further development of this topic will be the recreation of this system *in vitro*, where it will be even possible to measure some tension forces by using optical tweezers. However, the first vivid impression often determines the further course of thought, and even when studying the affect of the actomyosin system on centrosome centering in asymmetric cells with a geometry determined by micropatterning, cell biologists until recently always had the radial star of microtubules before their eyes first and foremost. The network of actin microfilaments remained invisible to Theodor Boveri, and his drawings from almost a century and a half ago largely predetermined the imbalance in the attitude of biologists to the two cytoskeletal systems involved in centrosome centering.

In conclusion, let us return to where this chapter began, namely to the words of Boveri, published by him back in 1887, "the centrosome represents the dynamic centre of the cell… around which all other cellular components arrange themselves symmetrically" (Scheer., 2014), and compare them with the text of the most recent work on this topic, "microtubules were involved in guiding cell reorganization around the centrosome, rather than in the physical movement of the centrosome in the cytoplasm" (Schaeffer et al., 2025). Having accumulated over the past century and a half a vast body of knowledge about the structure of cells and the centrosomes in particular, cell biology has moved up one turn of the spiral in its development, recalling some old sayings but putting a completely new meaning into them. A careful analysis of the new data shows that they do not essentially contradict the previous ones, but rather clarify them so much that this sharply limits the scope of application of old knowledge, just as the Theory of relativity once radically turned physics upside down, significantly supplementing (but not canceling) Newtonian mechanics. The most remarkable thing in this whole story is that when we had already decided that we were at the finish line, as a result of new discoveries, several new untrodden paths suddenly opened up before us, which are waiting for their researchers. Perhaps this is the most wonderful thing in science.

Chapter 10

The Centrosome in Generative Cells

Rustem E. Uzbekov[1,*], PhD, HDR, IR and Tomer Avidor-Reiss[2], PhD

[1]Laboratory of Cell Biology and Electron Microscopy, Faculty of Medicine, University of Tours, Boulevard Tonnellé, France;
Faculty of Bioengineering and Bioinformatics, Lomonosov Moscow State University, Moscow, Russia
[2]Department of Biological Sciences; Department of Urology, College of Medicine and Life Sciences, University of Toledo, Toledo, OH,USA

As noted earlier, in most somatic cells, the number of centrioles corresponds to the ploidy of the cell. In diploid cells during the G1 phase of the cell cycle, there are two centrioles, while in tetraploid cells during the G2 phase, there are four centrioles. Our studies of parasitic wasp cells, where haplo-diploid sex inheritance occurs (fertilized eggs develop into diploid females and unfertilized eggs into haploid males), demonstrate that this ratio is not genetically determined; haploid males have the same number of centrioles as diploid females (Uzbekov et al., 2024a). In most eukaryotes, the correspondence between ploidy and the number of centrioles is also disrupted when haploid germ cells are formed during gametogenesis. The significant differences in the fate of centrioles during spermatogenesis and oogenesis necessitate separate consideration. Note that because generative cells involve centrioles with distinct structures, we will use the terms centrioles or canonical centriole when referring to the standard cylindrical centriole observed in somatic cells, and we will use the term atypical centriole for centrioles with different structures.

* Corresponding Author's Email: rustem.uzbekov@univ-tours.fr

In: The Centrosome
Editor: Rustem E. Uzbekov
ISBN: 979-8-89530-815-8

10.1. Spermatogenesis

The correspondence between the number of centrioles and ploidy is preserved in spermatogonia, where cells reproduce through mitotic division. After spermatogonial division, some cells move away from the basal membrane and become primary spermatocytes. Before the onset of the first meiotic division, these cells contain a tetraploid set of chromosomes. However, the doubling of the number of centrioles from two to four does not occur in parallel with DNA replication (as in somatic cells). Instead, it happens during the prophase of the first meiotic division (Rattner and Phillips, 1973). For a long time, it was believed that the number of centrioles in secondary spermatocytes and spermatids could vary significantly across eukaryotes of different taxonomic groups. Nevertheless, new evidence contradicts this claim. Initial studies found that the poles of the second meiotic division possess one centriole, as described for some invertebrates, particularly insects (Friedlander and Wahrman, 1970; Krioutchkova and Onishchenko, 1999). More recently, several insects have been observed to possess an atypical centriole in addition to a canonical centriole, bringing the total number of centrioles in the spermatids to two.

Two centrioles are found in most studied animals, including mollusks and vertebrates (Fawcett, 1958, 1970; Gall, 1961; Krioutchkova and Onishchenko, 1989). Similar to insects and some fish, one of the centrioles in males and certain fish has an atypical structure, indicating that they possess two centrioles, not just one, as previously claimed (Fishman et al., 2018; Khanal et al., 2024; Turner et al., 2022). In some cases, the second centriole forms as an additional centriole duplication that occurs during the telophase of the second meiotic division (Schreiner and Schreiner, 1905, see Figure 1.3.) or during the formation of spermatids (Khire et al., 2016; Khire et al., 2015), which leads to the appearance of haploid germ cells with two centrioles. During spermiogenesis, i.e., the formation of mature spermatozoa from spermatids, the centriolar apparatus undergoes significant modifications referred to as centrosome reduction (Manandhar et al., 1998)or centriole remodeling (Avidor-Reiss and Turner, 2019). The molecular basis of these modifications has made some progress in insects (Khire et al., 2016; Khire et al., 2015). The mechanism of centriolar apparatus formation can have several variants. In rare cases, both centrioles, oriented parallel to each other, form flagella (Baccetti et al., 1969). However, in most cases, only one of the centrioles forms a flagellum; this centriole is called distal since it is located further from the nucleus. In some eukaryotes, spermatozoa do not have flagella, in which case the centrioles may either persist, as in lungfish (Boisson et al., 1967), or disappear altogether, as in helminths (Favord, 1961; Justine et al., 1985). The second centriole, closer to the nucleus, is called the proximal centriole. In some eukaryotes, this centriole disappears completely during spermatogenesis (Gall, 1961) or is transformed into an electron-dense plate at the proximal end of the distal centriole (Krioutchkova et al., 1994a). However, while the centriole structure may disappear, many centriole proteins are maintained at the centriole site (Khanal et al., 2024; Subbiah et al., 2024).

On the contrary, in other eukaryotes, including humans, the distal centriole that gives rise to the flagellum is remodeled, while the proximal centriole is preserved (Figure 10.1.) (Fishman et al., 2018). Thus, the presence of only one canonical centriole in a mature spermatozoon results from the resorption or transformation of one of the centrioles to centrioles with atypical structure, not the absence of centriole replication between two meiotic divisions.

It should also be noted that during spermatid maturation, a specific structure called the centriolar adjunct forms at the proximal centriole's distal end (Figure 10.2.). Unlike the flagellum, this structure does not consist of 9 MT doublets but of 9 incomplete MT triplets; the centriolar adjunct also lacks central MTs (formula 9 x 2.5 + 0), making it similar in morphology to the primary cilium. Because the centriolar adjunct is formed in the cytoplasm without being surrounded by the cell membrane, it is considered a cytoplasmic cilium (Avidor-Reiss and Leroux, 2015). The centriolar adjunct disappears in most mature spermatids and spermatozoa species but not in humans (Holstein and Roosen-Runge, 1981; Garanina et al., 2019; Zamboni and Stefanini, 1971). The incomplete disappearance of the centriolar adjunct in human spermatozoa may cause male sterility (Garanina et al., 2019).

The distal centriole of the spermatozoon is directed with its proximal (closest to the nucleus) end towards the second centriole. The proximal centriole has two "open" ends (Figure 10.1.), so it is logical to assume that of these two centrioles, the proximal one is the maternal one, and the flagellum, in this case, grows from the daughter distal centriole. We tested this assumption in our study of spermatogenesis in pigs. In these animals, the centrosome in early spermatids is not located near the nucleus but migrates to the cell membrane, where flagellum growth is initiated (Figure 10.1.). As can be seen in the photograph, the two centrioles maintain a mutually perpendicular orientation at this stage, and the flagellum grows precisely from the daughter centriole.

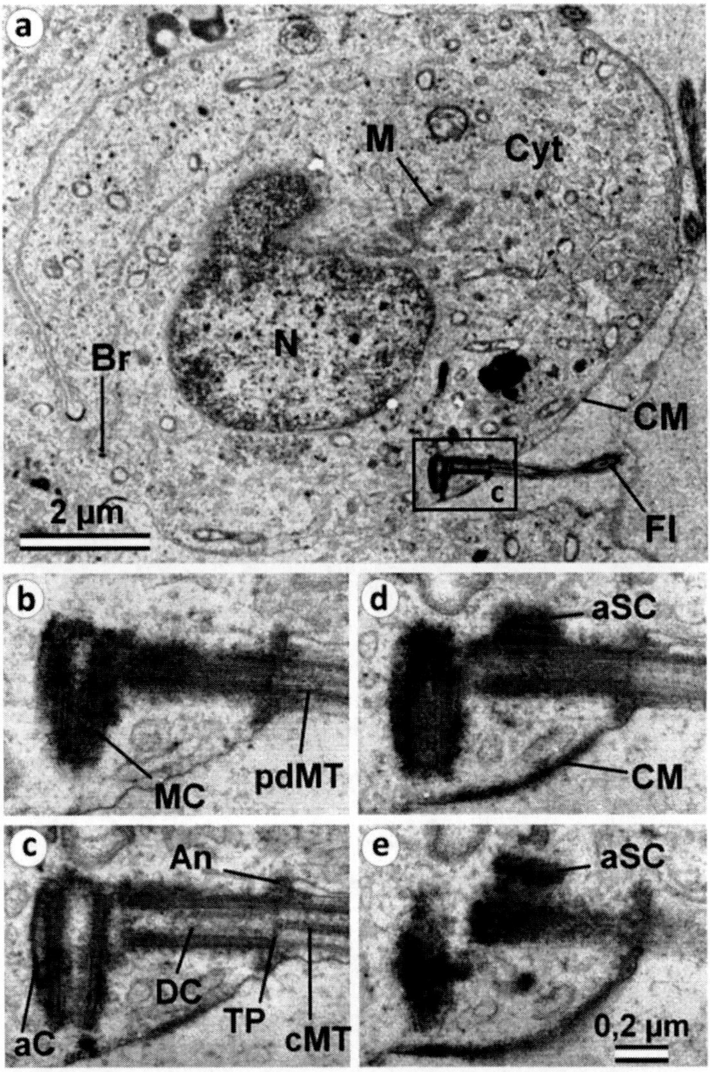

Figure 10.1. Ultrastructure of porcine (Sus domesticus) early spermatid. a - general view; b-e – four serial sections of the centrosome region (indicated by frame in a). aC – anlage of capitulum; An – annulus; aSC – anlage of segmented columns; Br – cytoplasmic bridge between two adjacent spermatids; CM – cell membrane; cMT – central MT of axoneme; Cyt – cytoplasm; DC - daughter centriole; Fl – flagella; M – mitochondrium; MC - mother centriole; pdMT – peripheral doublets of MT of axoneme; TP – terminal plate between internal volumes of centriole and flagellum. Scale bar 2 μm for a; 0.2 μm for b-e. Photo by R. Uzbekov from Alieva et al., 2018.

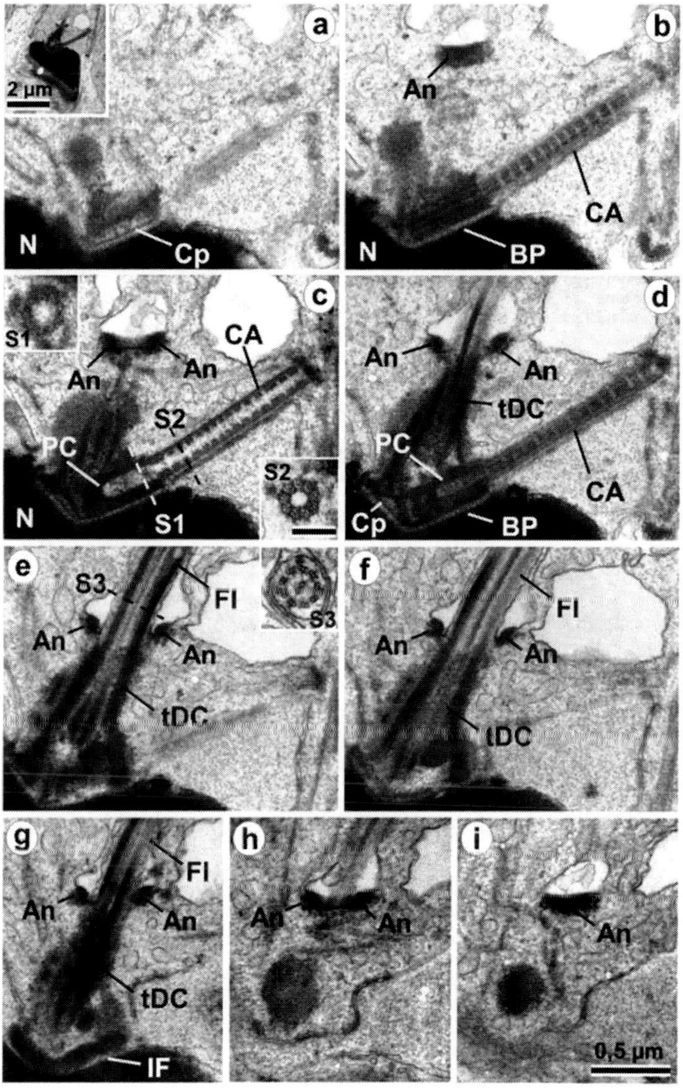

Figure 10.2. Ultrastructure of the porcine middle spermatid centrosome and basal part of flagella. Insertion in a - general view of the cell; b-i - nine serial sections of the centrosome region. An – annulus; BP – basal plate; CA – centriolar adjunct; Cp – capitulum; Fl – flagella; IF - implantation fossa; N – nucleus; PC – proximal centriole; tDC – transformed distal centriole on the base of flagella. S1 – a cross-section of centriole (cut level is shown by the dotted line in photo c); S2 – a cross-section of centriolar adjunct (cut level is shown by the dotted line in photo c), S3 – a cross-section of the flagellum (cut level is shown by the dotted line in photo e). Scale bar: insertion in a - 2 µm; b-i - 0.5 µm.

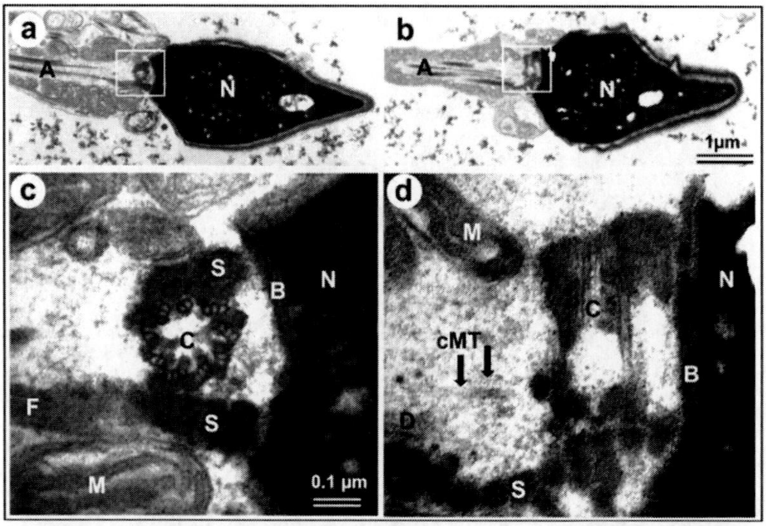

Figure 10.3. Ultrastructure of human spermatozoa. a, b - general view; c, d - enlarged images of the sections (indicated by frames in a and b) of the sperm cervix perpendicularly (c) and longitudinally (d) oriented proximal centriole. A – axoneme; B - basal plate; C - proximal centriole; D - end of outer MT doublets of axoneme; F - outer dense fibril (dense filament); M – mitochondria; N – nucleus; S - segmented columns. The arrows denote two central MT of the axoneme. Scale bar 1 μm (a, c); 0.1 μm (b, g).

Additional evidence for the correct identification of the origin of the two centrioles is the anlage of the capitulum on the mother centriole (Figure 10.1c); subsequently, the capitulum in the mature spermatoid is localized between the proximal centriole and the nucleus (Figure 10.2d), adjoining the basal plate in the implantation fossa of the nucleus. The anlage of segmented columns (Figure 10.1d) is associated with the distal centriole in the early spermatid.

Thus, in the early spermatid, both centrioles form structures consisting of MTs: the distal daughter centriole forms the flagellum of the spermatozoon, and the proximal mother centriole forms the centriolar adjunct. Since, as we described earlier, motile cilia in the cells of the ciliated epithelium are formed during one cell cycle, they are also formed "by definition" on the daughter centrioles (from the moment of its appearance, the procentriole must go through two mitotic divisions to mature into a mother centriole fully, i.e., it becomes mature only in the third cell cycle in its life) (Avidor-Reiss and Gopalakrishnan, 2013). Comparing this fact with the formation of a motile sperm flagellum also on the daughter centriole in the spermatid, we hypothesized that the structure of the axoneme depends on the maturity of the centriole, which gives it a basis: mature mother centrioles form sensitive primary cilia and a centriolar adjunct without central MTs, while daughter centrioles form motile cilia and flagella (Uzbekov and Avidor-Reiss, 2020). At later stages of spermatogenesis, the centrosome shifts into the cell, where it contacts and binds to the nucleus (Figure 10.2., 10.3.).

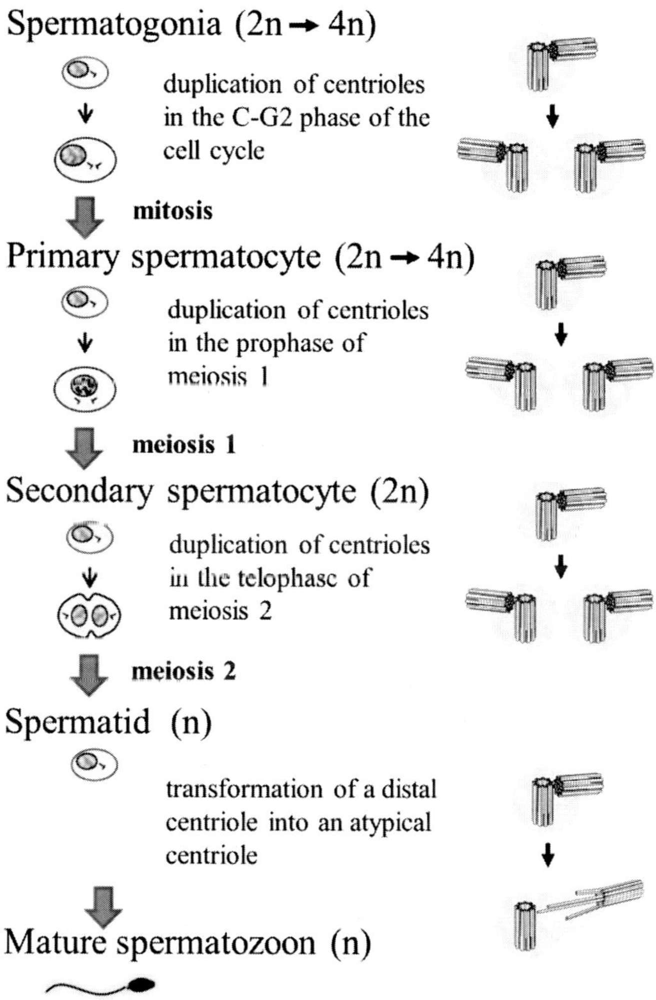

Figure 10.4. Transformation of the centrosome structure during spermatogenesis in humans. A row of differentiating cells is displayed on the left, and the centrosome structure in the corresponding cells at each stage of differentiation is illustrated on the right. The scale is not respected.

In all studied mammalian spermatozoa, the distal daughter centriole, which gives rise to the sperm flagellum, completely loses its morphology (Zamboni and Stefanini, 1971; Sato and Oura, 1984; Manandhar et al., 2000; Fishman et al., 2018). Detailed analysis on ultrathin serial sections perpendicular to the axis of the flagellum showed that in primate and human spermatozoa, the two central microtubules (MT) of the axoneme continue up to the surface of the proximal centriole, and there are no centriolar MT triplets in the expected distal centriole region (Manandhar et al., 2000; Leung et al., 2021). Moreover, the doublets of the MT flagellum diverge from each other and form a peculiar funnel expanding towards the nucleus. At the same time, this structure retains many proteins typical of centrioles and centrosomes - CETN1/2, POC523, CEP63, CPAP, POC1B, FAM161A, WDR90, CEP135, WDR62, Nek9, Potin, and Reptin (Fishman et al., 2018; Amargant et al., 2021; Khanal et al., 2021; Turner et al., 2023) in connection with which it was proposed to call the distal part of the sperm flagellum an atypical centriole (Fishman et al., 2018).

In describing the early stages of embryogenesis, we will revisit the atypical centriole and its role in forming centrosomes. Figure 10.4. illustrates the transformation of the centrosome during spermatogenesis in humans.

10.2. Oogenesis

In oogenesis, no additional doubling of centrioles between the two meiotic divisions has been described in contrast to spermatogenesis. Thus, in some eukaryotes (crustaceans and mollusks), at the beginning of the first meiotic division, the oocyte has two centrioles at each pole, and at the start of the second, one centriole (Ruthmann, 1959; Longo and Anderson, 1969). Moreover, centriole doubling does not occur before the first meiotic division in several eukaryotes. This leads to the fact that already, during the first meiotic division, there is only one centriole at each pole. During the second meiotic division, the only maternal centriole remaining in the oocyte is located at the outer pole of the spindle. This centriole is incorporated into the second polar body after meiosis is completed. The inner pole in the second division may contain centrioles from the basal body of the spermatozoon (Krioutchkova et al., 1994a, b) or be acentriolar, i.e., contain two electron-dense granules that serve as foci for the MT "asters" (Krioutchkova, Onishchenko, 1999).

This centriole goes into the second polar body after the completion of meiosis. The inner pole in the second division may contain centrioles from the basal body of the spermatozoon (Krioutchkova et al., 1994a, b) or be acentriolar, i.e., contain two electron-dense granules that serve as foci for the MT "asters" (Krioutchkova, Onishchenko, 1999).

In vertebrates (clawed frogs, mice, rabbits, sheep, cows, monkeys, and humans), centrioles in oocytes disappear even before the beginning of the first meiotic division. Therefore, the spindle poles of both meiotic divisions do not contain centrioles (Figure 10.5.), and the spindles themselves are not convergent but have a barrel shape (Szollosi et al., 1972; Le Guen and Crozet, 1989; Crozet 1990; Sathananthan et al., 1996; Sathananthan, 1997; Sathananthan, 1998). Figure 10.6. shows a diagram of the changes in centrosome structure during human oogenesis. In insect oogenesis, which has been studied in detail in Drosophila, the centrioles of all 16 cells formed due to the division of one cell migrate along cytoplasmic bridges into one of them, which becomes the oocyte. Later, during oocyte growth, before the onset of the first meiotic division, all centrioles disappear, possibly via the action of PLK1 (Raff, 2004; Pimenta-Marques et al., 2016). The spindle (as in the first meiotic division) is formed by the polymerization of MTs around chromatin and their organization through the coordinated work of various motor proteins (Raff, 2004).

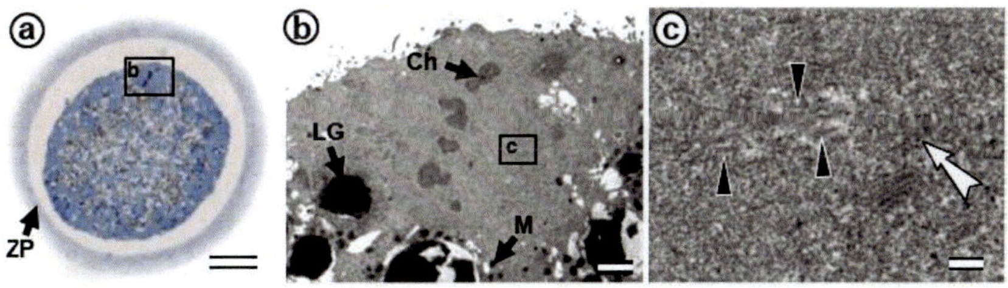

Figure 10.5. Spindle of the second meiotic division of a bovine oocyte. a - light microscopy, semithin section stained with toluidine blue; b - transmission electron microscopy, ultrathin section of the spindle region at low magnification; c - spindle pole region at high magnification. The spindle has a barrel-shaped form (arrow), MTs (shown with arrowheads) are directed toward the spindle pole, which has no centrioles. Ch - chromosomes, LG - lipid granule, M - mitochondrion, ZP - zona pellucida. Scale bar: a - 20 μm, b - 2 μm, c - 0.2 μm.

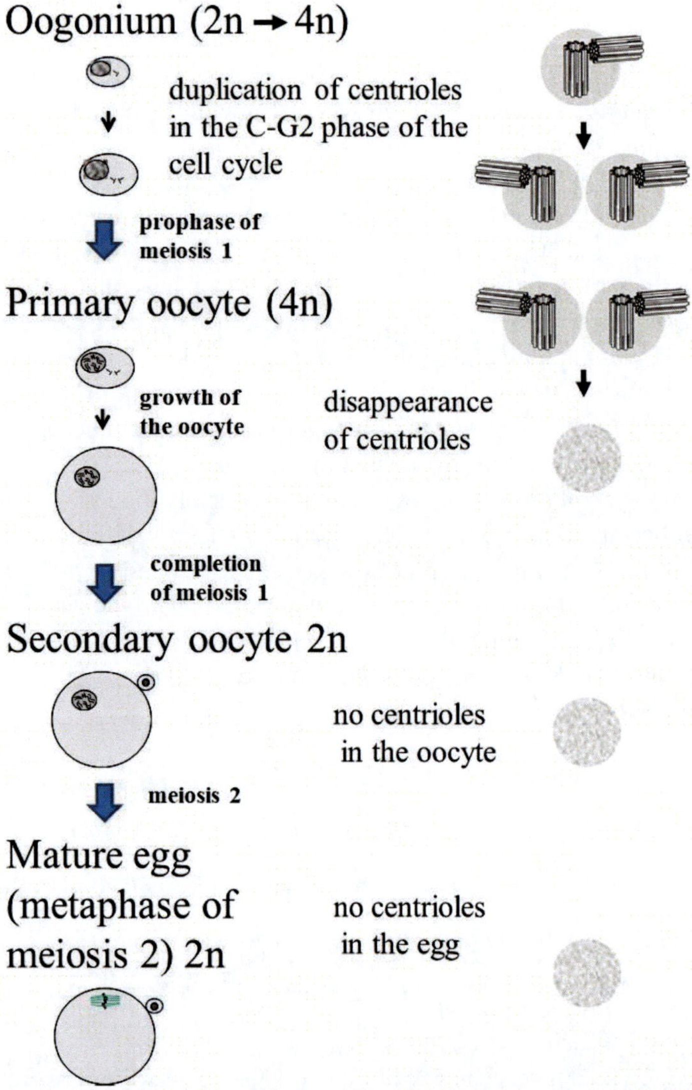

Figure 10.6. Transformation of centrosome structure during oogenesis in humans. A row of differentiating cells is depicted on the left, with the centrosome structure in the corresponding cells at each stage of differentiation illustrated on the right. The scale is not maintained.

10.3. Early Embryogenesis

10.3.1. Two Previously Unresolved Problems in the Inheritance of Centrioles in Mammals

As already noted, unlike other mammals, in rats, mice, and other murins, not only the distal centriole but also the proximal centriole is reduced in mature spermatozoa (Gueth-Hallonet et al., 1993; Manandhar et al., 1998; Khanal et al., 2024). As a result, after fertilization, the zygote of these rodents contains no centrioles at all, "either from the mother or the father," and the first cleavage divisions of the embryo occur without the participation of centrioles (Szollosi et al., 1972). According to various authors, centrioles appear at the stage from 8 to 32 or even 64 blastomeres (Maro et al., 1991; Gueth-Hallonet et al., 1993; Abumuslimov et al., 1994).

It turns out that centrioles appear in the cells in fundamentally different ways in evolutionarily close species such as mammals.

Regarding the question of centriole inheritance, in addition to the difference between murine and other mammals, there remains another unresolved issue concerning the disparity in centriole number between embryonic cells and somatic cells. As previously noted, one canonical centriole is introduced during fertilization into the zygote of a cow or a human - the proximal centriole of the spermatozoon. However, in somatic cells, there are four centrioles located in pairs at each of the poles of the division spindle before division (Figure 4.1.). How does the transition from one canonical centriole to four occur?

10.3.2. Three Hypotheses on the Mechanism of Centriole Inheritance in Humans

Three versions of the transformation mechanism of a single centriole from a spermatozoon in a zygote into four centrioles of somatic cells have been proposed. The most straightforward and logical explanation was that the proximal centriole duplicated twice before the zygote's first division.

This hypothesis was put forward by the famous, now deceased, scientist from Sri Lanka, A. Henry Sathananthan, who worked for a long time in Australia (Sathananthan et al., 1991, 1996). However, he and his colleagues could not detect two pairs of centrioles at the poles of the zygote's first division. And the hypothesis remained only a hypothesis.

The second version of the explanation of the transition from one centriole in the zygote to four centrioles before the first mitosis was proposed in the article by Schatten and Sun (2009). The authors of this paper suggested that the distal centriole is "recovered," and four centrioles emerge from a single duplication cycle on the proximal and recovered distal centrioles. However, this hypothesis was not supported by ultrastructural studies either.

A third hypothesis was proposed in the paper by Fishman et al., (2018). Based on immunofluorescence data, the authors suggested that the distal centriole can serve as a template for procentriole formation without restoring its canonical nine-triplet structure. They referred to this structure as an atypical centriole. This model also assumed that only one centriole duplication cycle occurs in the zygote before the first cleavage division.

So how and when does the transformation of a single canonical centriole in the zygote into four centrioles in the somatic cell occur? To answer this question, we examined the centrosome ultrastructure through a complete series of ultrathin sections of zygotes and early bovine embryos (Uzbekov et al., 2023).

10.3.3. How Does It All Happen?

It turned out that the proximal centriole does not double in zygotes. In all the zygotes we studied, up to the prometaphase stage of the first mitosis, the proximal centriole of the spermatozoon did not have a procentriole nearby (Uzbekov et al., 2023).

Moreover, in late anaphase-early telophase, there were no canonical centrioles at the poles of the first mitotic division (Figure 10.7.). The function of polymerization and anchoring of microtubules at the spindle poles was performed at each pole by round fibrogranular formations, for the name of which Rustem Uzbekov proposed to use the old term introduced at the end of the 19th century by one of the discoverers of the centrosome, Marie Joseph Edouard Van Beneden, to designate the structures that he discovered using a light microscope at the poles of the mitotic spindle - "polar corpuscles" (van Beneden 1876). These formations were, in all likelihood, double structures in which centrosomal proteins were concentrated, previously discovered immunocytochemically at the light-optical level in the laboratory of Professor Tomer Avidor-Reis (Fishman et al., 2018).

What is the origin of the polar corpuscles' material? We assume that they include, in particular, proteins of the striated columns, which were previously located in the "neck" section of the sperm (Avidor-Reiss, 2024). In addition, the "striated body" structure, which is found near the apical end of the sperm flagellum in the zygote and then in the blastomeres of the embryo at the two-cell stage (Uzbekov et al., 2023), probably originates from the same material.

When and how the proximal centriole of the sperm disappears remains to be determined; it can only be noted that this centriole was partially destroyed in one of the zygotes we studied.

Kalbfuss and Gönczy (2024) discuss various situations where cell centrioles are disassembled. In the comments to this article (Uzbekov and Avidor-Reiss, 2024), we suggested that the fate of the centrioles in the zygote may depend on the age of the sperm centriole.

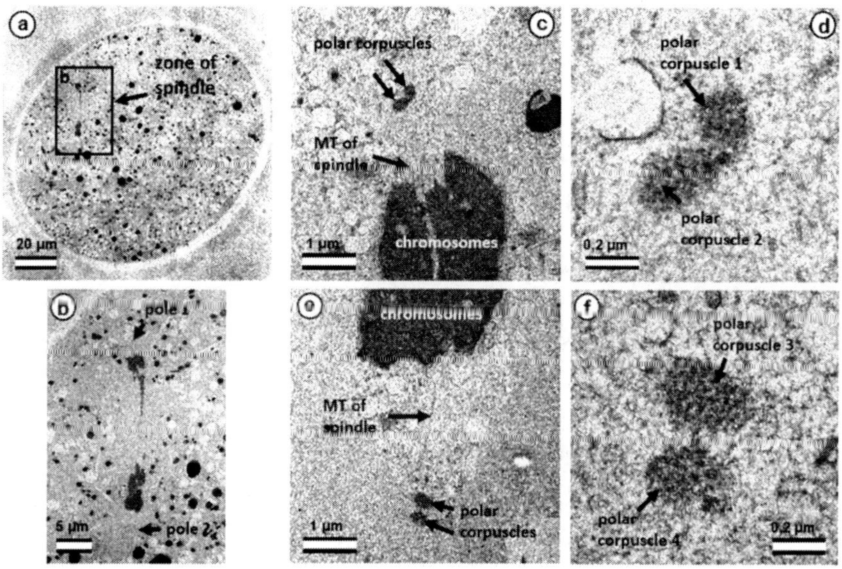

Figure 10.7. Ultrastructure of the bovine zygote at the late anaphase-early telophase stage of the first embryonic division. a– zygote at low magnification; b – mitotic spindle region; c – upper half-spindle; d – pole of the upper half-spindle at high magnification; e – lower half-spindle; f – pole of the lower half-spindle at high magnification. Each spindle pole has two polar corpuscles and no centrioles. The lower pole is associated with the axoneme of the sperm flagellum. A comprehensive series of sections for both poles is published in Uzbekov et al., (2023). Photographs from Uzbekov et al., 2023 with modifications.

As mentioned earlier, in the sperm of humans and most mammals, a single proximal centriole is preserved, which is the mother centriole. Unlike the daughter centriole, which appeared in the previous cell cycle, the mother centriole can have an age ranging from two to tens of cell cycles.

Throughout its life, this mother centriole could accumulate specific markers of elimination. Since centrioles are eliminated in the oocyte cytoplasm during oogenesis, the components responsible for centriole elimination may enter the zygote after fertilization, and if there are elimination markers on the sperm centriole, the centriole is resorbed.

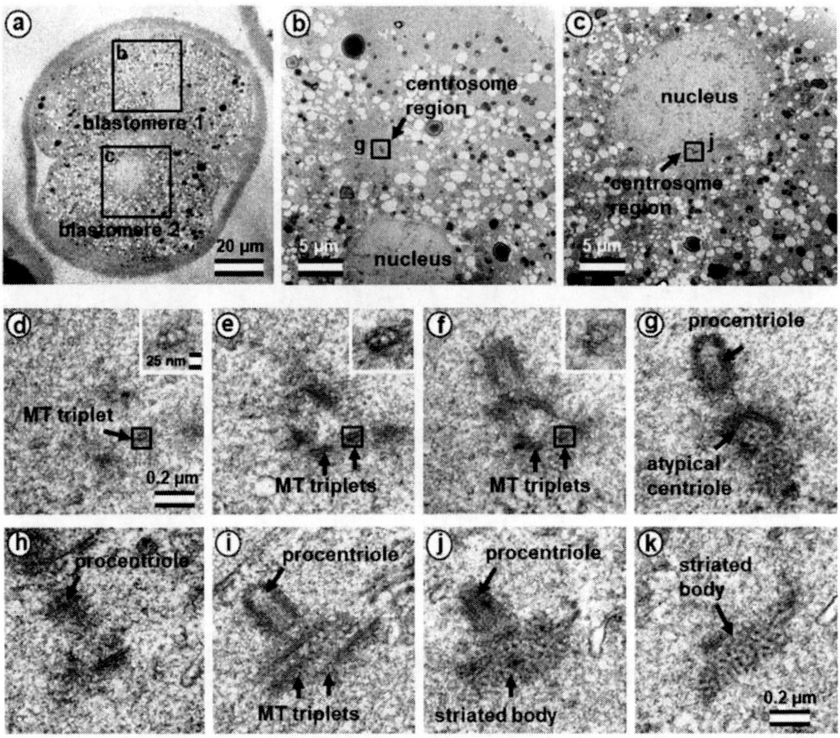

Figure 10.8. Ultrastructure of atypical centrioles in an embryo at the 2-blastomere stage. a - general view of the embryo at low magnification; b - centrosome region in the upper blastomere; c - centrosome region in the lower blastomere; d-g - four serial sections (from 9) through the centrosome region of the upper blastomere, the insets show an enlarged image of the MT triplet in a cross-section; h-k - four serial sections (from 12) through the centrosome region of the lower blastomere. A complete series of sections of both centrosomes are published in Uzbekov et al., 2023. Photos from Uzbekov et al., 2023, with modifications.

Our hypothesis suggests a possible different fate of sperm centrioles in the zygote (Uzbekov and Avidor-Reiss, 2024). It explains why various authors previously found centrioles only in a part of cells at the early stages of embryogenesis.

All centrioles identified in two-cell embryos differed significantly from the sperm's proximal centriole. They either had microtubule (MT) triplets that did not form a centriolar cylinder or exhibited a structure with MT triplets of unequal lengths, resulting in the total length of the centriole significantly exceeding that of the sperm's proximal centriole (Figure 10.8.). Considering the ultrastructural data, we proposed to extend the term "atypical centriole," previously introduced by Professor Avidor-Reiss (Fishman et al., 2018), to encompass all structures in early embryonic development that, possessing biochemical characteristics and functional features similar to centrioles, performed functions typical of centrioles, despite lacking a canonical structure - a centriolar cylinder of 9 triplets of microtubules of equal length (Uzbekov et al., 2023; Uzbekov and Avidor-Reiss, 2024).

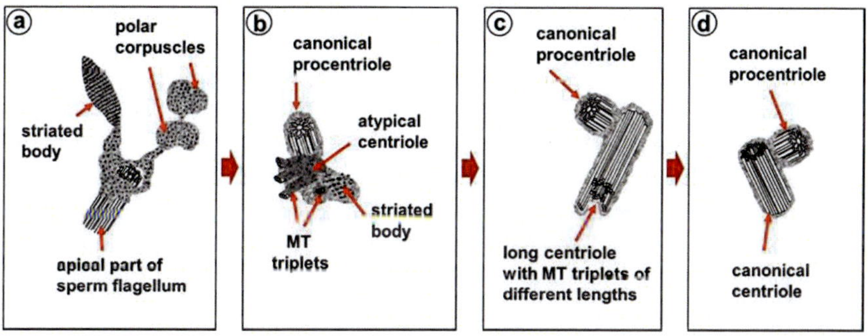

Figure 10.9. Stages of canonical centriole formation in early bovine embryogenesis. a – polar corpuscles in the pole of the first division of the zygote, only one pole associated with the apical end of the sperm flagellum is shown, the second pole also contains two polar corpuscles; b – atypical centriole with randomly oriented triplets of MT with a procentriole of canonical structure, centrosome in the embryo at the stage of 2 blastomeres; c – abnormally long centriole with MT triplets of different lengths with a procentriole of canonical structure, centrosome in the embryo at the stage of 4-16 blastomeres; g – centrosome containing a canonical centriole with a procentriole in somatic cells and at later stages of embryonic development. From Uzbekov et al., 2023, with modifications.

10.3.4. Why Are Atypical Centrioles Formed in Blastomeres at Early Stages of Development, and How Does the Transition from Atypical Centrioles to Canonical Centrioles Occur?

Thus, atypical centrioles can exhibit different structures. In the first stage, these appear as polar corpuscles located at the poles of the zygote's first mitotic division (Figure 10.7.). Subsequently, atypical centrioles form, featuring randomly oriented triplets of microtubules (Figure 10.8.). Following this, centrioles arise that already contain a centriolar cylinder. However, their length significantly exceeds that of typical centrioles found in somatic cells, and the triplets of microtubules forming the walls of the centriolar cylinder vary in length (Figure 10.9.).

When does the transformation of atypical centrioles into centrioles of canonical structure occur? Our experiments have shown that this process has several different variants. In the two-cell embryos we studied, the structure of the centrosomes varied significantly. The same can be said for the embryos consisting of 3 and 4 cells (Uzbekov et al., 2023).

The process of centriole formation results from the interaction of dozens of different proteins, and the construction of the correct shape of the cylinder does not occur all at once since it critically depends on the presence in the cytoplasm and the quantitative ratio of the proteins that constitute the centriole. Currently, methods of artificial activation or suppression of the activity of various genes have become widespread. The purpose of such studies is to determine the structural or functional significance of the corresponding proteins. In our study, we applied the opposite logical approach - based on the change in the structure of the centrioles, we assumed a deficiency in the expression of specific centrosomal or centriolar proteins.

The nine-ray symmetry of centrioles is based on the interaction of the lateral regions of the SAS6 protein dimers (Kitagawa et al., 2011; van Breugel et al., 2011), which form an angle of 40 degrees with each other. Thus, nine dimers form a closed structure, previously discovered in an ultrastructural study and called the "cartwheel structure" (Gibbons and Grimstone, 1960; Gall, 1961). We hypothesized that the formation of atypical centrioles with randomly located triplets may result from a deficiency of the SAS6 protein in embryos at the 2-4 blastomerestages. This hypothesis was supported by gene expression analysis in early bovine embryos. Indeed, in the zygote and at stages 2 and 4 blastomeres, SAS6 expression was at a very low level, and at stage 8 blastomeres, it increased approximately 10-fold (Uzbekov et al., 2023). We made a similar assumption to explain other deviations in the structure of centrioles from the canonical structure. The length of MT triplets of centriolar cylinders depends on the ratio of several proteins in the cell, particularly CPAP and CP110 (Schmidt et al., 2009; Tang et al., 2009). Local disturbances in the protein concentration ratios can lead to a violation of the length of MT triplets.

10.3.5. Functional Properties and Fate of Atypical Centrioles

In two-cell and four-cell embryos, we found that short procentrioles of the canonical structure were located on atypical centrioles consisting of randomly arranged triplets of microtubules. This observation provided significant information about the processes occurring in embryos at early stages. Previously, it was believed that only after the formation of the "cartwheel" did the growth of the first nine singlets (single MTs), then doublets (double MTs), and finally, MT triplets of procentriole begin (Sullenberger et al., 2020; Laporte et al., 2024). Our work suggests that the formation of MT triplets of the centriolar cylinder is possible without forming a full-fledged "cartwheel structure" (Uzbekov et al., 2023).

Another fundamentally important suggestion in our work is that an atypical centriole can serve as a template for forming a canonical procentriole. In this case, as in somatic cells, only one procentriole is laid down on each maternal centriole. Additionally, the centrosome, which contains the atypical centriole, is the center of microtubule (MT) organization in the blastomere. Thus, morphologically, the atypical centriole ideally performs the functions of the canonical centriole (except for the formation of the cilium or flagellum) (Uzbekov et al., 2023).

What happens next with these atypical centrioles during early embryonic development? We examined the entire volume of two embryos (7 and 14 blastomeres) through serial sections. In each of the embryos, one atypical centriole with triplets not organized into a centriolar cylinder was found in two blastomeres. All other blastomeres in these two embryos contained centrosomes that included only canonical centrioles.

Atypical procentrioles can form from both canonical and atypical centrioles. For example, the atypical centrioles of the polar corpuscles appeared to form from the canonical proximal centrioles and atypical distal centrioles of the sperm, as observed in the zygote. Additionally, the atypical centrioles did not transform into canonical centrioles and remained atypical in their structure after the zygote had divided into blastomeres. These atypical centrioles were even observed at the 7- and 14-cell stages. Unlike the zygote, in embryos at the stage of 2 blastomeres, even on atypical centrioles, only canonical procentrioles are formed; therefore, as we have shown, the number of atypical centrioles in the whole embryo at late stages does not exceed two (Uzbekov et al., 2023).

In a 7-cell embryo (divisions in the embryo are not synchronous; one of the blastomeres did not have time to divide by the time of fixation), we observed four centriolar cylinders in each blastomere for the first time. How did these centrioles arise? This question has not yet been finally resolved. We assume that, as in mice, centrioles at the stage of 2-4 blastomeres occur *"de novo,"* i.e., they form from polar corpuscles precursors that we observed at the poles of the first division of the embryonic cleavage.

Finally, we studied the formation of centrioles in parthenogenetically activated bovine embryos. With such activation, development occurs without the participation of a spermatozoon, meaning there are no proximal or atypical distal centrioles in the zygotes. However, these zygotes and subsequent blastomeres successfully divided into the blastocyst stage. At the 4-blastomere stage, we found a single centriole in one blastomere (only in one of the eight blastomeres from two 4-cell embryos after thoroughly studying the total volume of these embryos through serial ultrathin sections). This serves as direct evidence that cows also possess a "backup" program for *de novo* centriole construction in their genome, which is activated at the 4-blastomere stage if development occurs without the participation of a sperm (Uzbekov et al., 2023). It is important to note that single centrioles were also initially detected during *de novo* formation in mouse embryos (Abumuslimov et al., 1994), and only then did the blastomeres acquire somatic-type centrosomes.

Our experiments are ongoing, but we can already draw the first intermediate conclusions. We have found no fundamental differences between murine and other mammals in the mechanisms of somatic-type centrosome formation. In both cases, centrioles are formed *de novo* due to the activation of a genetic program. The only difference lies in the stage of development at which this program is activated in different animals and the presence of atypical centrioles in non-murine species. In cows, and probably also in humans, the activation of the centrosome construction program occurs very early, already at the stage of 2-4 blastomeres. At the early stages, due to the correct ratio of centriolar proteins not yet being established, atypical centrioles are formed, which, however, are functionally similar to canonical centrioles and, over a series of cell generations, continue to serve as a matrix for the formation of canonical procentrioles. As the embryo develops, the centrioles gradually acquire an increasingly canonical form (Figure 10-9).

We are currently conducting a detailed study of the protein composition of centrosomes in early embryos to enhance the morphological understanding of centrosome formation during the initial stages of embryonic development with a biochemical characterization of centrosomes.

Chapter 11

Anomalies of the Centrosome and Its Derivatives in Human Diseases

Lana Gorgidze[1], PhD
Emmanuelle Blanchard[2], PhD, HDR
and Rustem E. Uzbekov[3,*], PhD, HDR, IR

[1]Critical Care Department. National Medical Research Center for Hematology, Moscow, Russia
[2]Faculté de Médecine de Tours;
Centre Hospitalier Régional Universitaire de Tours Bretonneau,
INSERM U1259 Morphogenèse et Antigénicité du VIH et des virus des hépatites, University of Tours, Tours, France
[3]Laboratory of Cell Biology and Electron Microscopy, Faculty of Medicine, University of Tours, Boulevard Tonnellé, France;
Faculty of Bioengineering and Bioinformatics, Lomonosov Moscow State University, Moscow, Russia

Diseases associated with centrosome dysfunction are intrinsically linked to the critical role that the centrosome plays within the cell. These diseases can be broadly categorized into two primary groups.

The first category includes disorders related to defects in ciliogenesis, collectively referred to as ciliopathies (Reiter and Leroux, 2017). Ciliopathies can be either congenital, resulting from genetic mutations in genes encoding structural components of cilia and flagella, or induced by environmental factors that disrupt cellular and organismal homeostasis. In multicellular organisms, cilia and flagella mediate both motility and sensory functions; thus,

* Corresponding Author's Email: rustem.uzbekov@univ-tours.fr

In: The Centrosome
Editor: Rustem E. Uzbekov
ISBN: 979-8-89530-815-8

structural and functional defects in these organelles primarily impair these critical processes.

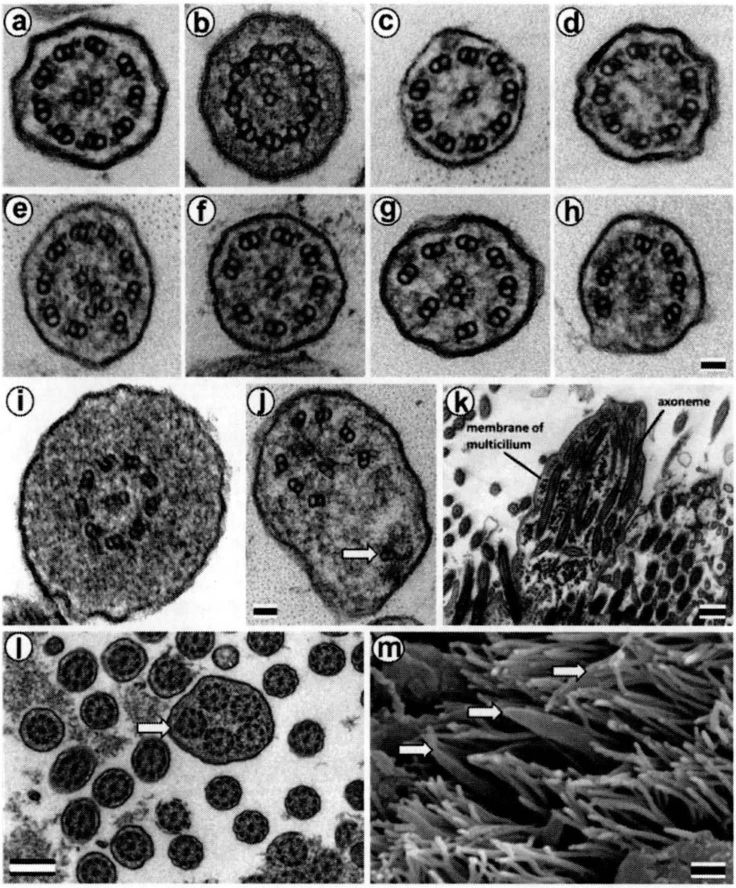

Figure 11.1. Ultrastructural abnormalities in cilia morphology. (a) Normal axoneme; (b) Displaced central MTs; (c) Absence of one of the central MTs; (d) Absence of both central MTs; (e) Duplicated central MTs; (f) Loss of one peripheral MT doublet; (g) Missing peripheral MT doublet with displacement of an adjacent doublet; (h) Combined absence of central MTs and three peripheral doublets; (i) Excess cytoplasm surrounding a normal axoneme; (j) Excess cytoplasm with missing and displaced peripheral doublets (arrow); (k) Oblique section of a multicilium; (l) Cross-section of a multicilium with five axonemes sharing a common membrane (arrow); (m) Scanning electron microscopy (SEM) image of multiple multicilia (arrows). (a–j) Olfactory epithelium cilia; (k–m) Oviduct epithelium cilia. Scale bars: 50 nm (a–j), 500 nm (k), 300 nm (l, m). Adapted in part from Bragina et al., 2018 (panels d, k, l, m).

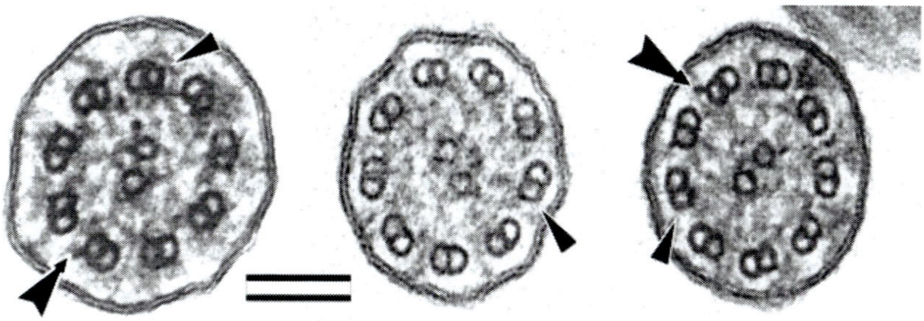

Figure 11.2. Cross-sections of cilia exhibiting ultrastructural defects: missing (arrowheads) or truncated (arrows) outer/inner dynein arms. Scale bar: 100 nm.

Many ciliary abnormalities can be detected via ultrastructural analysis. Most notably, these include defects in the arrangement of peripheral MT doublets and the central pair of MTs (Figure 11.1.). Further examination often reveals dysregulation of dynein arms (outer and inner) (Figure 11.2.), which frequently leads to reduced or absent ciliary beating. The second category encompasses diseases linked to aberrant cell cycle progression and division. Over a century ago, Theodor Boveri proposed a connection between centrosome anomalies and disease, hypothesizing that supernumerary centrosomes could promote defective mitotic spindle assembly, chromosomal instability, and oncogenesis (Boveri, 1914). This hypothesis has since been substantiated, with extensive research elucidating the molecular mechanisms underlying these defects. Beyond cancer, centrosome dysfunction contributes to a spectrum of other proliferative disorders.

This chapter examines both categories of centrosome-related diseases, focusing on their etiology and the molecular mechanisms driving pathogenesis.

11.1. Genetic Diseases Associated with Defects in Cilia and Flagella Formation

Ciliated epithelium is a hallmark of multiple human tissues, where it performs specialized functions essential for homeostasis. In the respiratory tract, bronchial cilia clear foreign particles, while ventricular cilia in the brain facilitate cerebrospinal fluid circulation. In the female reproductive system, fallopian tube cilia generate directional fluid flow that guides sperm selection and post-fertilization embryo transport to the uterus. Conversely, defective sperm flagella are a leading cause of male infertility. Nearly all cell types harbor primary (sensory) cilia, which are critical for signal transduction, particularly in specialized organs such as the cornea and vestibular system. Disorders arising from impaired motile or sensory cilia function are collectively termed ciliopathies, many of which were clinically defined prior to the identification of their genetic bases. Below, we discuss major ciliopathy syndromes linked to structural or functional ciliary defects.

11.1.1. Kartagener Syndrome

Kartagener syndrome, a classical manifestation of primary ciliary dyskinesia (PCD), presents with situs inversus in approximately 50% of affected individuals - a condition characterized by mirror-image reversal of internal organ positioning, including dextrocardia (right-sided heart) and left-sided liver localization (Nonaka et al., 1998; Khan et al., 2024). Patients invariably develop chronic respiratory complications, particularly recurrent bronchitis and sinusitis, due to impaired mucociliary clearance in the airways (Gohram and Merselis, 1959; Zariwala et al., 2001). The syndrome follows an autosomal recessive inheritance pattern and results from dysfunctional motile cilia in bronchial epithelium that exhibit impaired or absent beating. During embryogenesis, the coordinated beating of nodal cilia establishes left-right body asymmetry through directional fluid flow; disruption of this process in Kartagener syndrome leads to random organ positioning (Brueckner, 2001; Zariwala et al., 2001; Essner et al., 2002). Mutations in genes encoding axonemal dynein chains underlie this pathological condition (Brueckner, 2001; Bartoloni et al., 2001; Susalka et al., 2002). Ultrastructural abnormalities include: (1) axonemal disorganization with disrupted microtubule arrangements (Satir et al., 1980; Zariwala et al., 2001); (2) complete or partial absence of dynein arms (Afzelius et al., 1975; Antony et al., 2021); and (3) aberrant orientation of basal feet (Domaratskiĭ et al., 2005).

11.1.2. Meckel-Gruber Syndrome (MKS)

First described by Johann Friedrich Meckel in 1822 (Hartill et al., 2017), MKS is an autosomal recessive disorder characterized by encephalocele, postaxial polydactyly, polycystic dysplastic kidneys, and structural abnormalities of the heart and liver (Zhang et al., 2020, 2022). The condition is invariably lethal in the perinatal period due to severe multi-organ malformations (Williamson et al., 2017). Although centrioles in MKS retain normal ultrastructure, they fail to properly anchor to the cell membrane as a result of deficiencies in meckelin and MKS1 proteins (Dawe et al., 2007; Leightner et al., 2013). To date, mutations in at least 21 genes have been associated with the pathogenesis of this syndrome (Zhang et al., 2020).

11.1.3. Bardet-Biedl Syndrome (BBS)

Bardet-Biedl syndrome (BBS) represents a genetically heterogeneous ciliopathy, with 21 causative genes identified to date (Seo et al., 2011). The core molecular defect involves impaired assembly or function of the BBSome complex, composed of seven BBS proteins (BBS1, BBS2, BBS4, BBS5, BBS7, BBS8, BBS9) and BBIP10, which mediates selective protein trafficking to the ciliary membrane (Nachury et al., 2007; Loktev et al., 2008; Jin et al., 2010). Mutations in these genes leads to pleiotropic clinical features including: obesity (typically childhood-onset), cognitive impairment, postaxial polydactyly, renal malformations, progressive retinal degeneration (rod-cone dystrophy) (Green et al., 1989; Forsyth and Gunay-Aygun, 1993; Blacque and Leroux, 2006).

11.1.4. Autosomal-Dominant Polycystic Kidney Disease (ADPKD)

Developmental cysts represent epithelial-lined cavities that arise in multiple organs as consequences of systemic or hereditary disorders. ADPKD, the most prevalent lethal autosomal dominant disorder in the United States and Europe, results from mutations in PKD1 and PKD2 genes encoding polycystin-1 (PC1) and polycystin-2 (PC2) (Barr and Sternberg, 1999; Yoder et al., 2002; Luo et al., 2023). Both PC1 and PC2 localize to cilia, with ciliary dysfunction contributing to the characteristic cystic kidney phenotypes observed in mouse models (Pazour et al., 2000). Conditional PKD1 knockout studies in mice have established PC1's essential role in tubular morphogenesis and tissue maintenance (Kim and Walz, 2007). Furthermore, the KHD1 gene product, which is mutated in ADPKD, similarly demonstrates primary cilia localization (Ward et al., 2002; Onuchic et al., 2002).

11.1.5. Senior-Løken Syndrome (SLSN)

Age-related retinal degeneration has been linked to mutations in nephrocystin-5 (NPHP5) (Otto et al., 2005). Mutations in nephrocystin genes (NPHP1–NPHP11) cause nephronophthisis, an autosomal recessive kidney disorder characterized by fibrosis and corticomedullary cyst formation. Nephronophthisis represents the most frequent genetic cause of end-stage renal disease in individuals under 30 years of age (Hildebrandt and Zhou, 2007; Hildebrandt et al., 2011; Chen et al., 2024). NPHP5 interacts with RPGR, a gene associated with X-linked retinitis pigmentosa, and localizes to the cilia of retinal photoreceptors (Chang et al., 2006).

Ciliary dysfunction can also arise from environmental factors, including smoking and toxin exposure. Unlike genetically driven defects, these cases display heterogeneous phenotypic manifestations both within and across cells, with disease severity correlating with the duration and intensity of exposure.

11.2. Genetic Diseases Linked to Centrosomal Dysregulation

Diseases caused by mutations in centrosomal protein genes can be either congenital or acquired. In both cases, dysregulated cell division and cell cycle progression play a central role in disease pathogenesis. These processes rely on structural and regulatory proteins localized at the centrosome, which are essential for spindle assembly and cell cycle progression.

11.2.1. Primordial Dwarfism Syndromes (MOPD)

Primordial dwarfism represents a group of rare genetic disorders characterized by severe proportional growth restriction manifesting both prenatally and postnatally (Jurca et al., 2024). These conditions include: Seckel syndrome (SCKL1-10), Microcephalic osteodysplastic primordial dwarfism (MOPD types I/III and II), Meier-Gorlin syndrome (ORC1, ORC4, ORC6 mutations) and Silver-Russell syndrome (SRS; imprinting disorders of 11p15 or maternal UPD7). With the exception of SRS, all syndromes feature severe microcephaly (head circumference <-3 SD). While most cases follow autosomal recessive inheritance, some variants demonstrate autosomal dominant patterns and genomic imprinting mechanisms (particularly in SRS) (Widdowson, 1970; Khetarpal et al., 2016).

11.2.2. Seckel Syndrome (SCKL)

SCKL, the prototypical primordial dwarfism disorder, involves mutations in at least 10 genes encoding proteins critical for: CENP (CPAP) (Al-Dosari et al., 2010), CEP152 (Kalay et al., 2011), CEP63 (Sir et al., 2011), PLK4 (Martin et al., 2014), Ninein (NIN) (Dauber et al., 2012), and CEP215 (CDK5RAP2) (Yigit et al., 2015). These mutations disrupt the balance between cell proliferation and apoptosis, leading to reduced body mass and microcephaly.

11.2.3. Microcephalic Osteodysplastic Primordial Dwarfism Type II (MOPD II)

MOPD type II is the most common form of primordial dwarfism and is characterized by severe prenatal-onset growth restriction (birth weight/length typically <-4 SD), facial and skeletal anomalies, and treatment-resistant anemia (Hall et al., 2004). Affected individuals typically grow to about 110 cm and have a life expectancy of approximately 30 years (Eslava et al., 2021). Caused by biallelic PCNT mutations disrupting pericentrin function (Rauch et al., 2008). Defective mitosis induces apoptosis, leading to reduced cell counts and impaired tissue and organ growth. Pericentrin deficiency leads to mitotic spindle disorganization (Unal et al., 2014), prolonged mitotic checkpoints and reduced progenitor cell pools (Rauch et al., 2008).

11.3. Centrosome Anomalies in Tumors

The structure, protein composition, and functional activities of centrosomes in tumor cells differ significantly from those in normal cells. These alterations can affect both the number of centrosomes and centrioles per cell, as well as the structural integrity of the centriolar cylinders themselves (Pihan et al., 1998; Nigg and Raff, 2009; Song et al., 2023). Centrosome amplification is a common feature in many human cancers, raising the question of how extra centrosomes arise. Although multiple pathways contribute to centrosome amplification, the predominant mechanisms in cancer cells remain poorly understood (Nigg, 2002; Godinho et al., 2009). It is likely that dysregulation of centrosomal gene expression - whether through upregulation or downregulation - plays a key role in these abnormalities.

Excessive centriole duplication during the S phase can result from overexpression of centrosome duplication regulators, such as PLK4/SAK, or mutations in oncogenes and tumor suppressors, including *p53* and *BRCA1* (Fukasawa, 2007; Bettencourt-Dias and Glover, 2007, 2009; Cottee et al., 2017; Feng et al., 2024). Additionally, prolonged cell cycle arrest in the G2 phase can promote centrosome amplification (Bourke et al., 2007; Loncarek et al., 2010). Another mechanism involves cytokinesis failure or cell fusion, leading to tetraploidization (Kirsch-Volders et al., 2024). Normally, tetraploidization triggers p53 stabilization, inducing cell cycle arrest, senescence, or apoptosis, thereby preventing tetraploid cell proliferation. However, if tetraploid cells bypass the G1 tetraploidy checkpoint, uncontrolled proliferation, micronuclei formation, aneuploidy, and ultimately oncogenic transformation may occur, driving cancer progression (Kirsch-Volders et al., 2024).

Dysregulation of cell cycle-related gene expression is a hallmark of many cancers. Consequently, elucidating the mechanisms underlying aberrant regulation of key cell cycle genes provides broader insights than focusing on individual cancer types.

11.3.1. Polo-like Kinase 4 (PLK4/SAK)

A serine/threonine kinase family localized to centrioles, plays an essential role in controlling centriole duplication during the cell cycle (Bettencourt-Dias et al., 2005; Habedanck et al., 2005). PLK4 overexpression induces centrosome amplification, while its knockdown reduces centriole numbers. Due to its pivotal function, PLK4 has emerged as a promising therapeutic target for malignancies including malignant rhabdoid tumors, medulloblastomas, prostate cancer, and certain brain tumors (Sredni et al., 2017a, 2017b; Bailey et al., 2018; Suri et al., 2019).

11.3.2. The p53 Protein

A key transcription factor regulating cell cycle progression, localizes to centrosomes and is ubiquitously expressed. It becomes activated in response to diverse cellular stressors, including DNA damage, spindle defects, nucleotide depletion, hypoxia, heat shock, elevated nitric oxide levels, and ionizing radiation, thereby preventing malignant transformation (Fukasawa,

2007). p53 dysfunction is frequently associated with centrosome amplification, multipolar mitosis, and cancer development (Fukasawa, 2007; Feng et al., 2024). Its activity is tightly modulated through post-translational modifications and protein-protein interactions.

11.3.3. The p53-associated Protein MDM2

An E3 ubiquitin ligase (Dou et al., 2013), acts as an oncogene by targeting p53 for degradation. MDM2 overexpression is observed in multiple cancer types, where it drives centrosome amplification - particularly in head and neck squamous cell carcinomas (Carroll et al., 1999; Patel et al., 2008).

11.3.4. BRCA1 and BRCA2

Are tumor suppressor genes essential for repairing double-strand DNA breaks through homologous recombination repair (HRR) (Yang et al., 2024). BRCA1 mutations are strongly associated with hereditary breast and ovarian cancer syndrome (HBOC) (Kobayashi et al., 2013). Cells carrying BRCA1/BRCA2 mutations exhibit epigenetic effects on surrounding tissues, including suppression of proto-oncogenes and disruption of cytokine regulation. These alterations contribute to epithelial-mesenchymal transition, angiogenesis, and estrogen level dysregulation (Tay et al., 2024). BRCA1 mutations confer up to 85% lifetime risk of breast cancer and 50% risk of ovarian cancer, with additional elevated risks for colorectal and prostate cancers (Torr et al., 2024). Both proteins localize to centrosomes: BRCA1 is present throughout the cell cycle (Sankaran et al., 2006), while BRCA2 appears during S phase and dissipates by mid-mitosis (Nakanishi et al., 2007).

BRCA1 regulates centrosome duplication and function through multiple protein complexes, including BRCA1-BARD1 (Sankaran et al., 2006), BRCA1-γ-tubulin (Deng, 2002), BRCA1-GADD45 (Shao et al., 2006), and BRCA1-Nlp (Jin et al., 2009). While BRCA2 mutations generally confer lower breast cancer risk than BRCA1 mutations, they are more strongly associated with male breast cancer and show reduced ovarian cancer risk (Houlahan et al., 2024; Dean et al., 2024). Beyond DNA repair, BRCA2 participates in transcription regulation (Milner et al., 1997), cytokinesis (Daniels et al., 2004), and cell proliferation control (Tian et al., 2005). During S phase through early mitosis, BRCA2 localizes to centrosomes where it forms a complex with ROCK2 and nucleophosmin (B23) (Nakanishi et al., 2007; Malik et al., 2016). Disruption of this complex leads to centrosome amplification and multinucleation - key features of breast cancer cells (Wang et al., 2011).

11.3.5. Aurora A Kinase (AURKA) in Mitotic Regulation and Oncogenesis

Aurora A kinase (AURKA) is a pivotal regulator of mitosis, centrosome function, and cell cycle progression (Glover et al., 1995; Roghi et al., 1998; Uzbekov et al., 2001). In bovine embryos, this protein initially exhibits diffuse cytoplasmic distribution, becoming concentrated at spindle poles only at the 8-cell stage (Uzbekova et al., 2008). AURKA forms complexes with multiple partners to regulate centrosome localization and cell cycle transitions. Its expression and centrosomal localization peak during metaphase (Roghi et al., 1998; Giet et al., 1999b), followed by rapid ubiquitin-mediated degradation within one hour post-mitosis (Farruggio et al., 1999; Arlot-Bonnemains et al., 2001). Notably, AURKA reappears at centrosomes during late S-phase, preceding other mitotic centrosomal proteins (Uzbekov et al., 2024b), which is essential for centrosome separation (Giet et al., 1999a). AURKA mutations cause severe mitotic defects, including delayed mitotic entry, monopolar spindle formation, and impaired centrosome maturation (Sasai et al., 2008; Tatsuka et al., 2009). Clinically significant AURKA overexpression occurs in multiple cancers - particularly bladder (Fraizer et al., 2004), breast (Wang et al., 2006), ovarian (Mendiola et al., 2009), skin (Torchia et al., 2009), and colon cancers (Koh et al., 2017) - where it frequently drives centrosome amplification.

11.3.6. CDK-Cyclin Complexes in Cell Cycle Dysregulation and Cancer

Cell cycle dysregulation in cancer is characterized by elevated cyclin-dependent kinase (CDK) activity, increased cyclin expression, and reduced levels of endogenous CDK inhibitors like CIP/KIP proteins (Otto and Sicinski, 2017; Roskoski, 2019). Specific CDKs regulate critical cell cycle checkpoints (Leal-Esteban and Fajas, 2020), with CDK1, CDK2, CDK4, and CDK6 being frequently hyperactivated in tumors (Icard et al., 2019). While CDK inhibition represents a promising therapeutic strategy (Sanchez-Martinez et al., 2019; Tadesse et al., 2019), developing selective inhibitors remains challenging due to the broad substrate specificity of CDK1 and CDK2 (Tadesse et al., 2020; Spring et al., 2020).

In normal cells, p27 inhibits CDK2 to prevent uncontrolled proliferation (Hauck et al., 2008). Cancer cells subvert this regulation through CDK2-mediated p27 phosphorylation, which targets it for ubiquitin-dependent degradation (Shin et al., 2002; Starostina and Kipreos, 2012). CDK2 also phosphorylates nuclear hormone receptors (androgen, estrogen, and progesterone receptors), enhancing their transcriptional activity - a mechanism particularly relevant in hormone-dependent cancers (Jorda et al., 2018; Pierson-Mullany and Lange, 2004; Yin et al., 2018). Beyond proliferation, CDK2 regulates apoptosis; its suppression reduces FOXO1 phosphorylation at Ser249, enabling FOXO1 activation and subsequent upregulation of pro-apoptotic factors (FasL, TRAIL, Bim) (Huang et al., 2006; Huang and Tindall, 2007b). Since CDK2 is rarely mutated in human cancers yet remains highly active, selective CDK2 inhibitors are essential for targeting specific cancer subtypes.

CDK-cyclin complexes also regulate centriole duplication. The cyclin D/CDK4/6 complex phosphorylates pRB, preventing its interaction with the transcription factor EF2. This leads to activation of cyclin E and A synthesis, initiating centriole duplication (Hinchcliffe et al., 1999; Lacey et al., 1999; Meraldi et al., 1999). Thus, centriole duplication and DNA replication are coordinated by a shared cytoplasmic mechanism.

Centrosomal targeting of cyclin E depends on its 20-amino acid centrosomal localization sequence (CLS) (Matsumoto and Maller, 2004). This domain is essential for centrosomal anchoring and DNA synthesis initiation. Wild-type CLS peptides as opposed to mutant variants, competitively inhibit endogenous cyclin E/A centrosomal binding, blocking DNA synthesis (Matsumoto and Maller, 2004). CDK2-cyclin E overexpression drives premature S-phase entry, causing polyploidy and centrosome amplification (Fukasawa, 2008), while CDK2 inhibition induces centrosome declustering, multipolar spindle formation, and apoptotic death (Chen et al., 2023).

11.3.7. TRIM37 in Centrosome Regulation and Oncogenesis

TRIM37, an E3 ubiquitin ligase belonging to the TRIM protein family (comprising >70 human members), plays critical roles in centrosome biology and cancer pathogenesis (Domínguez-Calvo et al., 2021). Loss-of-function TRIM37 mutations cause MULIBREY nanism, a growth disorder with elevated cancer predisposition (Domínguez-Calvo et al., 2021; Gazzin et al., 2023). Mechanistically, TRIM37 depletion induces formation of aberrant Centrobin-PLK4 complexes that function as ectopic microtubule-organizing centers, disrupting mitotic fidelity (Domínguez-Calvo et al., 2021). Conversely, TRIM37 overexpression promotes CEP192 degradation, impairing centrosome maturation and increasing mitotic errors (Domínguez-Calvo et al., 2021). Notably, TRIM37 hyperexpression sensitizes cells to PLK4 inhibitors like centrinone (Meitinger et al., 2020). Clinically, TRIM37 dysregulation occurs in ovarian fibrothecomas (Karlberg et al., 2009), pancreatic cancer (Jiang et al., 2016), gliomas (Tang et al., 2018), and endometrial carcinomas (Zhou et al., 2024).

11.4. Centrosome Amplification in Human Malignancies

The link between centrosome abnormalities and cancer was first proposed by Theodor Boveri through sea urchin embryo studies (Boveri, 1914). In particular, an increased number of centrosomes - Centrosome amplification (CA), promotes mitotic chromosome missegregation, genomic instability, and aneuploidy - hallmarks of cancer progression. CA is now recognized across diverse malignancies - solid tumors: breast (Lingle et al., 1998; Fukasawa, 2007; Sutrisno et al., 2024; Torr et al., 2024), prostate (Pihan et al., 1998; Parvin et al., 2020), colorectal (Koh et al., 2017; Zhang et al., 2010), ovarian (Mendiola et al., 2009; Karlberg et al., 2009), pancreatic (Jiang et al., 2016) and gliomas (Klein et al., 2004; Tang et al., 2018; Zhu et al.) as well as in hematologic malignancies (Giehl et al., 2005; Krämer et al., 2005; Trachana et al., 2007; Krem et al., 2015).

11.4.1. Centrosomal Dysregulation in Breast Cancer Pathogenesis and Therapy

Breast carcinomas demonstrate significant centrosomal abnormalities, with 75% of malignant tumors exhibiting enlarged and amplified centrioles. The percentage of affected tumor cells varies substantially (1-100%), indicating intratumoral heterogeneity (Godinho and Pellman, 2014). These structural aberrations primarily result from pathological expansion of the pericentriolar material (PCM), a sophisticated protein network regulated through: centriolar control mechanisms, cytosolic αβ-tubulin availability, key regulatory kinases (PLK1, CHK1), structural organizers (pericentrin, CPAP) (Conduit et al., 2010; Chan, 2011; Antonczak et al., 2015). Clinical studies reveal that approximately 50% of breast tumors display centrosomal morphological defects, correlating with significantly worse overall and relapse-free survival outcomes (Woodruff et al., 2014). Molecular profiling identifies a centrosome-associated gene signature (CETN2, TUBG1, PCNT2, PLK4, CCNE1) whose overexpression predicts poor prognosis across all molecular subtypes (Denu et al., 2016).

Therapeutic strategies targeting centrosomal defects show particular promise due to: their near-ubiquitous occurrence in breast cancer; direct contribution to chromosomal instability.

Potential intervention approaches include: modulation of centriole biogenesis pathways, correction of ciliary dysfunction, regulation of phosphorylation cascades, targeting of overexpressed centrosomal components (Schatten, 2013; Pannu et al., 2015).

11.4.2. Prostate Cancer

Prostate cancer is one of the leading causes of morbidity and mortality in men. A common feature of prostate cancer is the mutation of the *p53* gene, which occurs in more than 50% of cases (Piemonte et al., 2021). Abnormal *p53* expression is also associated with advanced stages and higher grades of the disease (Navone et al., 1993, 1999). In addition, chromosomal instability, structural abnormalities, and aberrant mitoses are frequently observed in prostate cancer cells (Osman et al., 1999; Pihan et al., 2001). *p53* inactivation leads to centrosome amplification and structural heterogeneity, whereas reintroduction of wild-type *p53* into PC3 cells induces G2/M phase arrest and reduces centrosomal abnormalities (Pihan et al., 2001). Centriolar overduplication in prostate cancer cells is driven by overexpression of the serine/threonine kinase PLK4, with a strong correlation between elevated PLK4 levels and higher Gleason scores (a marker of tumor aggressiveness and metastatic potential) (Ouyang et al., 2001). PLK4 inhibition suppresses tumor cell proliferation, viability, and colony formation (Ouyang et al., 2001).

11.4.3. Colon Cancer

Colorectal cancer ranks as the third most prevalent cancer and the second leading cause of cancer-related deaths globally. Colorectal carcinomas serve as an ideal model for studying mitotic chromosome segregation mechanisms, as they can be categorized into two groups: tumors with a diploid genome and those exhibiting significant aneuploidy (Forozan et al., 1997; Singh et al., 2022). Aneuploid cell lines display severe centrosomal defects, including numerical and structural aberrations, with centrosome counts ranging from 3 to 7. These centrosomes often vary in size and morphology, and some are mislocalized from the nucleus (Ried et al., 1999). Aurora-A kinase is overexpressed in primary colorectal tumors, intestinal cancer stem cells, and differentiated tumor cells (Ghadimi et al., 2000; Zhang et al., 2010; Koh et al., 2017). Aurora-A has been functionally linked to centrosome amplification in stem cells, as its suppression reduces centriole overduplication. These findings suggest that Aurora-A-targeted therapies may selectively eliminate cancer stem cells in colorectal tumors (Ghadimi et al., 2000).

11.4.4. Ovarian Cancer

Epithelial ovarian cancer is the leading cause of mortality among gynecological malignancies. The overall survival rate for patients with serous ovarian carcinoma has remained stagnant over the past two decades, primarily due to persistent chromosomal instability, a hallmark of this disease (Cammareri et al., 2010). Centrosome amplification via centriole overduplication is a recurrent and heterogeneous feature of serous ovarian carcinoma, strongly associated with chromosomal instability and genomic subclonality (Karlberg et al., 2009; Mendiola et al., 2009; Morden et al., 2021; Torr et al., 2024; Sutrisno et al., 2024). In ovarian cancer, Aurora-A kinase is frequently amplified in cell lines (Mendiola et al., 2009; Sauer et al., 2023) and overexpressed during epithelial transformation (Hu et al., 2005). More than 400 distinct mutations in *BRCA1* and *BRCA2* genes have been identified, with familial-specific variants reported (Chung et al., 2005; Hoogerbrugge and Jongmans, 2016). *In vitro* studies demonstrate that centrosome amplification, recapitulated across ovarian cancer cell lines, correlates with multidrug resistance - particularly to paclitaxel, a first-line therapy for serous ovarian carcinoma. These findings suggest that centrosome amplification may serve as both a driver of tumor progression and a predictive biomarker for treatment response (Morden et al., 2021).

11.4.5. Pancreatic Cancer

Pancreatic ductal adenocarcinoma (PDAC) carries one of the most dismal prognoses among all malignancies. The majority of PDAC cases exhibit both numerical and structural centrosome abnormalities, though the precise impact of centrosome amplification on disease progression and patient outcomes remains unclear. The exceptionally high mortality rate in pancreatic cancer stems largely from the absence of reliable prognostic markers and effective therapeutic targets - a stark contrast to malignancies like breast and lung cancer where precision medicine has made significant advances (Chappuis et al., 1999). Centrosome amplification has been well-documented in pancreatic adenocarcinoma cells. Comparative studies reveal striking differences in centrosome profiles (both in number and size) between tumor tissue and normal pancreatic tissue (Mittal et al., 2015). Immunofluorescence analyses demonstrate that centrosome abnormalities (defined as excessive or structurally aberrant centrosomes) correlate strongly with nuclear anomalies, including bi or multinucleation and giant nuclei formation in pancreatic cancer cell lines (Sato et al., 1999). These findings support the hypothesis that centrosome amplification contributes to the aggressive nature of pancreatic cancer. Molecular analyses have identified multiple genes with dysregulated expression in primary PDAC tumors that are known to drive centrosome amplification, including: cell cycle regulators (CCNA2, CCND1, CCNE2, CDK1), centrosomal components (CEP63, CEP152, STIL), kinases and oncogenes (AURKA, NEK2, PIM1, PLK4, dNDR1), transcription factors (E2F1-3, LMO4, MYCN) and other regulators (MDM2, PIN1, RAD6). Notably, elevated expression of nine genes (AURKA, CCNA2, CCNE2, CDK1, CEP152, NEK2, PIM1, PLK4, and STIL) correlates with significantly worse overall survival (Zhu et al., 2005). These results strongly suggest that centrosome amplification may enhance the metastatic potential of PDAC and serve as a negative prognostic indicator (Chappuis et al., 1999).

11.4.6. Gliomas

Gliomas represent the most aggressive and treatment-resistant brain tumors, accounting for approximately 80% of primary brain malignancies (Diwanji et al., 2017; Fitzmaurice et al., 2017). These tumors are characterized by rapid proliferation, invasive growth patterns, and frequent centrosome amplification, often manifesting as multipolar mitoses (Telentschak et al., 2015). Despite multimodal therapy (surgical resection followed by radiation and chemotherapy), recurrence rates remain near-universal, with median survival below two years for high-grade gliomas (Giangaspero and Burger, 1983). Unlike normal cells, which typically undergo apoptosis in response to centrosome amplification, glioma cells evade mitotic catastrophe by clustering extra centrosomes into functional bipolar spindles (Levine et al., 2017; Liu and Pelletier, 2019). This process depends critically on kinesin motors, particularly KIFC1 (HSET in humans), which facilitates centrosome aggregation (Kleylein-Sohn et al., 2012; Rath and Kozielski, 2012). Clinically, elevated KIFC1 expression correlates with poor prognosis, while *in vitro* studies demonstrate that KIFC1 knockout suppresses glioma cell proliferation and invasiveness, highlighting its therapeutic potential (Xue et al., 2024).

11.4.7. Hematological Malignancies Associated with Centrosome Defects

Centrosome aberrations are increasingly recognized as hallmarks of hematological neoplasms. While these diseases are classically driven by oncogenic chromosomal translocations, recent evidence implicates centrosome protein dysregulation across diverse subtypes: acute leukemias (AML, ALL), myelodysplastic syndromes (MDS), lymphoid malignancies (Hodgkin's lymphoma, select NHL subtypes), chronic lymphoproliferative disorders (CLL, multiple myeloma).

11.4.7.1. B-Acute Lymphoblastic Leukemia (B-ALL)

B-ALL is a malignant clonal hematopoietic disorder characterized by uncontrolled proliferation of lymphoblasts. Hyperdiploidy represents the most frequent cytogenetic abnormality in both pediatric and adult B-ALL cases. Notably, 24% of studied patients exhibited centrosomal abnormalities, including enlarged, amorphous, and elongated centrosomes with irregular (uniform or sickle-shaped) edges (Krämer et al., 2005). Immunostaining for three centrosomal markers (α-tubulin, γ-tubulin, and pericentrin) in the same patient revealed distinct anomaly patterns, suggesting that centrosomal defects contribute to chromosomal instability and B-ALL pathogenesis (Krämer et al., 2005).

11.4.7.2. Burkitt Lymphoma

This aggressive B-cell malignancy is defined by the t(8;14) translocation involving *c-MYC* on chromosome 8 and, typically, immunoglobulin heavy-chain loci on chromosome 14 (Kerketta et al., 2017). While 30–50% of Burkitt lymphoma cells display centrosomal abnormalities (numerical/structural) and aberrant mitotic figures, metaphase analyses show no concurrent chromosomal instability (Klein, 1983). This results suggest that centrosomal anomalies do not universally correlate with ongoing chromosomal instability; multipolar metaphases alone cannot predict chromosomal stability status; tumors can maintain chromosomal stability *despite* severe centrosomal defects. These observations underscore that the functional consequences of centrosomal aberrations depend on cellular context, necessitating tumor-specific analyses of their role in genomic destabilization (Klein, 1983).

11.4.7.3. Hodgkin Lymphoma

Is unique among malignant neoplasms due to the rarity of tumor cells (approximately 1% of the total cell population), which drive tumor formation by recruiting reactive, non-clonal lymphocytes, eosinophils, and fibroblasts. A hallmark pathological feature of classical Hodgkin lymphoma is the presence of multinucleated Reed-Sternberg cells. These cells originate from germinal center B-lymphocytes and arise through cytokinesis defects in Hodgkin's mononuclear cells (Duensing et al., 2003). Centrosomal aberrations have been reported in two cases of classical Hodgkin lymphoma, potentially linked to the high chromosomal instability observed in Reed-Sternberg cells. However, after normalization for nuclear size, the difference in centrosome area between Reed-Sternberg cells and microenvironmental cells was not statistically significant (Krem and Horwitz, 2013). The *KLHDC8B* gene has been identified as a key candidate mitotic regulator in Hodgkin lymphoma. *KLHDC8B* knockout studies demonstrate that its loss induces centrosome amplification and aneuploidy - phenotypic hallmarks of Reed-Sternberg cells (Martín-Subero et al., 2003).

11.4.7.4. Multiple Myeloma

Is a hematologic malignancy that remains incurable despite advances in therapeutic approaches (Krem et al., 2012). This disease is characterized by clonal plasma cell proliferation in the bone marrow and exhibits remarkable clinical and genetic heterogeneity, featuring both structural chromosomal abnormalities and specific gene mutations (Munshi and Anderson, 2013). Multiple myeloma cells frequently display centrosome defects and amplification (Maxwell et al., 2005; Favasuli et al., 2024). Among recurrently mutated genes in myeloma, *DIS3* mutations occur in approximately 10% of cases and significantly impact disease progression (Morgan et al., 2012; Todoerti et al., 2022). Functional studies in *Schizosaccharomyces pombe* have established *DIS3*'s roles in chromosome segregation (Kinoshita et al., 1991; Bolli et al., 2014), cell cycle control (Ohkura et al., 1988; Noguchi et al., 1996), and spindle assembly (Smith et al., 2011). Similar cell cycle regulatory functions of *DIS3* have been confirmed in *Drosophila melanogaster* (Murakami et al., 2007; Robinson et al., 2015). Importantly, *DIS3* knockdown in multiple myeloma cells induces centrosome amplification and subsequent formation of multipolar mitotic spindles (Snee et al., 2016).

11.4.7.5. Myelodysplastic Syndromes (MDS)

Are a heterogeneous group of malignant oligoclonal hematopoietic stem cell disorders characterized by impaired growth and differentiation of hematopoietic progenitors, associated with peripheral blood cytopenias and an increased risk of transformation to acute myeloid leukemia (Kryukova et al., 2016; Sperling et al., 2017; Shastri et al., 2017). Abnormal centrosomes are known to be a common and early detectable sign in patients with MDS (Valent et al., 2007). ESPL1/separase, a cysteine endopeptidase, is a key player in chromosome segregation and centrosome duplication (Nolte et al., 2013; Kumar, 2016; Zhang and Pati, 2017). Proper temporal and spatial activation of separase's proteolytic activity ensures chromosomal accuracy and proper semi-conservative duplication of centrioles (Prinzhorn et al., 2016; Nigg, 2007). Pearson's correlation analysis showed a positive correlation between the SAD (separase activity distribution) value and the occurrence of abnormal centrosome phenotypes in MDS patients, and SAD values increased in all MDS patients during the course of the disease (Ruppenthal et al., 2018). It is also known that the transformation from MDS to Acute Myeloid Leukemia (AML) coincides with the appearance of a small fraction of bone marrow cells showing significant levels of intracellular separase proteolytic activity. Hematopoietic cells obtained from such patients showed an increased frequency of additional centrosomes (Ruppenthal et al., 2018). Similar observations were recorded for chronic myeloid leukemia, where clonal evolution and the progression time from chronic phase to blast crisis correlated with increased separase proteolytic activity in patients with the BCR-ABL e14a2 fusion type of chronic myeloid leukemia, during treatment with the tyrosine kinase inhibitor imatinib (Haass et al., 2015).

The topic of the functional activity of centrosomes in cells of different types and its manifestation in diseases at the organismal level is so vast that it cannot be fully covered not only in one chapter but even in an entire book. However, summarizing the research results presented in this chapter, it can be concluded that even the disruption of the expression of a single centrosomal protein can have an impact on the human body comparable to the loss of an entire chromosome. This is because in the centrosome intersects with numerous signaling and regulatory pathways, the disruption of which initially leads to changes at the cellular level, followed by tissue, organ, and organismal levels. As noted at the beginning of this chapter, the most noticeable manifestation of such deviations from the norm occurs in ciliopathies and oncological diseases. However, these diseases are far from exhaustive in listing deviations from the normal development of the human body, and every year, the research of scientists and doctors leads to the discovery of new cause-and effect relationships between disruptions in centrosomal protein expression and the manifestation of these disorders at the level of the human body.

11.4.7.6. Myelodysplastic syndromes (MDS)

Represent a heterogeneous group of malignant oligoclonal hematopoietic stem cell disorders characterized by defective growth and differentiation of hematopoietic progenitors, resulting in peripheral blood cytopenias and an elevated risk of progression to acute myeloid leukemia (Kryukova et al., 2016; Sperling et al., 2017; Shastri et al., 2017). Centrosomal abnormalities are recognized as a common and early detectable feature in MDS patients (Valent et al., 2007). The cysteine endopeptidase ESPL1/separase plays a critical role in chromosome segregation and centrosome duplication (Nolte et al., 2013; Kumar, 2016; Zhang and Pati, 2017). Precise temporal and spatial regulation of separase proteolytic activity is essential for maintaining chromosomal stability and ensuring semi-conservative centriole duplication (Prinzhorn et al., 2016; Nigg, 2007). Pearson correlation analysis revealed a positive association between separase activity distribution (SAD) values and the incidence of abnormal centrosome phenotypes in MDS patients, with SAD values increasing progressively during disease evolution (Ruppenthal et al., 2018). Notably, the transition from MDS to acute myeloid leukemia (AML) is marked by the emergence of a small subpopulation of bone marrow cells exhibiting elevated intracellular separase activity. Hematopoietic cells from these patients demonstrate a higher frequency of supernumerary centrosomes (Ruppenthal et al., 2018). Analogous findings have been reported in chronic myeloid leukemia, where clonal evolution and progression from chronic phase to blast crisis correlate with enhanced separase activity in BCR-ABL e14a2-positive patients undergoing imatinib therapy (Haass et al., 2015).

The functional complexity of centrosomes across cell types and their disease manifestations is too extensive to be comprehensively addressed in a single chapter or even an entire monograph. However, the evidence presented in this chapter underscores that dysregulation of even a single centrosomal protein can exert organism-wide effects comparable to chromosomal loss. This profound impact stems from the centrosome's integration with multiple signaling and regulatory networks, where primary cellular disturbances cascade to tissue, organ, and systemic levels.

As introduced earlier, the most clinically apparent consequences of such dysregulation manifest in ciliopathies and malignancies. Yet these represent only a fraction of potential developmental aberrations, with ongoing research continually revealing novel causal relationships between centrosomal protein dysfunction and their phenotypic expression at the organismal level.

Chapter 12

The Insect Centrioles: An Overlooked but Promising Frontier

Maria Giovanna Riparbelli[1], PhD
Romano Dallai[1], PhD
David Mercati[1], PhD
Rustem E. Uzbekov[2], PhD, HDR, IR
Denise Bonente[1], PhD
and Giuliano Callaini[1,*], PhD

[1]Department of Life Sciences, University of Siena, Siena, Italy
[2]Laboratory of Cell Biology and Electron Microscopy, Faculty of Medicine, University of Tours, Boulevard Tonnellé, France;
Faculty of Bioengineering and Bioinformatics, Lomonosov Moscow State University, Moscow, Russia

Centrioles are evolutionarily conserved small organelles (Azimzadeh, 2014; Avidor-Reiss and Turner, 2019) that serve dual functions: they are essential for centrosome assembly and for the nucleation of ciliary structures. Given that the centrosome is the primary microtubule-organizing center of the cell, centrioles play a critical role in regulating centrosome dynamics throughout the cell cycle. For instance, centrosome misregulation may result in the assembly of multipolar spindles, a process linked to the emergence of cellular aneuploidy, which is a hallmark of cancer. Besides, elongation of the A- and B-tubules of the centriole wall forms the structural framework of cilia and

* Corresponding Author's Email: giuliano.callaini@unisi.it

In: The Centrosome
Editor: Rustem E. Uzbekov
ISBN: 979-8-89530-815-8

flagella. Most of the immotile cilia, the primary cilia, are involved in cell signalling, whereas motile ciliary structures are involved in fluid and cell movement; consequently, centriole abnormalities may lead to infertility, tissue homeostasis defects, and debilitating diseases as ciliopathies. Therefore, the deep understanding of the centriole life and its structure may represent a useful tool to elucidate the complex dynamic of this organelle and its functional role during the cell cycle.

The majority of the studies on centrioles are focused on a few model organisms such as *Caenorhabditis elegans*, *Drosophila melanogaster*, and on vertebrate cells (Gupta and Kitagawa, 2018). Other systems are only sporadically examined. Therefore, additional analysis of the dynamics and structure of centrioles in different organisms can provide useful insights for the general understanding of these organelles.

The Hexapoda are the largest group within the Arthropod phylum and with more than one million described species represent the most species-rich group of organisms (Beutel et al., 2013). Remarkably, the insects display a great variability in sperm which axoneme structure, ranging from the conventional 9 x 2 + 2 model to atypical number and disposition of doublets (Dallai, 2014). Since the centriole acts as the primary organizer for axoneme nucleation, it is reasonable to expect that this great variability in the sperm axonemes is associated with a remarkably variability of the centriole structure. Therefore, the analysis of the centriole in germ cells and somatic cells of the insects could provide news insights to better understand the process of centriole assembly and its functions.

12.1. The Structure of the Insect Centriole

As is typical in most organisms, centrioles exhibiting nine-fold symmetry represent a conserved feature across the majority of insect species.

However, the number of microtubules (MTs) that form the centriole wall is heterogeneous. Indeed, there are insect species with centrioles consisting of doublet MTs as for instance the basal entognathan Collembola and Diplura (Figures 12.1. a,b) and other species having centrioles consisting of triplet MTs, as the pterygotan Zoraptera (Figure 12.1c), Thysanoptera (Figure 12.1. 1d-f) and Hymenoptera (Paoli et al., 2013).

A trivial observation might be that centrioles with triplets, a condition shared with other arthropods, represent the ancestral model of the insect centrioles. However, several data do not confirm this hypothesis highlighting a distribution of the two centrioles patterns that does not correlate with the evolutionary position of the species analysed (Dallai and Afzelius, 1999; Riparbelli et al., 2010). Centrioles with doublets observed in basal hexapods are scattered among several insect orders. Moreover, centriole geometry varies not only among different species but also through various tissues of the same organism. In *Drosophila*, for example, somatic cells present centrioles with nine doublets, whereas germline cells have centrioles with nine triplets. This observation raises questions on the functional meaning of this structure variation. A phenomenon of particular interest, is the male gametogenesis that culminates in the formation of mature sperm characterized by a 9×2+9+2 axoneme; in this case the accessory tubules are thought to arise from modifications of the C-tubule of the centriolar microtubule triplets or from the B-tubule of the axonemal doublets. This suggests a plausible correlation between the centriolar triplets and axoneme assembly (Gottardo et al., 2013; Basiri et al., 2014). Indeed, the axonemes growth is affected in *Uncoordinated* (*Unc*) *Drosophila* mutant spermatocytes in which the C-tubule dynamic is defective (Gottardo et al., 2013)

Moreover, the spermatids of Collembola, that are basal hexapods, display centrioles with nine doublets, and the sperm axoneme lacks accessory tubules (Fanciulli et al., 2013). By contrast, Diplura, also showing centrioles with nine doublets, are able to form accessory tubules in their sperm (Dallai et al., 2011) (Figure 12.1. a, b). The same occurs in the ephemeropteran *Chloeon* (Figure 12.1. g–i).

On the other hand, the *Drosophila* female germ cells display centrioles that despite having triplets do not organize ciliary structures (Mahowald and Strassheim, 1970). In addition, the sensory organs of *Drosophila* and other insects contain a distinct axoneme that is derived by a distal centriole which wall consists of nine doublets. However, it is important to note that the number of MTs that form the centriole wall in somatic cells is actually uncertain for most insect species, just as the architecture of spermatid centrioles is unclear. In fact, most of the published data concerns the structure of sperm rather than the process of spermatogenesis.

The most dramatic variation of the centriole geometry was found during the spermatogenesis of the proturan *Acerentomon* (Riparbelli et al., 2009; Dallai et al., 2010) and the dipteran *Sciara* (Phillips, 1967) where a small centriole with nine-fold symmetry represents the focal point for the assembly of a large daughter with a number of doublet MTs ranging from 14, in *Acerentomon* (Figure 12.1. j), to 60-70, in *Sciara*. The large centrioles of these species appear to lack a conventional cartwheel. Since it is generally assumed that spindle assembly abnormal 6 (Sas6) organizes the cartwheel with nine-fold symmetry, it is unclear how the centrioles of these insects can be assembled. Remarkably, the fibers emerging from the large hub of *Acerentomon* and the cylindrical sheets of *Sciara*, have similar length than the radial spokes of *Drosophila* somatic and germ cell centrioles that have a conventional nine-fold symmetry (Riparbelli et al., 2009). The possibility is that different isoforms of Sas6 could exist, working in different contexts, or that Sas6 might be dispensable for centriole assembly in different insect species.

A peculiar and interesting condition is the growth of centrioles during the process of spermatogenesis. In particular, the centrioles that drive the mitotic spermatogonial mitoses are short (Figure 12.2a), but during the prophase of the first meiosis the centrioles increase several time their length (Figure 12.2. b, c). This process has been described in *Drosophila melanogaster* (Riparbelli et al., 2012; Fritz-Niggli and Suda, 1972), Neuroptera (Friedländer, 1980), Coleoptera (Dallai et al., 2017), Trichoptera (Klein and Wolf, 1997), Lepidoptera (Yamashiki and Kawamura, 1988; Wolf and Kyburg, 1989; Gottardo et al., 2023; Friedländer and Wahrman, 1970) and could be shared by all the insect species. Moreover, very elongated centrioles with nine-fold symmetry and extended cartwheel have been observed in the coccinellid *Adalia* (Dallai et al., 2017), the neuropteran *Chrysopa* (Friedländer, 1980), and in the *Drosophila bifurca* (Riparbelli et al., 2024).

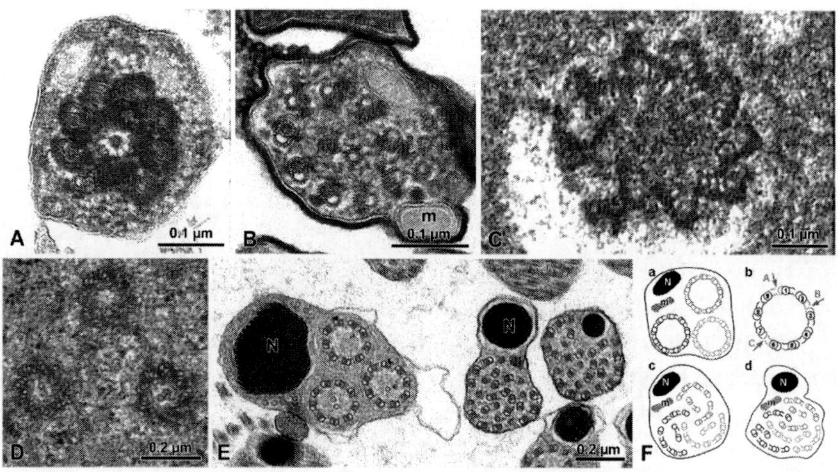

Figure 12.1. (Continued).

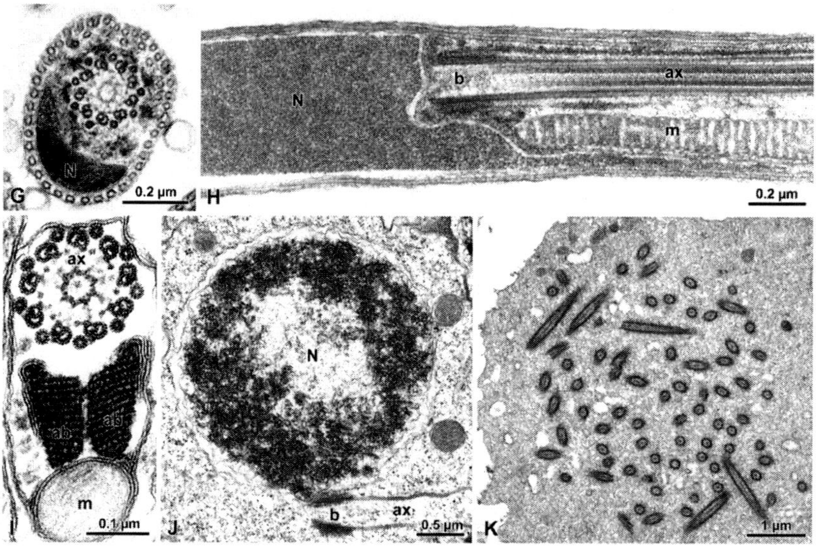

Figure 12.1. (A) Cross section passing through the centriole (basal body) in the dipluran Metajapyx braueri sperm. Note the presence of microtubule doublets. (From Dallai et al., 2011). (B) Cross section of the proximal region of the sperm flagellum of Japyx solifugus showing the presence of the accessory tubules along the right side. m, mitochondria. (From Dallai et al., 2011). (C) Cross section of the centriole (basal body?) in the spermatid of the zorapteran Zorotypus caudelli. Note the presence of the microtubule triplets. (From Dallai et al., 2011). (D) Cross section through spermatid of the thysanopteran Haplothrips simplex showing the three centrioles each consisting of microtubule triplets. (From Paccagnini et al., 2006). (E) Cross section through the thrips Agrostothrips meridionalis showing three axonemes in the same cytoplasm of an elongating spermatid (left side) and the fusion of the microtubule doublets in the mature sperm (right side). N, nucleus. (From Paccagnini et al., 2007). (F) Schematic drawing of the process of microtubule doublets fusion in the Thrips spermiogenesis, starting from the three axonemes of the spermatid elongating (a) up to in the final mature sperm (d). Arrows in (b) indicate the sites of rifts occurring in each axoneme. The final result will be the formation of three dyads, three triads and three tetrads of MTs (c). N, nucleus; m, mitochondria. (From Paccagnini et al., 2007). (G) Cross section of the sperm basal region of the flagellum in the mayfly Ecdyonurus venosus. Note the lack of the two central tubules. N, nucleus. (H) Longitudinal section of the basal region of the Ecdyonurus venosus sperm flagellum. ax, axoneme; b, basal body; m, mitochondria; N, nucleus. (I) Cross section through the sperm flagellum of Ecdyonurus venosus. Note the mitochondria (m), the accessory bodies (ab) and the axoneme 9x2+9+0 (ax). (From Baccetti et al., 1969). (J) Cross section of the early spermatid of the proturan Acerentomon microrhinus. Note the spherical nucleus (N) and the basal body (b) producing the axoneme (ax). (From Dallai et al., 2010). (K) Cross section of an early spermatid of the isopteran Mastotermes darwiniensis showing numerous centrioles in the cytoplasm. (From Riparbelli et al., 2009).

These observations raise questions on the regulatory mechanisms that control the centriole length. It is well-established that centriole length is regulated, at least in *Drosophila*, through the balance of the polymerization and depolymerization of the centriolar MTs. It has been reported that centriole growth in *Drosophila* spermatocytes is promoted in a dose-dependent manner by the Anastral spindle 1 protein (Ana1) (Saurya et al., 2016) and is reduced in Unc (Gottardo et al., 2013), and centrobin (Cnb) (Reina et al., 2018) mutants. Conversely, mutation in the kinesin-13 Klp10A that depolymerizes MTs, leads to longer centrioles in both cultured cells and spermatocytes (Delgehyr et al., 2012; Gottardo et al., 2013). Interestingly, Klp10A spermatocytes display short cilium like regions (CLRs) with remnant triplet MTs, whereas Unc mutants have elongated CLRs without C-tubule remnants. This suggests that the dynamics of the C-tubule could have unknown roles in the elongation of the centriole in *Drosophila* spermatocytes. However, reduced Klp10A expression results in severe structural defects of sensory type I neurons and triggers a dramatic centriole elongation, despite the centrioles lack of C-tubules (Persico et al., 2019).

A final interesting consideration concerns the unusual structure described in larval tissues of the male wasp *Anisopteromalus* with nine-fold symmetry but lacking MTs (Uzbekov et al., 2018a). Surprisingly, adult tissues of these species display typical centrioles with triplet MTs and nine-fold symmetry.

12.2. Asymmetry between Parents

In contrast to vertebrate centrioles in which the parents are easily recognized by the presence of distal and subdistal appendages in the mother centriole, and thanks to the transient presence of the cartwheel in the daughter, the parent centrioles of the insects are morphologically indistinguishable and display a distinct cartwheel along their life. However, these centrioles are usually orthogonally oriented and the daughter is placed at the basis of the mother (Figure 12.2a).

In addition, it has been reported in *Drosophila* that several proteins have an asymmetric localization on the parent centrioles. A transient enrichment of the centriole duplication factor Ana2 has been observed in the daughter centrioles of primary spermatocytes (Stevens, 2010). Moreover, the spindle assembly abnormal 4 (Sas4) and Sas6, two conserved proteins mainly involved in centriole assembly and elongation, are enriched at the basal regions of the daughter centrioles during male gametogenesis, whereas they accumulate to the mother centrioles in somatic tissues (Persico et al., 2020) (Figure 12.3.). This observation supports a subtle molecular diversity among germline and somatic cell centrioles. Remarkably, Centrobin (Cnb) that localizes specifically to the daughter centriole (Januschke et al., 2013), is essential for C-tubules to attain the right position within the centriole wall (Reina et al., 2018). In wild-type centrioles, the center of the C-tubules is slightly shifted away from the midline connecting the A- and B-tubules, whereas in Cnb mutants, the C-tubules are dramatically displaced. Moreover, replacement with a Sas4 allele that has decreased tubulin interaction results in the loss of the C-tubules (Zheng et al., 2016). These observations highlight unresolved questions about how C-tubules are assembled.

Interestingly, mutations in *Unc* and *Klp10A* revealed a different growth of the parent centrioles with the daughter being shorter (Gottardo et al., 2013). This is particularly evident in germline stem cells (GSCs) and during the spermatogonial mitoses (Gottardo et al., 2016a). Peculiar microtubule-based structures positive for Sas-4 and Klp10A extend into the peripheral cytoplasm of the *Drosophila* polar spermatocytes at the onset of the first meiosis. These structures are associated with one of the two mother centrioles and have a short lifespan (Riparbelli et al., 2018). These findings reveal a characteristic asymmetry between the parent centrioles at the onset of male meiosis and also uncover unexpected different functional properties between the mother centrioles of the same cells.

A structural asymmetry between mother and daughter centrioles was described in *Drosophila* GSCs. Here, the old mother centriole, which stays in the apical cytoplasm of the male GSC is invariably composed of triplets, whereas its daughter always consist of mixed doublets and triplets (Gottardo et al., 2015b).

This structural asymmetry might reflect a correlation between the architecture of the parent centrioles and the unequal distribution of centrosomal proteins. Moreover, the mother centriole is bound to the cell membrane by specific links which may contribute to the maintainance of its apical positioning during centrosome separation.

Keeping the focus on the variability in centriole length, it is important to note that an unusually long mother centriole has been reported during early *Drosophila* oogenesis. This centriole is firstly inherited by the cystoblast after the asymmetric division of the GSC and persists during the process of oocyte selection; thus it represents a marker for oocyte identification at the beginning of its formation and during its ensuing developmental stages (Riparbelli et al., 2021).

One last noteworthy observation is that the centrioles of auditory and olfactory sensilla of *Drosophila* and other insects are aligned in tandem (Jana et al., 2018; Vieillard et al., 2016; Keil, 2012) with the distal centriole consisting of nine doublets and nucleating the axoneme, whereas the proximal centriole is composed of mixed singlets and doublets (Gottardo et al., 2015b). In addition, the tandem aligned centrioles lack a distinct cartwheel, a peculiar condition also found in the post-mitotic cells that form the hub region of the male stem cell niche (Gottardo et al., 2015b) (Figure 12.3.).

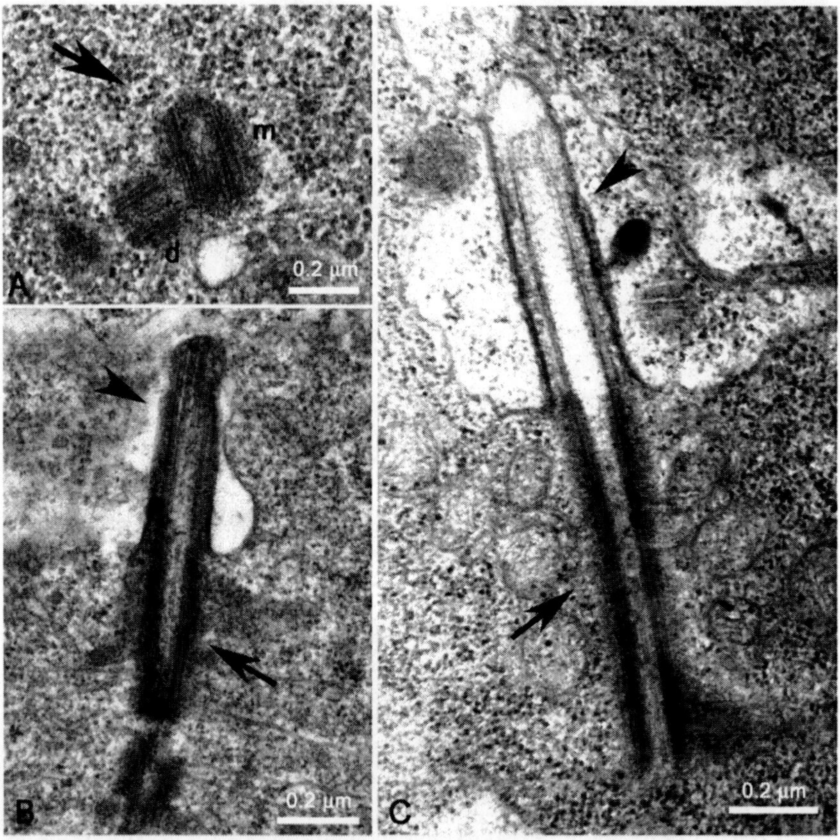

Figure 12.2. Centrioles (arrows) and Cilium Like Regions (CLRs) (arrowheads) in Drosophila melanogaster (a,b) and Drosophila bifurca (c). In young primary spermatocytes the centrioles are short (a), and reach their final dimension in mature spermatocytes of the primary prophase (b,c). Note the different length of the centrioles in Drosophila melanogaster (b) and Drosophila bifurca (c) primary spermatocytes. Mother (m) and daughter (d) centrioles are orthogonal.

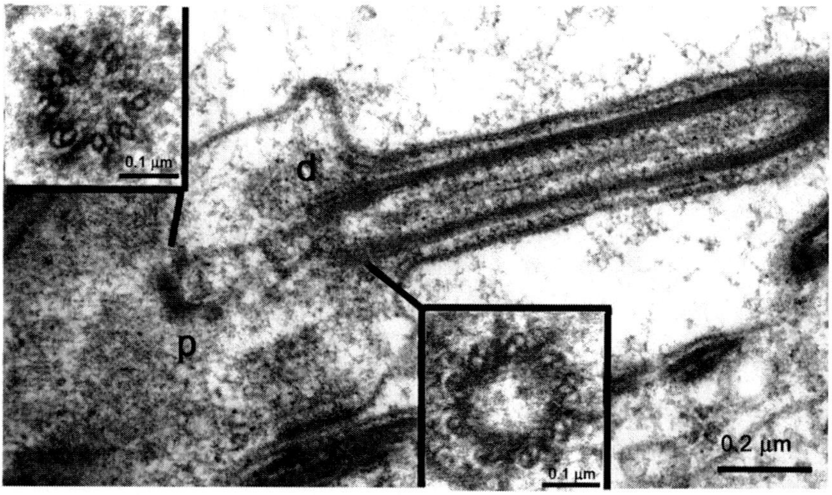

Figure 12.3. Basal region of an auditory sensillum of Drosophila showing distal (d) and proximal (p) tandemly aligned centrioles; the distal centriole consists of doublets and nucleates the axoneme, whereas the proximal one is composed of mixed singlets and doublets.

12.3. Centriole Number in the Insect Spermatids

The number of centrioles carried by the insect spermatids appears to be a trivial question. It might be obvious to hypothesize that these spermatids consist of a centriole pair as it usually occurs in other organisms. Indeed, the number of centrioles held by the mature sperm of most animal species results from two successive meiotic divisions with two centriole duplication cycles (Krioutchkova and Onishchenko, 1999; Avidor-Reiss et al., 2015).

Contrary to this general rule, several data suggested that the mature insect male germ cells contain only the centriole that nucleate the sperm axoneme, when the latter is present (Friedlanlder and Wahrman, 1971). Indeed, the spermatids of the fungus gnat *Sciara coprophila* (Phillips, 1967), *Bombyx mori* (Friedländer and Wahrman, 1971), and several Neuropterans species (Friedländer and Wahrman, 1966) display only one centriole. In addition, more recent ultrastructural data support the finding of only one centriole in the insect spermatids (reviewed in Dallai et al., 2016; Ross and Normark, 2015; Riparbelli et al., 2020; Uzbekov et al., 2017, 2018b).

The centriole duplication, at least in somatic cells, is strictly associated to the DNA replication cycle. Thus, the duplication of these organelles in the absence of DNA replication appears as a remarkable exception that rises questions on the regulation of the centrosome cycle in the absence of DNA duplication. However, the number of centrioles in insect spermatids is still uncertain with numerous previously published works supporting the possibility that the insect spermatids carry two centrioles. One centriole pair is present in the spermatids of the cricket *Pachyrhamma fasciferum* and the grasshopper *Stenobothrus viridulus* (Mathur, 1960). Moreover, electron micrographs have reported two centrioles in the spermatids of the beetle *Cicindela flavopunctata*, the mosquito *Culiseta ornata*, the neuropteran *Hesperoleon sp.* and the cockroach *Blatella germanica* (Breland et al., 1968).

How is it possible to explain these contrasting findings? The images representingcentriole pairs are typical from spermatids due to the presence of mitochondria clusters looking like nebenkern. However, similar mitochondria disposition was also found in secondary spermatocytes (Friedlanlder and Wahrman, 1971). Moreover, electron microscope observations in the spermatids of several species of Coccinellidae (Dallai et al., 2018; Dias et al., 2020) and Hymenoptera (Uzbekov et al., 2017, 2018b) failed to reveal centriole pairs. However, normal spermatids with two orthogonal centrioles are found in the spermatids of the isopterans *Reticulitermes* and *Zootermopsis* (Baccetti et al., 1981), and the thricopteran Hydroptilidae (Dallai and Afzelius, 1995), whereas the spermatids of Mallophaga and Anoplura (Baccetti et al., 1969), the related Psocoptera (King and Ahmed, 1989) and the gall-midge Lestodiplosis (Baccetti and Dallai, 1976; Dallai et al., 1993) show two parallel centrioles. The two centrioles probably exist before the reduction divisions and no subsequent replication of centrioles occurs. Indeed, the arrhenotokous parthenogenetic parasitoid *Gryon* lacks male meiosis and each spermatocyte gives rise to two spermatids, each containing a pair of orthogonally oriented centrioles (Paoli et al., 2013). However, the spermatids of *Gryon* assemble only one axoneme nucleated by the distal centriole, whereas the proximal one displays a reduced size. By contrast, both the centrioles of Mallophaga and Anoplura organize functional axonemes and the sperm is biflagellate (Baccetti et al., 1969).

The spermatids of the thrip *Halothrips*, that also lack a meiotic process, inherit a pair of centrioles, but surprisingly one of these two centrioles gives origin to a third one. All the centrioles nucleate single axonemes that during the spermiogenesis fuse together to form a single sperm flagellum consisting of a microtubule bundle of 27 doublet and singlet MTs (Paccagnini et al., 2007, 2010) (Figure 12.1. D-F).

Although the centrioles of the insects usually do not duplicate during the second male meiosis, short procentrioles consisting of nine singlets appeared at the basis of the centriole at the beginning of spermatid differentiation in *Drosophila* (Gottardo et al., 2015a) and the coccinellid *Adalia* (Dallai et al., 2017). Moreover, an unstructured cylinder is detectable in the spermatid of the beetle *Tribolium* (Fishman et al., 2017). In Drosophila, it is challenging to observe the short procentriole during axoneme elongation, due to its immersion in the pericentriolar material enveloping the basal body (Anderson, 1967). Moreover, the procentriole does not acquire the B- and C-tubules and is no longer visible in mature sperm despite accurate serial sectioning analysis (Gottardo et al., 2015a) suggesting that this organelle has a transient life and disappears at the end of spermiogenesis. The origin of these incomplete centrioles is unclear. Although the formation of these organelles involves in *Drosophila* typical centriolar proteins, such as Ana1, Bld10/Cep135, Sak/Plk4 and Sas6, usually required to initiate centriole formation (Stevens et al., 2010; Blachon et al., 2009, 2014; Mottier-Pavie and Megraw, 2009; Jo et al., 2019), their assembly process is unknow. Moreover, the assembly of the procentrioles at the beginning of spermatid differentiation opens new questions on the mechanism regulating the timing of their appearance. Finally, it is unclear the persistence of their neotenic aspect characterized by the lack of the A- and B-tubules that would complete the centriole wall. Hypothesizing a possible function, these transient organelles may act as recruiting centre for proteins needed during the assembly of the second centriole at fertilization (Blachon et al., 2014). This is consistent with the finding that Sak/Plk4 and Sas6, essential for centriole biogenesis, localize close to the spermatid centriole at the beginning of the axoneme elongation in *Drosophila* (Blachon et al., 2009). Moreover, the reduction and the definitive loss of the typical centrosomal proteins Ana1 and Asterless (Asl) (Khire et al., 2015) during *Drosophila* spermiogenesis is associated with the accumulation of Poc1B (Jo et al., 2019; Khire et al., 2016), a conserved centriolar protein essential for centriole elongation and stability (Pearson et al., 2009). This suggest, that despite the lack of a distinct structured procentriole in mature sperm, some precursors involved in the assembly of a daughter centriole at fertilization can be provided by the sperm in form of an atypical centriole (Avidor-Reiss et a., 2015; Blachon et al., 2014). If this is the case, the generally assumed inheritance of a single centriole at fertilization in insects needs to be reconsidered.

An extreme condition has been described in the termite *Mastotermes* in which the young spermatid inherits only one centriole at the end of meiosis.

However, new centrioles soon appear and their number highly increases before spermatids enter the elongation phase. Remarkably, about a hundred centrioles that nucleate as many independent flagella are assembled in a very short time (Baccetti and Dallai, 1978; Riparbelli et al., 2009; Dallai et al., 2010) (Figure 12.1. K).

On the other hand, the spermatids of the archaeococcid *Matsucoccus* inherit, at the end of spermatogenesis, two orthogonal centrioles, unable to organize a functional axoneme at the onset of spermiogenesis (Paoli et al., 2015). Likewise, centrioles that do not organize axonemes were found in differentiating spermatids of thricopteran Hydropsychidae (Friedländer and Morse, 1982), in the isopteran Hodotermitidae (Baccetti et al., 1981) and in the ephemeropteran Leptophebiidae (Gaino and Mazzini, 1991). Moreover, in some insect species the centrioles organize a distinct axoneme that degenerates during spermiogenesis resulting in aflagellate sperm cells with a rod immotile structure in place of the axoneme, as it occurs in the Aleyrodidae (Baccetti and Dallai, 1977; Bào et al., 1997; Dallai, 2014).

The finding of sperm with hundred centrioles opens important questions on how the egg cytoplasm may control the number of centrioles at fertilization to avoid the assembly of multipolar spindles that are detrimental to proper zygotic division. On the other hand, it is unclear how the egg could assemble a functional zygotic spindle in the absence of inherited centrioles following fertilization by aflagellate sperm devoid of centrioles such as the proturan *Huhentomon* and *Acerella* (Dallai et al., 1990) and the ephemeropteran Leptophebiidae (Gaino and Mazzini, 1991).

A remarkable aspect of the spermatogenesis of some Lepidopteran species is the assembly of several procentrioles close to the basal region of each parent centriole in primary spermatocytes of *Bombyx* (Yamashiki and Kawamura, 1998), *Ephestia* (Wolf and Kyburg, 1989) and *Pieris* (Gottardo et al., 2014) that have orthogonal or V-shaped engaged parent centrioles. This is a peculiar condition, as it is generally assumed that for centriole duplication during the cell cycle, the loss of their orthogonal configuration, or disengagement, is mandatory and begins at the onset of anaphase. Therefore, the separation between the parent centrioles represents the licensing factor enabling duplication during the following interphase. Thus, the appearance of procentrioles when mother and daughter centrioles are still engaged represents a unique condition, that overrides the general rule. The number of procentrioles increases during the meiotic progression, and up to four procentrioles were found at the base of each mother centriole in *Pieris brassicae* (Gottardo et al., 2014), suggesting that repeated duplications during the meiotic progression occurred. The procentrioles in *Pieris* look like intermediate structures, since they consist of a ring of dense material in which incomplete number of singlet tubules are embedded. The assembly of several procentrioles at the basis of the mother centriole, points to the absence of a single template that would ensure a centriole copy number control. The function of these organelles is unclear, but their presence may be redundant, since they fail to complete their wall and disappear at the end of spermatogenesis. Such a condition, in which engaged centrioles can support procentriole assembly, was also observed in mammalian cultured cells expressing high level of Plk1 (Shukla et al., 2015).

12.4. Ciliary Structures in the Insect Spermatocytes

A common characteristic of the insect centrioles during male gametogenesis is that both parents work as basal bodies for the assembly of the axonemes of ciliary structures (Figure 12.2. B,C) that are inherited by the spermatids and organize the sperm axoneme at the onset of spermiogenesis (Riparbelli et al., 2012). These ciliary structures resemble primary vertebrate cilia. However, in contrast to the latter, CLRs of the insect spermatocytes are organized by both the parent centrioles, they lack typical Y link structures in the transition region (TZ) between the distal end of the centriole and the forming axoneme, and persist during the meiotic divisions. Although the CLRs of *Drosophila* spermatocytes diverge from the conventional primary cilia, they share with vertebrate cilia some conserved TZ module proteins (Basiri et al., 2014; Jana et al., 2018; Vieillard et al., 2016; Pratt et al., 2016; Gonçalves and Pelletier, 2017; Wu et al., 2024) that are assembled thanks to intraflagellar transport (IFT)-independent mechanisms (Witman, 2003). Indeed, it is generally considered that the assembly of the CLRs in *Drosophila melanogaster* spermatocytes requires free diffusion of tubulin until the tip of the growing ciliary axoneme. This phenomenon could be involved in the formation of very short CLRs, of approximately 0.8-0.9 μm in length; however, it seems unlikely that this mechanism is responsible for the formation of longer CLRs, such as those of *Drosophila bifurca*, which measure approximately 4 μm in length (Riparbelli et al., 2024). In this case the simple tubulin diffusion seems only to drive axoneme elongation.

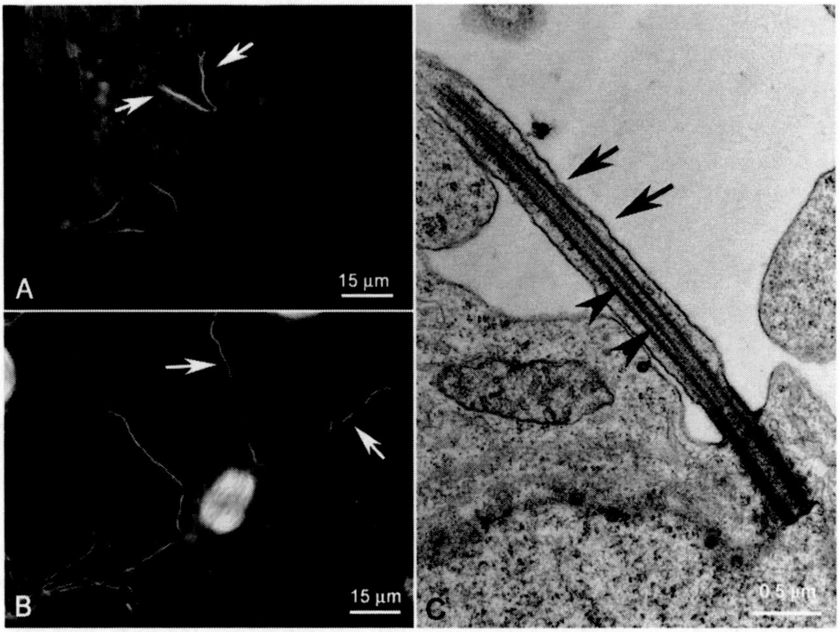

Figure 12.4. Cilium Like Regions (CLRs) (arrows) of prophase (A) and metaphase (B) primary spermatocytes of the lepidopteran Pieris brassicae; each pole of the cells displays a pair of CLRs that elongate further during the meiotic division. (C) Ultrastructural detail of a CLR in mature primary spermatocytes showing the presence of doublets (arrowheads) inside the axoneme.

Ciliary structures resembling the CLRs of *Drosophila* have been occasionally observed in the spermatocytes of different Hexapoda groups. These ciliary structures share a similar axonemal scaffold, but their shape and size can vary among different species. For example, the centriole of the proturan *Acerentomon* that consists of 14 doublet MTs form a large and short CLR (Riparbelli et al., 2009). Whereas the CLRs of some coleopterans resemble those of *Drosophila*, the CLRs observed in the spermatocytes of the crane fly *Nephrotoma* (Diptera) (LaFountain, 1976) and the caddisfly *Potamophylax* (Trichoptera) (Klein and Wolf, 1997) are very elongated.

Some lepidopterans, such as Bombyx (Yamashiki and Kawamura, 1998; Friedländer and Wahrman, 1970), Ephestia (Wolf and Kyburg, 1989) and Pieris (Gottardo et al., 2014) have extraordinarily elongated CLRs (Figure 12.4. A-C), which had already gained attention at the beginning of the last century (Henneguy, 1898; Meves, 1903; Bloodgood, 2009). Remarkably, distinct Y-links have been observed at the transition between the centriole and the ciliary region in Ephestia (Wolf and Kyburg, 1989) and Pieris (Gottardo et al., 2023) spermatocytes in which the CLRs are extraordinarily elongated and display dynein arms associated with the C-tubule. Moreover, the CLRs of Pieris contain thin filamentous structures of electron dense material (Gottardo et al., 2023) that look like the IFT trains that were first observed in Chlamydomonas (Petriman and Lorentzen, 2020; Scholey, 2003). This observation provides new insights into a potential role for IFT in insect spermatocytes and opens new questions about the differential involvement of IFTs in Drosophila spermatogenesis versus other systems.

12.5. Centrioles in the Insect Ommatidia

Eakin (1965) recognized two main types of photoreceptor cells in the animal kingdom, the ciliary form, found in chordates, which consist of photoreceptor cells with a 9 x 2 + 0 cilium, and the rhabdomeric form, typical of arthropods, devoid of ciliary structures. However, the significance of these models in terms of phylogenetic evolution of the photoreceptors is questionable (Salwini-Plawen, 1982). Indeed, these models are not so rigid as rudimentary ciliary structures have been occasionally observed in the rhabdomeric eyes of some onychophorans (Mayer, 2006), and arachnids (Muñoz-Cuevas, 1975).

Of particular interest is the case of some coleopterans where rudimentary ciliary structures have been reported in the retinula cells (Home, 1972, 1976; Wachmann et al., 1983). Such structures in form of short buds were organized by the distal of a pair of tandemly aligned centrioles as it occurs in the axoneme of the sensory organs. In particular, centrioles associated with rudimentary ciliary structures have been observed at the beginning of formation of the rhabdomeres in Leptinotarsa decemlineata; the ciliary structures soon disappear as rhabdomere differentiation proceeds leaving only the tandemly aligned centrioles, which in turn are no longer observed in the fully differentiated rhabdomeres (Wachmann et al., 1983). It is therefore evident that the presence of centrioles and rootlets in photoreceptor cells of rhabdomeric-type insects suggests a potential role for these structures in rhabdom formation.

On the other hand, centrioles aligned in tandem, but without evident ciliary buds, are reported in hymenopterans (Eisen and Youssef, 1980; Wachmann and Hennig, 1974). In this case, the absence of ciliary structures could be explained considering a sort of intermediate condition sporadically occurring during the development of the rhabdomeric photoreceptors. In wasp neurons, two centrioles are aligned with their distal ends in contact with each other (Uzbekov et al., 2024a).

The case of the *Drosophila* ommatidia may represent the terminal event of this process. Tandemly aligned centrioles are observed at the beginning of ommatidia formation in the *Drosophila* imaginal disc eye, but rootlets or ciliary buds are lacking (Figure 12.5.) suggesting that the program of rhabdomere differentiation might be different from other insect ommatidia in which there are rudimentary ciliary structures. However, the *Unc* gene product, a potential orthologue of mammalian *orofaciodigital syndrome 1* (*Ofd1*) (Baker et al., 2004), that is constrained to cell types that have ciliary structures, namely type I sensory neurons and male germ cells, is also present as evenly spaced spots on the eye imaginal discs of third instar larvae (Gottardo et al., 2016b).

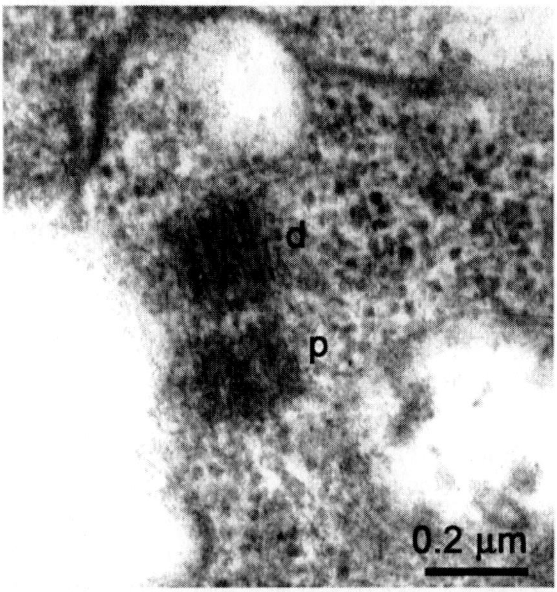

Figure 12.5. Ommatidia in the Drosophila imaginal disc eye. Detail of two tandemly arranged centrioles within the ommatidial cells: d, distal centriole; p, proximal centriole.

Interestingly, the spots of Unc are restricted to the R8 photoreceptor cell which is the founder cell for each ommatidium and promotes the sequential differentiation of the other photoreceptor cells (Tomlinson and Ready, 1987; Morante et al., 2007; Roignant and Treisman, 2009). However, the Unc signal appeared in association with mother centrioles, before the expression of the Boss Bridge of Sevenless (Boss) ligand that is specifically expressed on the apical surface of the R8 photoreceptors (Krämer et al., 1991). Given that *Sas4* mutants that develop without centrioles have normal eye imaginal discs (Basto et al., 2006), it is conceivable that the Unc-associated centriole does not reflect functional activities, and it is not engaged in the process of rhabdomere formation. Rather, the mother centriole to which Unc is associated, may reflect the conservation of an ancestral ciliary type program and could represent the remnant of a vestigial structure present in the fly ancestor. Thus, the *Drosophila* eye could recapitulate the evolutionary trend of the insect eye: loss of the ciliary structures, but maintenance of centrioles aligned in tandem with the distal centriole fated to become a basal body, but unable to nucleate a ciliary axoneme. Tandemly aligned centrioles seems to be characteristic of sensory organs. Indeed, these organelles are observed at the basis of the ciliary axoneme of the sensory organs. Tandemly aligned centrioles are also found at the basis of the axoneme of the mechanosensory bristles at the apex of each ommatidial unit. Remarkably, similar arranged centrioles lacking of axoneme are observed in the flanking sheath cells (Persico et al., 2021). The function, origin, and significance of these centrioles within the sheath cells are still unknown. Tandemly aligned centrioles and striated rootlets have been usually associated with the extension of the neurons in sensory organs. The finding of such structures in sheath cells represents, therefore, a remarkably exception. The distal centriole of the sheath cells is unable to nucleate a ciliary axoneme, suggesting that this property is exclusive of the sensory neurons and likely correlated with the presence of a dendritic extension. Tandemly aligned centrioles without ciliary axoneme have been also observed in the sheath cells of some orthopteran sensilla (Keil, 2012) and hymenopteran sensory organs (Uzbekov et al., 2024a).

Transient ciliary structures with tandemly aligned centrioles have been also described during the development of insect secretory organs (Sreng and Quennedey, 1976; Johnson and Berry, 1977; Keil, 2012; and references therein). These ciliary structure supports the formation of the cuticular duct of the glands and then disappear.

The morphological diversity of centrioles in insects reflects the vast species diversity, their more ancient evolutionary origin (compared, for instance, to vertebrates), and the different conditions of their habitats. Differences in centrosome structure among various insect representatives, combined with comprehensive taxonomic knowledge, enable the reconstruction of the evolutionary pathways of centrioles and centrosomes in living species and consequently allows for parallel analysis of genomes, which makes insects very useful research subjects. Currently, only a limited number of insect species have been extensively studied regarding centrosome structure, and future research will undoubtedly yield valuable insights into the mechanisms underlying centrosome evolution.

Chapter 13

The Role of Centrioles in the Functioning of the Cell and the Organism

Rustem E. Uzbekov[1,*], PhD, HDR, IR
and Alain Debec[2], PhD
[1]Laboratory of Cell Biology and Electron Microscopy, Faculty of Medicine, University of Tours, Boulevard Tonnellé, France;
Faculty of Bioengineering and Bioinformatics, Lomonosov Moscow State University, Moscow, Russia
[2]S Sorbonne Université, Université Paris Cité, Univ Paris Est Créteil, CNRS, IRD, INRAE, Institut d'écologie et des sciences de l'environnement de Paris, IEES, Paris, France

Unlike mitochondria, the endoplasmic reticulum, and other organelles, the centrosome is present in the cell (during the G1 phase of the cell cycle) as a single copy, similar to the nucleus. However, unlike the nucleus, it occupies merely ~0.1% of the cellular volume. This structural minimalism has long motivated biologists to remove the centrosome and observe the functional consequences of its absence.

Centrioles form the structural foundation of the centrosome, and their role within this organelle has been a focal point of biological inquiry. Diverse methods have been employed to eliminate centrosomes from cells. These include microirradiation, microsurgical approaches, and, more recently, strategies such as targeted inhibition of centriolar protein expression and the application of centrinone, a specific inhibitor of Plk4 kinase. This chapter synthesizes findings from these studies, which collectively elucidate the centrosome's role as a central organelle in orchestrating intracellular

* Corresponding Author's Email: rustem.uzbekov@univ-tours.fr

In: The Centrosome
Editor: Rustem E. Uzbekov
ISBN: 979-8-89530-815-8

processes, while distinguishing cellular activities that depend on centrioles from those that do not.

The centriole-free Drosophila cell line serves as a unique experimental model for studying centriole-independent cellular mechanisms and will be analyzed in the first section of this chapter.

13.1. The 1182-4 *Drosophila* Acentriolar Cell Line

13.1.1. Origin of the 1182-4 Cell Line

As with many influential discoveries, the generation of the acentriolar 1182-4 cell line occurred serendipitously. Initially, the laboratory of M. Gans produced ninety-five X-linked female sterile mutants in Drosophila melanogaster via EMS mutagenesis (Gans et al., 1975). Among these, seven mutants yielded lethal haploid embryos (Zalokar et al., 1975). The primary objective was to establish Drosophila haploid cell lines from such embryos. Most mutants arrested at early developmental stages, but mutant #1182 was chosen due to its relatively advanced embryonic progression. Six haploid cell lines were derived from dissociated 1182 embryos, with the 1182-4 line retaining the highest proportion of haploid cells (Debec, 1978). The acentriolar nature of these cells was identified subsequently.

13.1.2. The Path to Conceptualizing Centriole Loss

The 1182-4 line exhibited a high frequency of multinucleated and dikaryotic cells. Rare diploid cells displayed aneuploidies of chromosomes 2 and 3, anomalies absent in wild-type Drosophila cell lines. Notably, most mitotic spindles exhibited a barrel-shaped morphology. These observations suggested potential abnormalities in the mitotic apparatus. The absence of centrioles emerged as a plausible explanation, as barrel-shaped spindles are characteristic of acentriolar systems, such as animal meiotic spindles and the initial cleavage divisions of mouse embryos (Szöllösi et al., 1972). To confirm this hypothesis, ultrastructural analyses of 1182-4 cells were conducted to assess centriolar presence.

13.1.3. Ultrastructural and Immunofluorescence Studies of Acentriolar Mitosis

Initial ultrastructural studies utilized conventional electron microscopy on ~50–60 nm ultrathin sections (Debec et al., 1982). Both random cell profiles and serial reconstructions of mitotic cells were analyzed. No centrioles were detected at spindle poles or elsewhere in the cytoplasm. Instead, small foci of dense filamentous material, associated with MTs, were observed at the poles. These structures were interpreted as acentriolar MT-organizing centers (MTOCs) (Figure 13.1.).

A follow-up study employed high-voltage electron microscopy on 1 μm-thick serial sections of entire 1182-4 cells. Consistent with prior findings, centrioles remained undetectable, whereas control wild-type cells exhibited centrioles in every instance (Szöllösi et al., 1986) (Figure 13.2.).

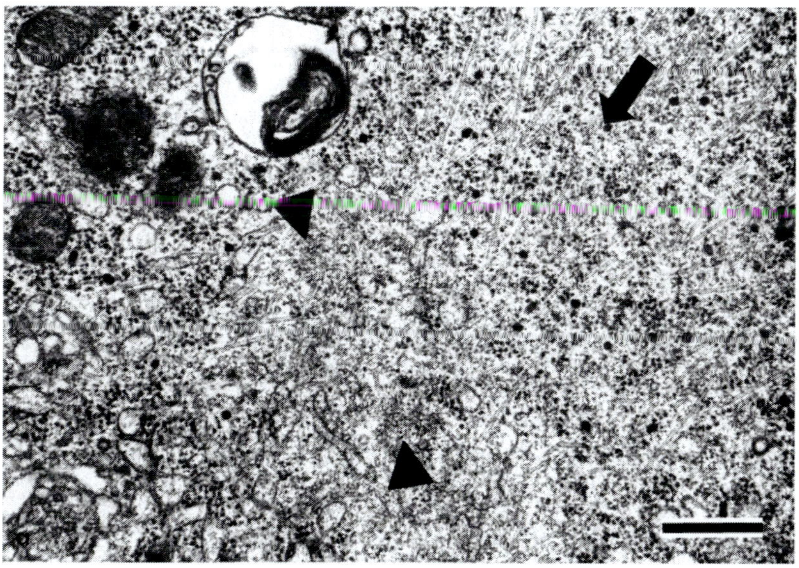

Figure 13.1. Ultra-thin section through a half-spindle from a 1182-4 cell. At the spindle pole, MTs converge toward multiple electron-dense areas (arrowheads) clustered in a cytoplasmic region devoid of other organelles. Arrow: spindle axis. Scale bar: 0.5 μm.

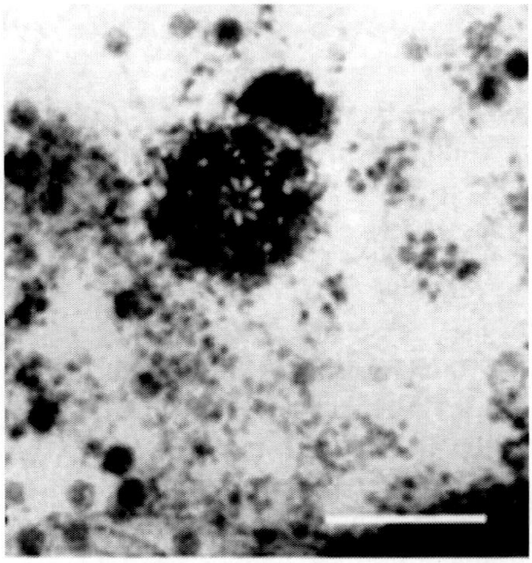

Figure 13.2. Thick section (0.5 μm) of a control Kc cell displaying a diplosome with the mother centriole in cross-section. Scale bar: 0.2 μm.

These two ultrastructural studies definitively established that 1182-4 cells lacked centrioles. This represented the first unambiguous demonstration of animal cells undergoing regular division in the absence of this organelle.

In subsequent years, the growing availability of antibodies specific to centrosomal components enabled comprehensive evaluation of entire cell populations and facilitated detailed analysis of acentriolar mitosis. Immunostaining for CP190, a centrosomal marker, revealed that four additional 1182-derived lines exhibited a proportion (36–88%) of acentrosomal cells. The absence of centrioles in 1182-4 cells, previously documented, was therefore concluded to be non-random and intrinsically linked to their unique origin (Debec and Abbadie, 1989).

13.1.4. Identification of the Maternal Haploid Gene

A pivotal discovery was the identification of the gene disrupted in the maternal haploid mutant as CG9203 (Delabaere et al., 2014; Tang et al., 2017). CG9203 encodes the Drosophila ortholog of human Spartan/DCV1, a conserved metalloprotease essential for processing DNA-protein crosslinks (DPCs). DPCs induce severe DNA damage and can obstruct DNA replication machinery, as demonstrated in human cells, yeast, nematodes, and flies (Stingele et al., 2017).

The CG9203 allele in mh1 flies carries a non-synonymous substitution (M151K) affecting a conserved residue within the SprT domain, likely abolishing protease activity (Delabaere et al., 2014). The mechanistic link between Spartan inactivation and centriole absence remains unclear (see Section 13.1.6 for hypotheses).

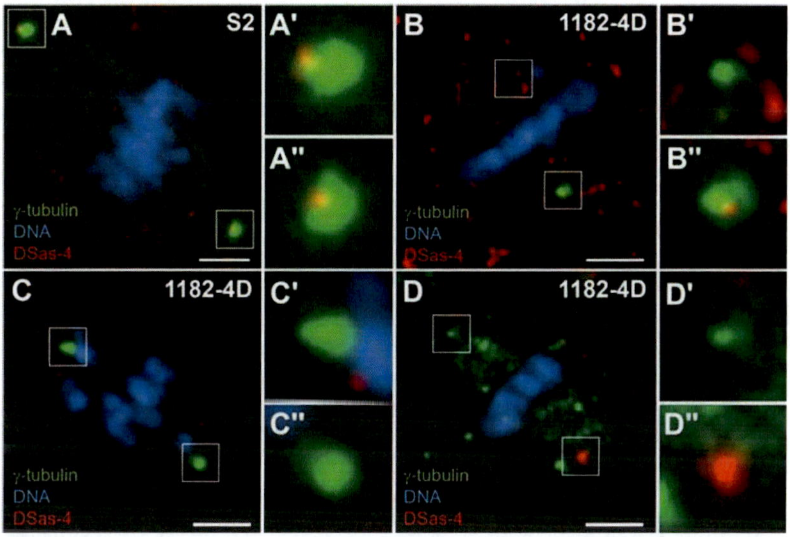

Figure 13.3. Molecular and structural characterization of spindle poles in control S2 cells and acentriolar 1182-4D cells. Courtesy of Sara Moutinho-Pereira and Helder Maiato. Immunofluorescence images showing γ-tubulin (green), DNA (blue), and DSas-4 (red). (A) Control S2 cell. (B–D) 1182-4D acentriolar cells. Insets display enlarged views of spindle poles. DSas-4 accumulated at the centrosomal core in S2 cells but was predominantly dispersed in the cytoplasm of 1182-4D cells. Scale bars: 5 μm.

13.1.5. Acentriolar Mitosis and Centriole Roles in Development

The 1182-4 cell line provided the first evidence of animal cells proliferating indefinitely without centrioles. However, this phenomenon was long viewed as a cell culture-specific anomaly.

A critical advance in understanding centrosome dispensability during Drosophila development was the discovery of centrosomin (cnn), a key pericentriolar material component. Despite non-functional centrosomes, cnn mutant flies exhibit normal morphology and behavior (Megraw et al., 2001).

Notably, centriole structure remains intact in cnn mutants, unlike in *DSas-4* mutants (discussed later).

Basto et al., (2006) further demonstrated that centriole loss in *DSas-4* mutants does not impair development or adult fly viability, though adults lack coordination due to absent cilia. Centrioles, however, remain essential during early embryogenesis (e.g., syncytial divisions) (Rodrigues-Martins et al., 2008).

A natural absence of centrioles as mitotic organizers was unknown until their discovery in planarians (Azimzadeh et al., 2012). In these organisms, centrioles are critical for ciliogenesis but dispensable for mitosis. Given Earth's biodiversity, future studies will likely identify additional animal models defying centriole requirement in development (Sullivan, 2015).

13.1.6. Hypotheses on the Acentriolar State of 1182-4 Cells

The persistent absence of centrioles in 1182-4 cells remains enigmatic. Multiple hypotheses have been proposed to explain this phenomenon.

13.1.6.1. Hypothesis 1: Loss of Basal Body Complex

One initial assumption posited that the paternal chromosome loss during early embryogenesis might also eliminate structures like the basal body complex. However, this explanation is likely incomplete. Loppin et al., (2001) observed catastrophic first mitoses in most mh eggs, yet embryos bypassing early arrest execute syncytial divisions with normal asters until cellularization. Crucially, centrioles are indispensable for early syncytial stages (Rodrigues-Martins et al., 2008).

13.1.6.2. Hypothesis 2: De Novo Centriole Biogenesis Block

Centrioles can regenerate *de novo* even after experimental ablation (e.g., irradiation, centrinone treatment) (see Sections 13.2–13.4). Structural continuity is unnecessary, as shown by Blachon et al., (2014), where basal body/PCL precursor loss in zygotes did not preclude later centriole formation. Thus, the acentriolar state may stem from a blockage in centriole biogenesis pathways. Genomic analyses of 1182-4 cells revealed no loss-of-function mutations in key centriole biogenesis genes (*SAK, Sas-6, Sas-4, Asl, Cep135/bld10, ana1, ana2, ana3, Cp110*) (Debec et al., 2020). Observed sequence variants were likely benign, though allelic identity of CG9203 with the original mh1 strain was confirmed.

13.1.6.3. Hypothesis 3: Spartan Inactivation and Replication Stress

Spartan/DCV1 dysfunction in mh1 cells may impair centriole duplication. Persistent DNA-protein crosslinks (DPCs) due to Spartan deficiency could induce chronic replication stress, triggering DNA damage responses that disrupt centriole cycles. Similar centrosome inactivation occurs in replication-stressed Drosophila embryos (Sibon et al., 2000).

13.1.6.4. Hypothesis 4: Diploid-Haploid Transition Disruption

Haploid-diploid transitions in non-physiological contexts (e.g., cell culture) might uncouple DNA replication and centriole duplication. While haploid male Drosophila and diploid wasp cells maintain normal centriole numbers (Uzbekov et al., 2018a, 2024a), the artificial haploidization of 1182-4 cells could disrupt coordination. Notably, centriole loss occurs in near-haploid human HAP1 cells (Yaguchi et al., 2018). However, this hypothesis is weakened by the acentriolar state of diploidized 1182-4 sublines.

13.1.6.5. Hypothesis 5: Wolbachia Endosymbiont Interference

Proliferation of Wolbachia pipientis - a MT-associated endosymbiont in Drosophila - might destabilize centrioles. Despite antibiotic treatment failing to restore centrioles, Wolbachia's spindle pole localization and similarities between mh defects and Wolbachia-induced incompatibility (Charlat et al., 2001) warrant consideration.

13.1.7. Unresolved Questions Regarding the Acentriolar Nature of the 1182-4 Cell Line

A critical consideration is that, unlike experimentally ablated centrioles (discussed in subsequent chapters), centrioles lost in 1182-4 cells cannot regenerate *de novo*.

These observations suggest two possibilities: (1) a critical element has been lost (e.g., the original haploid genome configuration or Wolbachia symbionts) or (2) a step in centriole biogenesis is irreversibly blocked. The latter scenario implicates Spartan dysfunction as the most plausible candidate. While the mechanism linking persistent DNA-protein crosslinks (DPCs) to centriole loss remains unclear, Spartan's multifunctional domains may play a role. Future studies must elucidate how Spartan inactivation disrupts centriole assembly.

13.2. Study of Centrosome Functions via Microsurgical Methods

Microsurgical approaches to generate centriole-free cells fall into two categories: (1) microneedle-mediated excision of a cell portion (Maniotis and Schliwa, 1991) and (2) centrifugal removal of cytoplasmic fragments perpendicular to the culture plane, optionally combined with MT- and/or actin-depolymerizing agents. The latter method yields karyoplasts (nucleated cell fragments with minimal cytoplasm) or cytoplasts (enucleated cell fragments lacking a nucleus) (Prescott et al., 1972), and under specific conditions, cells lacking both nucleus and centrosome (Karsenti et al., 1984a; Rodionov et al., 1999).

Early studies demonstrated that enucleation with cytochalasin retains centrioles in cytoplasts but not in karyoplasts (Goldman et al., 1973; Shay et al., 1974). Notably, some karyoplasts from mouse L929 cells regenerated cytoplasm and re-entered mitosis (Lucas et al., 1976, 1978). Subsequent work revealed *de novo* centriole biogenesis in karyoplasts (Zorn et al., 1979). Although centriole absence and restoration were inferred statistically from random sections rather than exhaustive serial reconstructions, this study pioneered the concept of centriole regeneration in somatic cells. Single centrioles emerged in karyoplasts within 24 hours, reaching normal numbers by 72 hours under thymidine-induced G1/S arrest. Thus, nucleated fragments can rebuild centrioles and resume cycling, unlike cytoplasts, which lack nuclei. A critical unanswered question concerns centrosome fate in cytoplasts.

Immediately after enucleation, cytoplasts have different shapes (Pittenger and Cleveland, 1985), but after several hours they take a shape similar to the shape of whole cells. Like whole cells, cytoplasts spread well on the substrate (Prescott et al., 1972), and in terms of functional parameters do not differ from whole cells (Prudovsky and Zelenin, 1978). For example, cytoplasts obtained from mast cells retain the normal orientation of the plasma membrane, ATP is synthesized in them, and labeled fatty acids are capable of combining into fat complexes (Kennerly, 1987). Cytoplasts retain many properties inherent in whole cells - the above-mentioned external shape, pinocytotic activity, the ability to move and synthesize proteins. This is one of the reasons why cytoplasts are a common model for embryological studies (Mohammed et al., 2008; Bowles et al., 2008). And not only experimental studies, but also practical developments, thanks to which it became possible to transfer nuclei of somatic cells into an enucleated ooplast and thus clone mammals (reviewed by Fulka and Fulka, 2007). It has also been shown that the bulk of cytoplasts retain most of the cytoskeleton components, including MTs (Caron et al., 1985b; Karsenti et al., 1984a; Gorgidze, Vorobjev, 1995; Komarova et al., 2002; Brodsky et al., 2007). Centrioles, if they are preserved after enucleation, are located in the center of the cytoplast (Goldman et al., 1975), their normal location and the structure of the centrosome are not disrupted due to the removal of the nucleus (Gorgidze, Vorobjev, 1995). Thus, cytoplasts are a classic object for studying the functions of the centrosome that are autonomous from the nucleus (Gorgidze, Vorobjev, 1995). Considering the possibility of obtaining both cytoplasts lacking a nucleus and cytoplasts lacking both a nucleus and a centrosome, they are a convenient and adequate model for studying the centrosome's own capabilities, independent of the nucleus, for example, those associated with its ability to duplication or nucleation of MTs.

When cells are enucleated by centrifugation after depolymerization of actin microfilaments using cytochalasin B, the centrosome mainly remains in the cytoplasts. According to various authors, a pair of centrioles is preserved either in the majority (Brown et al., 1980; Gorgidze and Vorobjev, 1992) or in virtually all cytoplasts (Goldman et al., 1975). In a small part of cytoplasts only one centriole is preserved - either the mother or the daughter (Gorgidze and Vorobjev, 1992). The structure of the centrosome in cytoplasts 2 hours after enucleation did not differ from the normal one, however, after 14-16 hours, free foci of MT convergence were found in the centrosome in some cytoplasts, which were not observed normally (Gorgidze and Vorobjev, 1992). In the absence of a nucleus, centrioles did not lose the ability to change their position in the cell volume and reorient themselves perpendicular to the substrate in response to the action of energy metabolism inhibitors (Gorgidze and Vorobjev, 1992), which is typical of centrioles in native cells (Alieva, Vorobjev, 1989d, 1992a, 1992b).

Immediately after enucleation, cytoplasts exhibit varied morphologies (Pittenger and Cleveland, 1985), but within hours, they adopt a shape resembling intact cells. Like whole cells, cytoplasts adhere effectively to substrates (Prescott et al., 1972) and functionally mirror intact cells (Prudovsky and Zelenin, 1978). For instance, mast cell-derived cytoplasts maintain plasma membrane polarity, synthesize ATP, and incorporate labeled fatty acids into lipid complexes (Kennerly, 1987). Cytoplasts retain numerous whole-cell attributes, including morphology, pinocytotic activity, motility, and protein synthesis, making them a standard model in embryological research (Mohammed et al., 2008; Bowles et al., 2008). Beyond basic studies, cytoplasts enable practical applications such as somatic cell nuclear transfer into enucleated ooplasts for mammalian cloning (Fulka and Fulka, 2007). Cytoplasts also preserve cytoskeletal components, including MTs (Caron et al., 1985b; Karsenti et al., 1984a; Gorgidze and Vorobjev, 1995; Komarova et al., 2002; Brodsky et al., 2007). Centrioles retained post-enucleation localize centrally, with centrosomal architecture intact (Goldman et al., 1975; Gorgidze and Vorobjev, 1995). Thus, cytoplasts serve as a classical system for studying centrosomal functions independent of nuclear input (Gorgidze and Vorobjev, 1995). The ability to generate cytoplasts lacking either the nucleus alone or both nucleus and centrosome renders them ideal for probing autonomous centrosomal activities, such as duplication or MT nucleation.

Centrosomes predominantly remain in cytoplasts after cytochalasin B-mediated actin depolymerization and enucleation. Most studies report retention of centriole pairs in the majority (Brown et al., 1980; Gorgidze and Vorobjev, 1992) or all cytoplasts (Goldman et al., 1975). A minor subset contains only one centriole (mother or daughter) (Gorgidze and Vorobjev, 1992). Centrosomal structure in cytoplasts matches controls at 2 hours post-enucleation, but by 14–16 hours, aberrant MT convergence foci emerge in some cases (Gorgidze and Vorobjev, 1992). Without nuclei, centrioles retain mobility and reorient perpendicularly to the substrate under metabolic inhibition, mirroring behavior in intact cells (Alieva and Vorobjev, 1989d, 1992a, 1992b).

It was also shown that in cytoplasts obtained by enucleation of L line cells, previously synchronized by selection in mitosis and located in the late G1 period, centrioles duplicated (Gorgidze and Vorobjev, 1994). Thus, centrioles in cultured fibroblasts can duplicate in the absence of RNA synthesis. Since procentrioles did not appear under conditions of complete suppression of RNA synthesis in whole cells (incubation of synchronized cells, starting from the end of the G1 period in a medium with actinomycin D for 16 hours), the authors suggested that the cell nucleus had not only a positive but also a negative effect on centriole duplication.

Cytoplasts are a convenient model for experiments related to the study of the ability of the centrosome to polymerize MTs and organize them into a radial system. It has been shown that these functions of the centrosome do not depend on the nucleus: in enucleated cell fragments, a radial network of MTs is located around the centrosome, similar to the MT system in intact cells, while in cytoplasts lacking a centrosome, MTs were present but located randomly (Karsenti et al., 1984a; Rodionov et al., 1999; Komarova et al., 2002; Chernobelskaya et al., 2004; 2009). At the same time, nuclear and enucleated cells react in the same way to MT-depolymerizing effects: free

MTs are disassembled first, and only then MTs associated with the centrosome (Chernobelskaya et al., 2004).

It has been shown that the level of tubulin synthesis in cytoplasts after washing off colcemid is significantly lower than in nucleated cells, despite the apparent similarity in the number and location of MTs (Caron et al., 1985b). Apparently, this is due to the fact that the level of tubulin mRNA in cytoplasts remains quite low after removal of the effect, whereas in intact cells the level of tubulin mRNA increases significantly (Caron et al., 1985a). Nevertheless, during the recovery process, the dynamics of growth in the number of centrosome-associated MTs, as well as their length, in cytoplasts lag only slightly compared to whole cells (Chernobelskaya et al., 2004).

The main difference between intact cells and cytoplasts is the differences in the dynamic properties of MTs: the behavior of most MTs in centrosome-containing cytoplasts is described by the parameters of dynamic instability, as is their behavior in intact cells, while in centrosome-less cytoplasts MTs are characterized by treadmilling (Rodionov et al., 1999). It should be noted that the switch to treadmilling occurs only in centrosome-less cytoplasts obtained from fibroblasts, while the behavior of MTs in centrosomal-less cytoplasts obtained from epithelial cells was described by the parameters of dynamic instability (Rodionov et al., 1999). The authors believe that the behavior of MTs in cytoplasts lacking a centrosome correlates with the degree of influence of the centrosome on the MT system in intact cells, and suggest that a certain factor stabilizing the minus ends of MTs is present in the cytoplasm of epithelial cells and cytoplasts obtained from them. Such a factor is absent in the cytoplasm of fibroblasts and cytoplasts obtained from them, probably being completely localized on the centrosome (Rodionov et al., 1999).

It was demonstrated that cytoplasts derived from enucleated L-line cells (synchronized in mitosis and arrested in late G1) retain centriole duplication capacity (Gorgidze and Vorobjev, 1994). Thus, centrioles in fibroblasts can duplicate without RNA synthesis. However, procentriole formation was absent in whole cells under complete RNA suppression (actinomycin D treatment post-G1 for 16 hours), suggesting the nucleus exerts both positive and negative regulatory effects on centriole duplication.

Cytoplasts serve as a robust model for studying centrosome-dependent MT polymerization and organization. Centrosomes in enucleated fragments assemble radial MT networks akin to intact cells, whereas centrosome-free cytoplasts exhibit disorganized MTs (Karsenti et al., 1984a; Rodionov et al., 1999; Komarova et al., 2002; Chernobelskaya et al., 2004, 2009). Nuclear presence does not influence centrosomal responses to MT-depolymerizing agents: free MTs disassemble prior to centrosome-anchored MTs in both intact cells and cytoplasts (Chernobelskaya et al., 2004).

Tubulin synthesis in cytoplasts post-colcemid washout is reduced compared to nucleated cells, despite comparable MT numbers and distribution (Caron et al., 1985b). This likely reflects low tubulin mRNA retention in cytoplasts versus mRNA upregulation in intact cells (Caron et al., 1985a). Nevertheless, MT regrowth dynamics (number and length) at centrosomes are only marginally delayed in cytoplasts (Chernobelskaya et al., 2004).

A key distinction between intact cells and cytoplasts lies in MT dynamics: centrosome-containing cytoplasts display dynamic instability like intact cells, while centrosome-free fibroblast cytoplasts exhibit treadmilling (Rodionov et al., 1999). Notably, centrosome-free epithelial cytoplasts retain dynamic instability, implying epithelial cytoplasm harbors factors stabilizing MT minus ends, absent in fibroblasts where such factors may be centrosome-localized (Rodionov et al., 1999).

In the study by Maniotis and Schliwa (1991), the centrosome was microsurgically removed from cultured BSC-1 cells along with a cytoplasmic fragment (cytoplast). The residual karyoplast retained sufficient cytoplasm for survival and growth. Both karyoplasts (centriole-free) and cytoplasts (containing centrioles) were monitored. Karyoplasts restored the perinuclear MT-organizing center (MTOC), astral MT organization, and a compact Golgi complex. These cells entered and likely completed S-phase, expanding beyond typical BSC-1 cell size. However, centrioles failed to regenerate even after >10 cell cycle equivalents, and mitosis was never observed. In contrast, karyoplasts retaining centrioles post-surgery progressed through the cell cycle and formed clones. The authors concluded that centriole removal uncoupled cell growth from proliferation, halting centriole biogenesis and duplication.

Subsequent work revealed that this interpretation overlooked external factors during prolonged live imaging (Uetaki et al., 2007). Critical among these was exposure to blue or intense green light. Minimizing such interference allowed centriole-ablated cells to undergo mitosis and re-enter S-phase. As this study combined microsurgery and laser ablation, its findings will be detailed in Chapter 13.3.

13.3. Centrosome Function Studies via Microirradiation

Ultraviolet (UV) irradiation, employed for localized ablation of biological structures since the early 20th century (Tschachotin, 1912; reviewed in Chachotin, 1959), induces photochemical reactions in proteins and nucleic acids (Vladimirov and Potapenko, 1989). In proteins, UV triggers S–S and peptide bond cleavage (Lvov et al., 1975; Burshtein, 1976), leading to denaturation (Vladimirov and Potapenko, 1989). Forer (1965) demonstrated that UV microirradiation locally disrupts the mitotic spindle, creating an Area of Reduced Birefringence (ARB) where spindle MTs lose ordered alignment. Later studies confirmed UV's ability to sever MTs *in vitro* and *in vivo,* with action spectra closely matching, validating direct MT targeting (Hughes et al., 1988).

Laser microirradiation of centrosomes during prophase - early metaphase in the 1970s–1980s revealed extreme cellular sensitivity at these stages (Peterson and Berns, 1980; Berns et al., 1981; Vorobjev et al., 1988). Cells failed to progress through mitosis post-irradiation, even with ultrastructurally intact centrosomes (Vorobjev et al., 1988). Resistance increased from late prometaphase onward (Berns et al., 1977; Vorobjev et al., 1988). Unlike laser ablation, UV microirradiation halted mitosis at metaphase, likely reflecting distinct mechanisms of damage between visible-light lasers and UV.

Uzbekov and colleagues (Uzbekov and Vorobjev, 1991a,b, 1992; Uzbekov et al., 1995; Uzbekov, 2004) utilized a UV microirradiation setup (280 nm, beam diameter ~1.6 μm). Their work highlighted divergent cell responses to centrosomal irradiation in metaphase versus anaphase. These differences arise because UV damages both regulatory centrosomal proteins and adjacent spindle MTs. Centrosome irradiation severs kinetochore MT bundles approaching the centrosome.

In metaphase cells, centrosomal microirradiation induced degradation of the irradiated half-spindle, prompting chromosome migration to the non-irradiated pole (Uzbekov and Vorobjev, 1991a). Chromosomes coalesced into a "rosette" around the intact pole, followed by failed cytokinesis and multinucleation upon interphase transition (Figure 13.4.). In contrast, anaphase irradiation displaced the centrosome to the cell periphery without blocking mitosis or cytokinesis, yielding two daughter cells (Uzbekov and Vorobjev, 1991b, 1992).

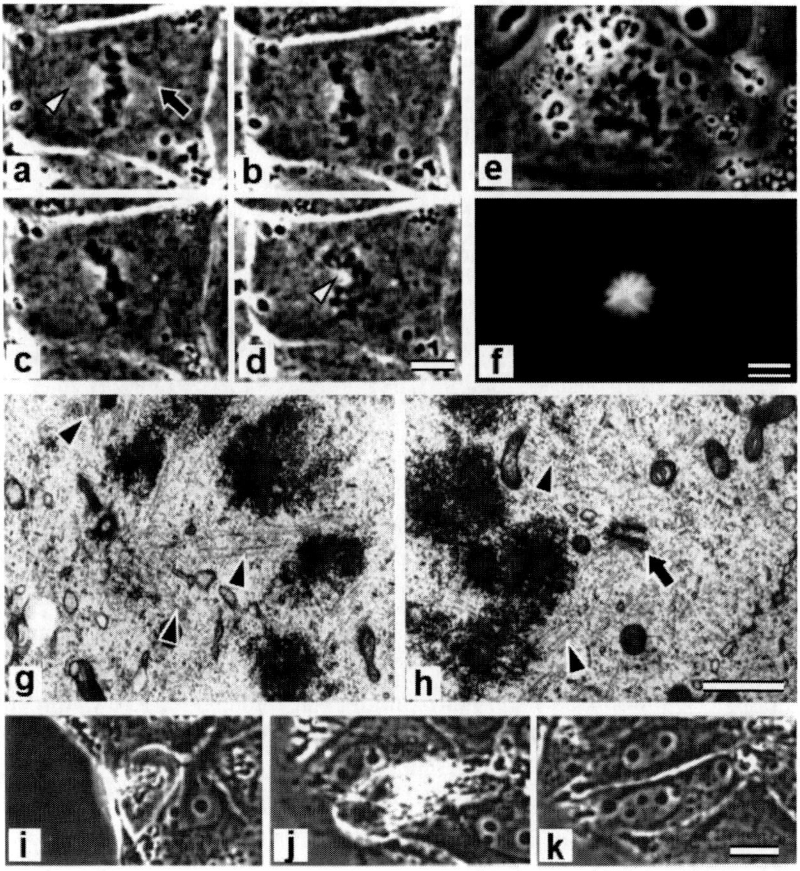

Figure 13.4. Ultraviolet microirradiation of the centrosome (spindle pole) in mitotic metaphase of the pig embryonic kidney (SPEV) cell line. (a–d) Live-cell imaging before (a) and 1 min (b), 3 min (c), 6 min (d) post-microirradiation of the right spindle pole (arrow). The non-irradiated pole (white arrowhead) becomes the focus of chromosome migration. (e–f) Phase-contrast (e) and α-tubulin immunofluorescence (f) 80 min post-irradiation, showing monopolar spindle formation. (g–h) TEM images 1 h 39 min post-irradiation: intact centrosome within the chromosome "rosette" and irradiated centrosome (arrow) exterior to the rosette. MT bundles (black arrowheads) are visible. (i–k) Phase-contrast time-lapse 2 h 55 min (i), 6 h 45 min (k), and 8 h post-irradiation, demonstrating failed cytokinesis and binucleation. Scale bars: a–f = 5 μm; g,h = 1 μm; i–k = 10 μm. Adapted from Uzbekov and Vorobjev, 1991a, with modifications.

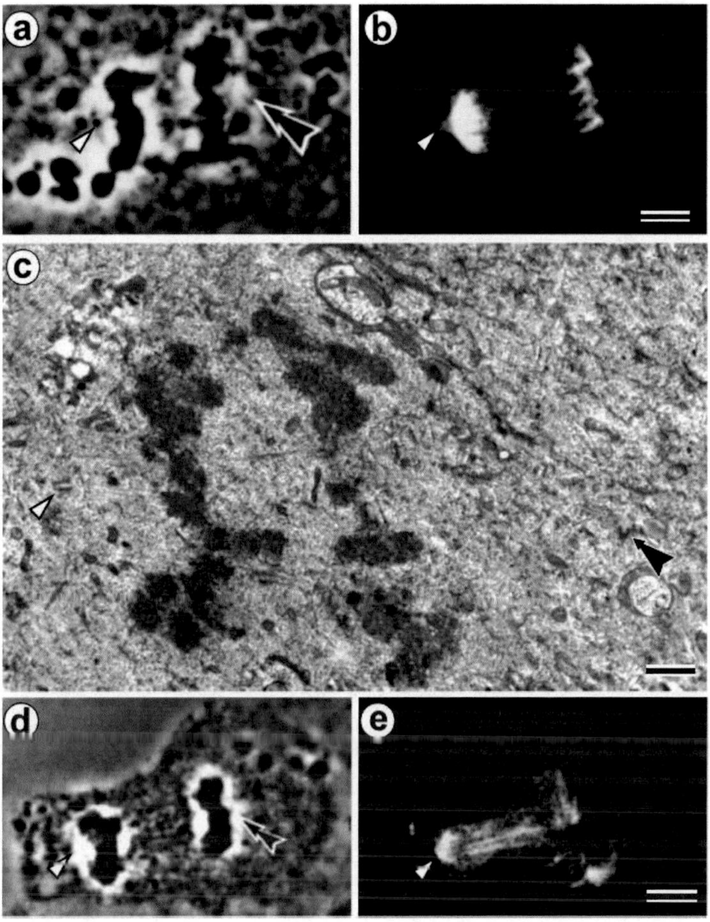

Figure 13.5. Ultraviolet microirradiation of the centrosome (spindle pole) during anaphase in pig embryonic kidney (SPEV) cells. (a–b) Phase-contrast (a) and α-tubulin immunofluorescence (b) 3 min post-irradiation of the spindle pole (arrow). The non-irradiated pole (white arrowhead) displays intact half-spindle MTs converging at the pole, with chromosomes migrating toward it. Irradiated pole MTs lose polarity, and chromosome movement slows. (c) TEM image 3 min post-irradiation: intact half-spindle MTs at the non-irradiated pole (white arrowhead); irradiated centrosome (arrow) detaches from spindle MTs, positioned twice as far from chromosomes. (d–e) Phase-contrast (d) and α-tubulin immunofluorescence (e) 15 min post-irradiation. Interzonal MTs remain organized, cytokinesis initiates midway between chromosome groups, and the daughter cell inheriting the irradiated centrosome retains most maternal cytoplasm. Scale bars: a,b,d,e = 5 μm; c = 0.5 μm. Adapted from Uzbekov and Vorobjev, 1991b; Uzbekov et al., 1995, with modifications.

Anaphase cells tolerated triple the metaphase-blocking irradiation dose without mitotic arrest (Figure 13.5.), indicating a critical threshold beyond which centrosome involvement becomes dispensable for mitotic completion.

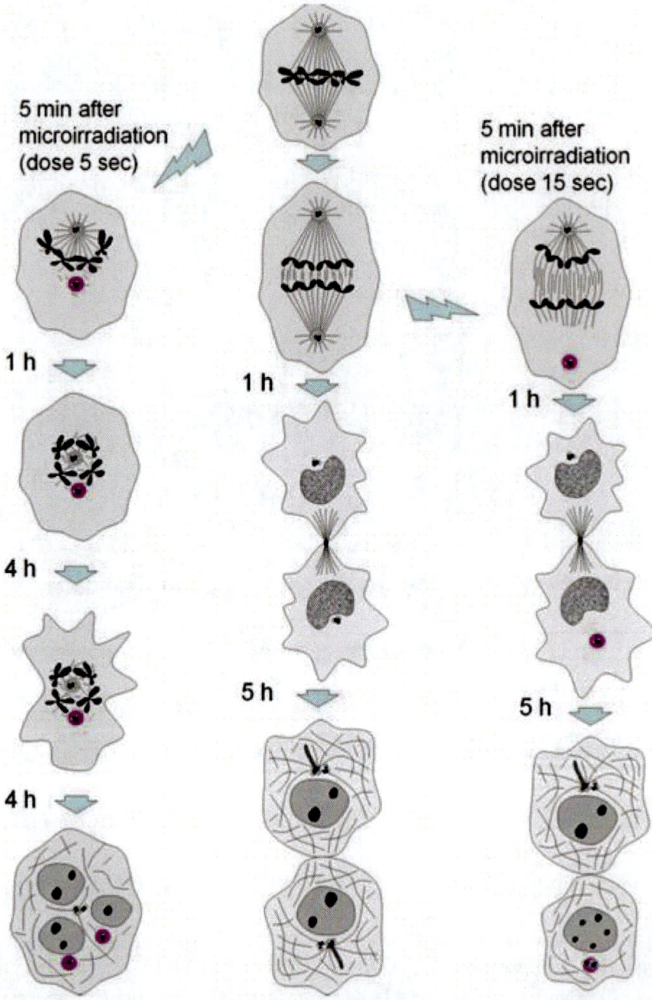

Figure 13.6. Summary diagram illustrating the effects of UV microirradiation of the centrosome during metaphase (left panel) and anaphase (right panel) on SPEV cell cycle progression. The irradiated lower pole is marked (red circle). Middle panel: control cell division. Time intervals between stages are indicated. Post-interphase, only non-irradiated centrosomes form primary cilia and initiate procentriole assembly.

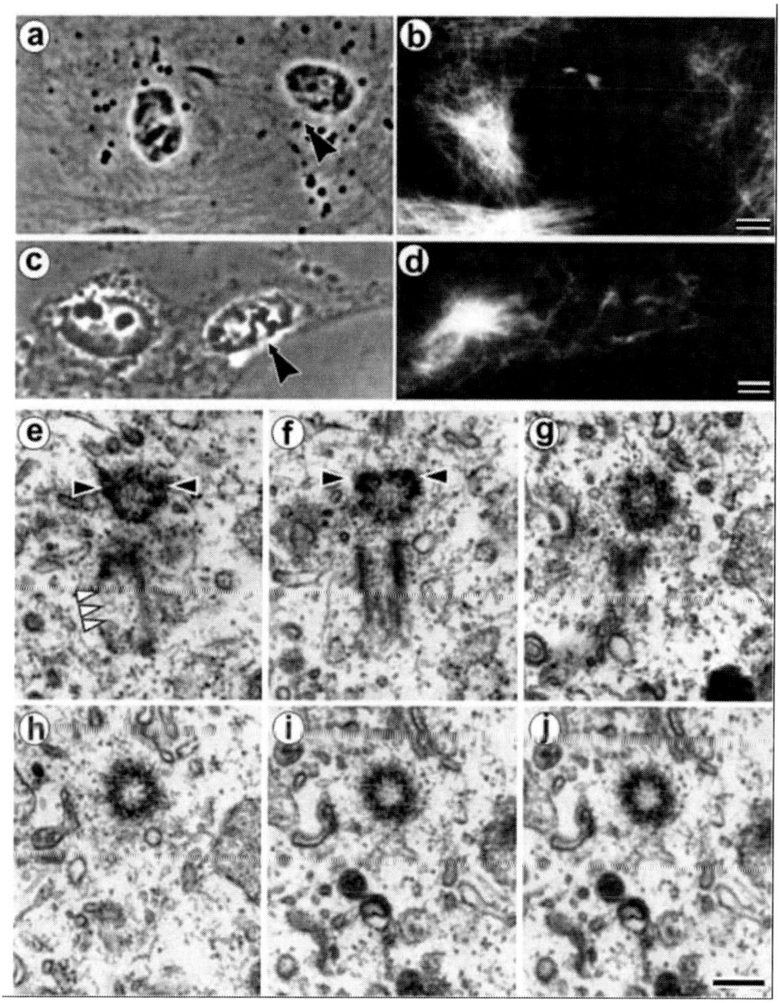

Figure 13.7. Differential centrosome activity in sister SPEV cells post-anaphase UV irradiation. (a–b) Phase-contrast (a) and α-tubulin immunofluorescence (b) 1 h 10 min post-irradiation. The irradiated centrosome (arrow) exhibits a sparse MT network compared to the intact centrosome. (c–d) Phase-contrast (c) and α-tubulin labeling (d) 3 h 12 min post-irradiation with nocodazole (0.15 μg/ml, 35 min). The intact centrosome forms a robust MT aster, absent in the irradiated cell (arrow). (e–j) TEM serial sections of the irradiated centrosome 3 h 52 min post-treatment. The daughter centriole shows wall damage (white arrowheads); the mother centriole lacks appendages and MT-nucleation capacity. Scale bars: a–d = 5 μm; e–j = 0.2 μm. Adapted from Uzbekov and Vorobjev, 1991b, 1992; Uzbekov et al., 1995.

As demonstrated by later laser microirradiation studies (Uetaki et al., 2007), the critical factor was not the centrosome's regulatory role but its structural-mechanical function in anchoring MTs at the spindle pole, including both kinetochore MTs connecting to chromosomes and astral MTs extending toward the cell periphery.

In metaphase, UV-induced MT destruction at the centrosome disrupted MT restoration and force balance, preventing sister chromatid separation and driving chromosome migration to the non-irradiated pole (Figure 13.4.). In anaphase, centrosomal UV microirradiation similarly severed MTs, but chromatid separation had already occurred. Mitosis completed via anaphase A (chromosome movement) and anaphase B (chromosome group separation), resulting in a larger daughter cell with the irradiated centrosome, as cytokinesis initiated midway between chromosome clusters (Figure 13.5. Uzbekov and Vorobjev, 1991b).

Figure 13.6. summarizes centrosome UV microirradiation outcomes in metaphase and anaphase. Partial kinetochore MT severing (cf. Spurck et al., 1990) selectively disrupted associated chromosome movement. Post-anaphase irradiation produced one daughter cell with an irradiated centrosome and one with an intact centrosome.

This model revealed long-term centrosome irradiation effects: cells with irradiated centrosomes failed to form radial MT systems (Figure 13.7. Uzbekov and Vorobjev, 1991b, 1992), ceased growth and S-phase entry (Neverova et al., 1996), and exhibited impaired primary cilium formation, centriole duplication, cell spreading, nucleoli development, and B23 protein accumulation (Uzbekov and Vorobjev, 1992; Neverova et al., 1996, 1999).

Notably, irradiated centrioles often retained close proximity and perpendicular orientation (Figure 13.7e,f), suggesting blockage during the centriolar disorientation phase. Partial centriolar wall damage was also observed (Figure 13.7e).

Centrosomal irradiation impacts not only centrioles but all proteins within the irradiated spindle pole region. In mammalian cells, critical mitotic regulators localize near spindle poles in regions far exceeding centriole dimensions (Alieva and Uzbekov, 2016).

These proteins include a significant fraction of the p34-cyclin B complex (Bailly et al., 1992), a key regulator of the cell cycle (Nurse, 1990). In anaphase, cyclin B degradation inactivates the p34 cdc2 (CDK1) kinase, which is later reactivated via site-specific phosphorylation to drive G1/S transition (Moreno and Nurse, 1994). UV microirradiation of the centrosome in anaphase may denature CDK1, preventing its phosphorylation and subsequent cyclin B degradation, thereby blocking cell cycle progression. Additionally, the B23 protein, critical for nucleolar organization and centriole replication (Okuda et al., 2000), localizes to the centrosome during mitosis and G1 (Zatsepina et al., 1995, 1999). According to Okuda et al., (2000), unphosphorylated B23 remains centrosome-associated post-mitosis until CDK2/cyclin E phosphorylates it at the restriction point, triggering centriole replication. UV-induced B23 damage likely disrupts this phosphorylation, halting centriole and cell cycles.

Khodjakov, Sluder, and colleagues advanced centrosome studies using laser microirradiation to ablate centrioles, tracked via GFP-γ-tubulin fluorescence in CVG-2 cells (Khodjakov et al., 1997; Khodjakov and Rieder, 2001; Khodjakov et al., 2002; Uetaki et al., 2007) (Figure 13.8.). Uetaki et al., (2007) combined microsurgery and laser ablation, revealing that external stressors critically influence centrosome-ablated cell fate. Their findings necessitated re-evaluation of prior interpretations by the authors and others (see below). As in microsurgical centrosome removal experiments in BSC-1 cells (Maniotis and Schliwa, 1991), laser-ablated centrosomes in early G1 phase failed to regenerate (Khodjakov and Rieder, 2001). However, centrosome ablation during S phase triggered *de novo* centriole formation within 24 hours (Khodjakov et al., 2002), mechanistically resembling centriole biogenesis during ciliogenesis and embryogenesis (Dirksen and Croker, 1966; Sorokin, 1968; Palazzo et al., 1992; Abumuslimov et al., 1994).

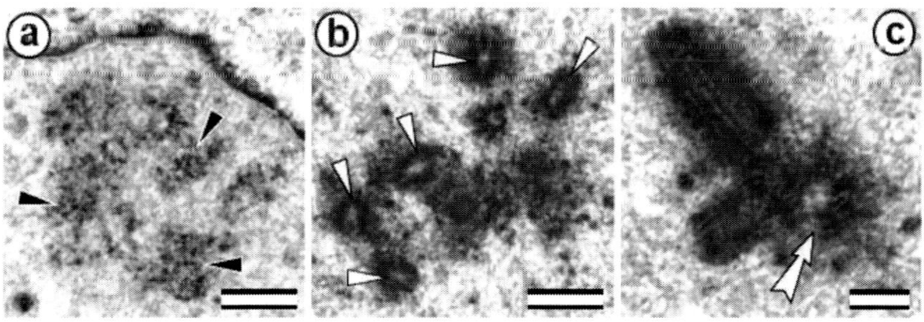

Figure 13.8. Centriole regeneration after laser ablation. (a) *De novo* γ-tubulin/GFP foci (black arrowheads) form PCM clouds without centrioles 8–9 h post-centrosome removal (0.25-μm section). (b) By 24 h, γ-tubulin foci contain multiple centrioles, many structurally abnormal (14 total in this cell; 0.25-μm section). (c) High-magnification view of centriolar defects: truncated blades and incomplete MT triplets (arrow). Scale bars: a,b = 0.5 μm; c = 0.2 μm. Images courtesy of A.L. Khodjakov (Khodjakov et al., 2002).

This restoration occurs post-restriction point, where centriole replication pathways are active and cell cycle checkpoints are bypassed (Hinchcliffe et al., 1999; Lacey et al., 1999; Meraldi et al., 1999), and allowing cycle progression without division.

Notably, BSC-1 cells post-centrosome removal regained normal size, assembled radial MT networks, reformed MT-organizing centers, and traversed S-phase over 10 cycles, yet remained mitotically inactive (Maniotis and Schliwa, 1991). Similar outcomes followed interphase centrosome excision in BSC-1 cells (Hinchcliffe et al., 2001). Earlier studies used green light (546 nm) for time-lapse imaging, presumed non-damaging. However, reduced illumination intensity (64-fold) via high-sensitivity cameras revealed centrosome-free cells exhibited heightened light sensitivity (Uetaki et al., 2007). Cells were maintained in sealed chambers post-ablation, avoiding oil-coated micromanipulation environments. Controls included cytoplasm-ablated (non-centrosomal) and intact cells under identical illumination. Centrosome-free cells arrested more frequently under elevated light, though most progressed cyclically at minimal exposure.

Initially, *de novo* centriole assembly post-ablation was observed only in transformed cells (Khodjakov et al., 2002; La Terra et al., 2005), suggesting transformation bypasses centrosome assembly constraints (Maniotis and Schliwa, 1991; Hinchcliffe et al., 2001; Khodjakov and Rieder, 2001). However, replication in non-transformed RPE1 and HMEC cells confirmed this as a universal mammalian somatic cell trait, independent of transformation (Uetaki et al., 2007).

Following centrosome removal during the G1 phase, variable numbers of GFP-fluorescent centrin foci - termed "pre-centrioles" by the authors - appeared during the first S phase. These foci gradually intensified in brightness over time, eventually forming morphologically recognizable centrioles. While the number of centrioles assembled *de novo* in RPE1 cells was variable, their maximum count (up to 6) remained lower than in HeLa cells (up to 14). Furthermore, a significant proportion of RPE1 cells lacking centrioles contained two centrioles after the second mitosis. This observation suggests either the existence of a cellular mechanism controlling centriole copy number or the *de novo* assembly of a single centriole that undergoes subsequent duplication (Uetaki et al., 2007).

During early embryogenesis, blastomeres typically retain no more than two centrioles post-mitosis under normal development. Similarly, parthenogenetically activated bovine embryos (Uzbekov et al., 2023) and early mouse embryos initially form only one centriole (Abumuslimov et al.,

1994). These findings support the hypothesis that a single centriole may initially form in RPE1 cells following centriole removal (Uetaki et al., 2007).

To localize GFP-centrin-labeled centrosomes, 488 nm illumination was employed. Earlier studies demonstrated the inhibitory effect of blue light on cultured cells (Gorgidze et al., 1991, 1998). In this case, similar to high-intensity green light exposure, illumination exerted an additive stress effect on centrosome-ablated cells (Uetaki et al., 2007). Cells lacking centrosomes but shielded from intense illumination progressed through the cell cycle without centriole formation. These cells traversed G1 and entered mitosis regardless of whether centrosome removal occurred early in G1 or they originated from mitotic divisions without centrosomes. The identical behavior of RPE1 and HMEC cells post-centrosome ablation indicates that these outcomes are not cell-type specific or dependent on the ablation method (e.g., microirradiation vs. microsurgery) (Uetaki et al., 2007).

Upon discovering that centrosome-free cells can cycle normally, Uetaki et al., (2007) proposed two hypotheses to reconcile earlier reports of cell cycle arrest following centrosome disruption:

1. Centrosome surveillance mechanism: Cells detect centrosome damage or functional impairment via a dedicated mechanism (Murray, 2001; Doxsey et al., 2005a,b). If residual centrosomal components act as signaling platforms for the p53 pathway, complete centrosome removal might evade detection, whereas partial damage triggers arrest.
2. Stress synergy: Centrosome dysfunction synergizes with external stressors (e.g., scattered UV during ablation) to activate p38-p53-mediated cell cycle arrest.

Revisiting UV microirradiation experiments during anaphase (Uzbekov, 1991b, 1992), cell cycle arrest in G1 could align with both hypotheses. Damaged centrosomes (Figure 13.7e) exerted a stronger inhibitory effect on cell cycle progression than complete centrosome absence, consistent with Hypothesis 1. Concurrently, UV scatter during anaphase irradiation likely introduced additional stress, supporting Hypothesis 2. Phototoxicity from imaging was ruled out, as orange light (non-toxic wavelength) was used for observations. Sister cells to irradiated ones exhibited normal centriole duplication (Uzbekov and Vorobjev, 1992) and S-phase entry (Neverova et al., 1996), confirming minimal illumination artifacts.

Summarizing the works employing microirradiation and microsurgery to study centrosome functions, these methods have significantly advanced our understanding of the centrosome as an integrated cellular organelle. It is now evident that the centrosome represents a polyfunctional, multiprotein regulatory complex, with dynamically changing protein ensembles throughout the cell cycle.

Data from microirradiation and microsurgery revive the hypothesis, first proposed in the 1970s (Friedlander and Wahrman, 1970; Pickett-Heaps, 1971), that centrioles act as "passengers" leveraging their mitotic spindle pole positioning to ensure equal distribution to daughter cells for cilia and flagella formation. Intriguingly, as discussed in this book (Figure 6.11.), certain viruses similarly localize to spindle poles to infect both daughter cells. This parallels the evolutionary strategy hypothesized for ancient flagellates, where centrioles may have adopted pre-existing duplication mechanisms rather than relying solely on *de novo* genetic programs.

Animal cells can regenerate lost centrioles by activating genes encoding centriolar proteins, ultimately restoring structurally normal centrioles. This was directly demonstrated in cultured cells lacking essential roles for primary or motile cilia (Uetaki et al., 2007). Notably, such cells initially form multiple centriolar cylinders but revert to the canonical "two centrioles per diploid genome" within several cycles.

Unlike artificial centriole removal in cultured cells, embryonic development naturally eliminates centrioles. During oogenesis, all centrioles are removed from oocytes, while spermatogenesis eliminates them entirely (e.g., mice) or partially (other mammals). Our studies in bovine embryos revealed that sperm-derived proximal centrioles are eliminated, leaving no centrioles at the first mitotic spindle poles (Uzbekov et al., 2023). Tracking centriole formation *in vivo* showed that, akin to cell culture models, canonical centriole numbers emerged after multiple cycles: excess centrioles were reduced in cultured cells (Uetaki et al., 2007), while blastomeres upregulated centriole numbers to match genomic content (Uzbekov et al., 2023). This supports the hypothesis of cellular mechanisms regulating centriole quantity.

Furthermore, centrioles form *de novo* in parthenogenetically activated embryos (Dirksen, 1961; Szollosi and Ozil, 1991; Uzbekov et al., 2023), confirming an intrinsic genetic program for centriole biogenesis. The normalization of centriole numbers to "one cylinder per genome copy" echoes historical hypotheses about quantity regulation. Testing this in haploid male wasp somatic cells revealed centriole numbers identical to diploid females

(Uzbekov et al., 2024a), demonstrating that ploidy-independent mechanisms control centriole homeostasis.

13.4. Genetic and Biochemical Methods for Acentriolar Cell Generation

All cells retain an evolutionary imperative for cilia and flagella formation, necessitating centriole biogenesis within their genetic program. This program, however, can be disrupted. In Drosophila, the DSas-4 protein is critical for centriole assembly. Mutants lacking DSas-4 lose centrioles during embryogenesis, and centrosomes become undetectable in third-instar larval cells (Basto et al., 2006). Although acentriolar cells exhibit delayed mitotic spindle assembly, larvae develop to adulthood. Postnatal lethality occurs due to sensory neuron cilia absence, underscoring centrioles' essential role in ciliogenesis (Basto et al., 2006). This study extended earlier findings of centriole-independent mitosis in Drosophila cell culture (Debec et al., 1982), demonstrating organismal viability without centrioles.

Another key regulator of centriole formation is Polo-like kinase 4 (Plk4/SAK), identified in proliferating cells, including prostate (Korzeniewski et al., 2012), breast (Li et al., 2016; Kazazian et al., 2017), lung (Kawakami et al., 2018), and neuroblastoma cancers (Tian et al., 2018). Plk4 knockdown suppresses proliferation and invasion while inducing apoptosis *in vitro* and impairs tumor growth *in vivo* (Bao et al., 2018). Localized to centriole proximal regions via Cep152 interaction (Zitouni et al., 2014), Plk4 levels are tightly controlled to ensure single procentriole formation per parental centriole during G1. Quantitative analysis revealed ≈50 Plk4 copies per centrosome in KE37 cells, representing a minor fraction of total cellular Plk4 (Bauer et al., 2016).

Plk4 regulation involves autophosphorylation-triggered degradation via SCFβ-TrCP ubiquitin ligase (Cunha-Ferreira et al., 2009; Guderian et al., 2010). This feedback loop maintains precise kinase levels, as both depletion and overexpression disrupt centriole homeostasis: Plk4 inhibition blocks duplication (Bettencourt-Dias et al., 2005; Habedanck et al., 2005), while overexpression induces centriole amplification (Peel et al., 2007; Kleylein-Sohn et al., 2007; Kuriyama, 2009).

Targeted Plk4 inhibition led to centrinone development, with two variants (centrinone: 100 nM EC50; centrinone-B: 500 nM EC50) preventing new procentriole formation without disassembling existing centrioles (Wong et al.,

2015). Crucially, centrosome loss differentially affects cell types: non-transformed cells undergo irreversible p53-mediated G1 arrest, independent of DNA damage, stress, or mitotic errors, whereas cancer cells continue proliferating (Wong et al., 2015).

In contrast, HeLa cancer cell lines proliferated indefinitely following centrosome loss (Wong et al., 2015), albeit at a reduced rate compared to controls. Centriole depletion proved fully reversible in cancer cells, with normal centrosome numbers restored within 10 days post-centrinone washout (Wong et al., 2015). Proliferation rates similarly rebounded to baseline levels. While mitotic duration increased significantly, other cell cycle phases remained unaffected. The overall population growth reduction stemmed from elevated cell death, likely due to mitotic errors in acentriolar cells (Wong et al., 2015). Notably, centrosome-depleted NIH/3T3 and HeLa cells retained DNA damage-induced G1 arrest, reversible by caffeine treatment (Wong et al., 2015).

Nine transformed cell lines with varying centriole amplification levels were analyzed. Centrinone-induced centriole removal reduced proliferation rates independently of the initial proportion of cells with excess centrioles, indicating no direct correlation. Intriguingly, centrinone withdrawal triggered transient Plk4 upregulation, causing a surge in centriole amplification. However, each line eventually reverted to baseline amplification levels via cell death from multipolar mitoses, with death rates reflecting genetic backgrounds (Wong et al., 2015).

Non-transformed human cell lines and primary cultures universally underwent irreversible p53-dependent G1 arrest within 1–2 cycles post-centriole loss, accompanied by p53/p21 upregulation. This arrest was independent of stress signaling, as p38 kinase activation - observed during DNA damage or chromosome missegregation - did not occur (Wong et al., 2015). The authors reconciled discrepancies with prior microsurgery/microirradiation studies (Uetaki et al., 2007) by proposing that G1 arrest requires 1–2 cycles post-centrosome removal. Immediate centrosome ablation (Uetaki et al., 2007) instead activated Plk4-mediated *de novo* centriole formation, circumventing arrest.

Centrinone also reduced centriole numbers in multiciliated Xenopus epithelial cells, confirming Plk4's role in differentiated cell centriole amplification (Wong et al., 2015). Unlike cancer cells, non-transformed cells failed to regenerate centrioles post-centrinone removal, despite remaining viable for >3 weeks. This paradox mirrors Joseph Heller's *Catch-22*: cells cannot enter S-phase (and thus synthesize centrioles) without centrosomes, nor can they restore centrosomes without S-phase entry, trapping them in G1 (Wong et al., 2015).

These findings suggest the Drosophila acentriolar cell line (Debec et al., 1982) represents a transformed lineage, mirroring cancer cell behavior post-centrinone treatment.

Acknowledgments

The authors thank Alexey L. Khodjakov (New York State Department of Health, Wadsworth Center) for providing Figure 13.8. micrographs and critical comments instrumental in preparing this chapter.

Chapter 14

Centrosome-Associated Splicesome Components as Regulators of Ciliogenesis

Massimo Pancione*, PhD

Department of Sciences and Technologies, University of Sannio, Benevento, Italy;
Department of Biochemistry and Molecular Biology, Faculty of Pharmacy, Complutense University Madrid, Madrid, Spain

One of the compelling unanswered questions in biology is how identical information encoded in the genome results in cell diversity. This complexity can be seen in multicellular organisms where the same genome gives rise to a variety of specialized cells and tissues. The process of differentiation allows cells to acquire specialized functions or adapt to different environments (Pelechano, 2017).

Primary cilia and centrosomes are emerging as critical components of stem cells. They play a critical role in the definition of cellular phenotypes in tandem with lineage-dependent cytoskeletal dynamics (Palla et al., 2022). Primary cilia are differentially expressed in different types of mesenchymal, adipose-derived stem cells, or during neuronal morphogenesis. Several signaling pathways, particularly Wnt and Hedgehog, known to be key regulators of cell proliferation and differentiation signals are located to the cilia (Clement et al., 2009; Egorova et al., 2011; Han et al., 2009; Bodle and Loboa, 2016; Sarkisian et al., 2015).

For more than 50 years, it was believed that the centrosome and primary cilia were uniquely composed of proteins. However, research, especially during the past 15 years, has clearly shown that a number of RNAs and mRNA processing proteins localize to the ciliary basal body and the centrosome

* Corresponding Author's Email: pancione.massimo@gmail.com

In: The Centrosome
Editor: Rustem E. Uzbekov
ISBN: 979-8-89530-815-8

(Alliegro et al., 2006; Ryder et al., 2020; Ito et al., 2020; O'Neill et al., 2022; Cerulo et al., 2023., Busselez et al., 2023). This chapter will explore the exchange of spliceosome and RNA molecules between the nucleus and centrosome and analyze how this interdependence is becoming increasingly recognized for the control of stem cell differentiation and pluripotency. The study of cytoplasmic spliceosome components implicated in the modulation of transcriptional networks, needed to regulate stemness and cell-type specific differentiation, will be the subject of the first part of this chapter.

14.1. RNA and Cytoplasmic Spliceosome Localized at the Centrosome

The genome is typically regarded as an instruction manual for building an organism. However, the majority of coding DNA in every plant and animal cell contains non-coding DNA sequences called "introns" that are a crucial part of eukaryotic genomes. The excision of the so-called spliceosomal introns from the "pre-mRNA" is facilitated by the nuclear spliceosome - a ribonucleoprotein complex consisting of RNA molecules and proteins (Zhang et al., 2017). In addition to spliceosomal introns with canonical GT-AG junctions, many eukaryotic organisms harbor introns with noncanonical splice sites, such as AT-AC. These introns are removed by another type of spliceosome called the U12 spliceosome (minor spliceosome) in eukaryotic cells (Bai et al., 2021).

Although introns are mostly excised in the nucleus, a number of cytoplasmic intron sequence-retaining transcripts (CIRTs) have been identified across mammalian genomes (Buckley et al., 2014; Pabis et al., 2024; Tyzack et al., 2021., Mikina et al., 2024).

Often CIRTs contain a variety of transposable elements, the most common of which are retrotransposons called short interspersed nuclear elements (SINEs). Despite the prevalence of transposable elements in cytoplasmic intron transcripts, their functional relevance is not known (Pabis et al., 2024; Mikina et al., 2024).

Potentially, retained intron sequences in cytoplasmic transcripts can form secondary structure, alter mRNA decay rates, and be potential sites of cytoplasmic mRNA splicing.

However, only sparse evidence exists that cytoplasmic intron sequences are target by cytosolic constituents of the spliceosome. Recent array-based technologies and subcellular spatial omics studies are providing previously unknown information on the subcellular cytoplasmic sites of spliceosomes (Busselez et al., 2023). An early example came from the study of Mitochondria (Herai et al., 2017). A long-standing prevailing concept was that mitochondria produce intronless transcripts resembling prokaryotes, which are thought to have originated. A study by Herai et al., 2017, challenged this view providing evidence that mitochondrial RNA transcripts (mtRNA) present canonical spliceosomal introns along with a dedicated spliceosome machinery. In addition to mitochondria, an exceptional variety of spliceosome proteins within the centrosome and ciliar-basal body has recently been discovered (Ito et al., 2020; O'Neill et al., 2022; Cerulo et al., 2023., Busselez et al., 2023). These extra-nuclear macromolecular complexes result particularly abundant within the centrosome of neural or cancer stem cells but their biological relevance is not known. The spliceosomes could be important for stem cell maintenance and cellular differentiation. In line with this, spliceosome inhibition by RNA interference or selective chemical inhibitors in pluripotent embryonic stem cells (ESCs) has been suggested as a key mechanism for the transition from pluri- to totipotency (Shen et al., 2021).

For years, the prevalent notion has been that the nuclear compartmentalization of the spliceosome is needed to prevent primary gene transcripts "pre-mRNAs" from engaging in translation before they are fully matured.

In 2007, a paper challenged this view, claiming that the excision of minor introns, U12-type introns, occurs in the cytoplasm after export of unspliced pre-mRNAs from the nucleus (König et al., 2007). The next year, this study was strongly contested. In fact, Pessa and colleagues confirmerd that probes for U11, U12, U4atac, and U6atac snRNAs strongly label the nucleoplasm of human HeLa cells (Pessa et al., 2008). Similarly, also probes specific for the major spliceosome, including U1, U2, and U6 snRNAs were found to exhibit a strong nuclear staining, as expected. Later, comparable results were obtained in Xenopus oocytes, supporting the large body of prior evidence for the nuclear location of both the major and minor spliceosomes (Friend et al., 2008). Rare exceptions that pre-mRNA splicing can occur in biological environments devoid of a nucleus were found in anucleate platelets or dendrites of neurons (Denis et al., 2005; Glanzer et al., 2005). In recent years, more refined and physiological approaches less prone to artifacts, including single molecule fluorescence *in situ* hybridization (smFISH), dual protein-mRNA localization assays, and spatial transcriptomics, have facilitated the identification of mRNA and RNA-binding proteins at subcellular resolution in several species including zebrafish, Drosophila, Xenopus, and humans (Zein-

Sabatto and Lerit, 2021; Chouaib et al., 2020; O'Neill et al., 2022). For example, centrosomal transcripts encoding for *Pericentrin (PCNT) and NINEIN (NIN)* appear to preferentially enrich at interphase/early mitotic centrosomes. An intriguing possibility is that centrosome-localized transcripts could be modulated "on-demand" before mitosis or in ciliogenesis, suggesting that RNA localization to the centrosome is a dynamic process (Safieddine et al., 2021; Busselez et al., 2023). Similarly, centrosomal mRNAs containing introns could be subject to cytoplasmic splicing, or when no longer needed can be readily shuttled away in a translationally-suppressed condition until they are required again, or destroyed (Busselez et al., 2023). However, presently the notion that pre-mRNA splicing events may occasionally occur, or be completed, in the cytoplasm, although possible, remains unproven. As we outline before, capturing the subcellular spatial information of RNA and RNA-binding proteins is providing new roles of spliceosome components localized into cytoplasm including to the centrosome. Here below, we present a variety of splicing factors localized at the centrosome that could regulate gene expression in higher eukaryotes in a different way than originally thought.

14.2. Functional Relevance of the Spliceosome Localized at the Centrosome

One of the first evidence that splicing exhibited a function at the centrosome came from the discovery that the deletion of exon junction complex (EJC) protein Magoh produced cilia defects linked to microcephaly in a mouse model (Silver et al., 2010). Additional evidence that EJC core proteins concentrate around centrosomes at base of primary cilia both in mouse neural stem cells (mNSCs) and human retinal pigment epithelial (RPE1) cells has been provided recently (Kwon et al., 2021). Similarly, a recent study on human induced pluripotent neural stem cells (IPS) and neurons has clearly demonstrated a dramatic reduction of the EJC/TREX complex protein interaction network in centrosomes during neuronal differentiation (O'Neill et al., 2022). The splicing factor SON, which interacts with EJC proteins as part of the mRNP complex, has been shown to be essential for centrosome assembly and ciliogenesis (Singh et al., 2012; Sharma et al., 2011; Stemm-Wolf et al., 2021). Thus, it is becoming clear that EJC contribute to transport/localize specific centrosome-related transcripts subject to translation at centrosomes.

In 2015, a number of Pre-mRNA processing factors (PRPF6, PRPF8, and PRPF31) that mediate the incorporation of U4/U6/U5 tri-snRNP subunits into the spliceosome, were found to localize to the ciliary basal body and the centrosome (Wheway et al., 2015). More recently, it was discovered that several centrosomal PRPF members are NSC-specific factors exhibiting high affinity for centriolar subdistal appendages proteins (O'Neill et al., 2022; Cerulo et al., 2023).

Interestingly, only a few PRPFs are found at mid-to-late mitosis centrosomes, suggesting that the majority of pre-mRNA processing factors preferentially enrich at interphase centrosomes (O'Neill et al., 2022).

In addition to the snRNPs, non-snRNP splicing factors like the Prp19 complex also play a role at the centrosome. Prp19-associated proteins including BCAS2, BUD31 and CTNNBL1 are localized in interphase centrosomes in a variety of human cells including neurons (O'Neill et al., 2022; Cerulo et al., 2023; Thul and Lindskog, 2018). In addition to a role in interphase centrosomes, the Prp19 complex contributes to spindle assembly (Hofmann et al., 2013). Although the Prp19 complex is clearly linked to centrosomes, its precise mechanism of action is not known.

Another class of RNA-binding proteins are the SR proteins, known as alternative splicing regulators. Most SR proteins are located exclusively in the nucleus, but some of them (mainly SRSF1, SRSF2, SRSF3, and SRSF7) can shuttle between the nucleus and the cytoplasm (Graveley, 2005). SR proteins compose a large network of the neuronal centrosome proteome undergo changes during stem cell differentiation (O'Neill et al., 2022; Cerulo et al., 2023). A direct involvement for SR proteins in the biology of cilia has recently emerged (Haward et al., 2021). This study documented that the forced nuclear retention of SRSF1 inhibits the transport and/or translation of mRNA to the cilia in a mouse model. Haward and coworkers concluded that the lack of cytoplasmic SRSF1 primarily affects mRNA translation that contributes to the biogenesis and function of cilia (Haward et al., 2021). The direct relevance of SR proteins, including SRSF1, at the centrosome and ciliary base still needs to be determined in mechanistic detail.

Surprisingly, also some centrosomal proteins can re-localize in the nucleus and regulate alternative splicing. The best described example is the centrosomal protein NEK2 (never in mitosis gene A-related kinase 2), a member of the NIMA-related family of serine/threonine protein kinases (Naro et al., 2014). NEK2 dissolves the centrosome linker in G2/prophase through phosphorylation of the linker components C-Nap1, rootletin and CEP68 (Remo et al., 2020). Thus, an unexpected mode of gene expression regulation in metazoans is emerging that involves a bidirectional nucleocytoplasmic shuttling between spliceosomes and centrosomes. The role of centrosome-associated spliceosome machinery in stem cell pluripotency and differentiation is discussed in the next paragraph.

14.3. Centrosome-associated Spliceosome Proteins as Orchestrators of Stem Cell Control and Cellular Differentiation

The centrosome-associated spliceosome machinery could be an active contributor to stem cell identity and cellular differentiation (Kwon et al., 2021; O'Neill et al., 2022). Alternative splicing is also regulated by DEAD-Box Helicase (DDX) families, many of which interact with the centrosome and are involved in pluripotency maintenance and cellular differentiation (Cerulo et al., 2023). For example, DDX5 and DDX17 expression is downregulated during myogenesis and epithelial-to-mesenchymal transdifferentiation, which leads to splicing program switching and actin cytoskeleton remodeling (Dardenne et al., 2014). In constrast to the neural stem cells, centrosome neurons show a high affinity for DDX5/DDX17 and heterogeneous nuclear ribonucleoproteins (hnRNPs) during differentiation (O'Neill et al., 2022). Changes in the transcriptional networks and the control of pluripotency vs cell-type specification may be linked by the accumulation of DDX and hnRNP proteins at the centrosome of differentiated neurons. Other RNA helicases that accumulate at the centrosome, such DDX53 and DDX3, have also been shown to play a role in regulating stem cell-like properties (Kerr et al., 2019; Kim et al., 2017; O'Neill et al., 2022; Thul and Lindskog, 2018). Similarly, the splicing factors RBM39, BCAS2 and BUD31 of the Prp19 complex, influence the control of stem and progenitor cells binding directly to centrosome proteins (Cerulo et al., 2023; Chou et al., 2015; Wang et al., 2019; Yu et al., 2019; Chen et al., 2022). Our group and others recently proposed that the induction of cellular differentiation by extracellular differentiating factors or centrosome-reducing factors could activate or deactivate the action of the centrosome-associated spliceosome modifying the target RNA substrates at specific subcytoplasmic sites (Ng et al., 2020, Cerulo et al., 2023). Upon external stimuli, centrosome-associated splicing factors can: a) re-localize in the nucleus, or b) be downregulated via proteasome, miRNA-mediated inhibition or c) be subject to cytoplasmic splicing. These events could favor alternative splicing events and contribute to differences in centrosome composition during cell-type specification. Therefore, the spliceosome arsenal around the centrosome could control gene expression pathways and cytoskeletal remodeling with a modality more intriguing than originally believed. The existence of a close relationship between the spliceosome and the centrosome is also evident in the striking phenotypic similarities between splicing and centrosome-related genetic disorders, particularly retinal degeneration, and brain and craniofacial developmental defects (Busselez et al., 2023). In line with this, a recent study uncovered another complex syndrome resulting from GPATCH11 mutations, which is involved in pre-mRNA splicing and transcription regulation, characterized by overlapping clinical features of spliceosomopathies and ciliopathies (Zanetti et al., 2024).

Conclusion

Extra-nuclear spliceosome components play a fundamental role in eukaryotic cellular physiology. The bidirectional exchanges between centrosome and spliceosome suggests that cytoplasmic splicing factors govern post-transcriptional regulation, including the maintenance of pluripotency. Our model suggests that cell-type specific splicing factors assemble on cytoplasmic mRNA in a variety of subcytoplasmic sites including Mitochondria, Golgi, vesicles, centrosomes and ciliary basal body. In particular, centrosome and ciliary basal body act as a cytoplasmic anchor for cell-type specific spliceosome components. This interaction may play a fundamental role during cell-type specification, functioning as a cytoplasmic site of alternative splicing and control of gene expression. Therefore, these cytoplasmic condensates could promote stem cell pluripotency and facilitate effective reprogramming of somatic cells. However, the precise function of RNA-processing complexes at the centrosome remains an ongoing field of investigation. The centrosome-associated spliceosome proteins are an intrinsic component of the intracellular microenvironment of eukaryotic cells that can no longer be ignored. Centrosomes have acquired various accessory structures to facilitate these disparate functions. These accessory structures including centriolar appendages and satellites may have co-evolved to serve the greater number of cell and tissue types of higher eukaryotes. In conclusion, the centrosome can be seen as a cytoplasmic site that must be coordinated with nuclear factors to promote the regulation of gene expression and cellular differentiation to serve the greater variety and complexity of cell types of higher eukaryotes.

Chapter 15

The Principal Postulates of Centrosomal Biology After 150 Years of Study: Version 2025

Rustem E. Uzbekov[1,*], PhD, HDR, IR
Claude Prigent[2], PhD, HDR, DRCE
and Tomer Avidor-Reiss[3], PhD

[1]Laboratory of Cell Biology and Electron Microscopy, Faculty of Medicine, University of Tours, Boulevard Tonnellé, France;
Faculty of Bioengineering and Bioinformatics, Lomonosov Moscow State University, Moscow, Russia
[2]L'équipe de recherche Cycle Cellulaire, Centre de Recherche en Biologie cellulaire de Montpellier, U'unité mixte de recherche 5237 CNRS;
University of Montpellier, Montpellier, France
[3]Department of Biological Sciences; Department of Urology,
College of Medicine and Life Sciences, University of Toledo, Toledo, OH,USA

The principal postulates of centrosomal biology were first formulated by us in 2007 (Uzbekov and Prigent, 2007). A revised version was published later (Uzbekov and Avidor-Reiss, 2020). In the past five years, new data have emerged and are now included in the updated version of the principal postulates.

---

* Corresponding Author's Email: rustem.uzbekov@univ-tours.fr

In: The Centrosome
Editor: Rustem E. Uzbekov
ISBN: 979-8-89530-815-8

15.1. The Centriole and the Basal Body Are Two Forms of the Same Organelle

The centrosome typically organizes and orients the mitotic spindle. The basal body nucleates and anchors the cilium during interphase. After mitosis, in the G1 or G0 phase of the cell cycle, the centrosome comprises two centrioles surrounded by pericentriolar material. By the end of interphase, just before the next mitosis in the G2 phase, the cell contains two centrosomes, each consisting of two centriolar cylinders. Each centriole is made up of nine microtubule triplets and is associated with subdistal appendages, distal appendages, striated roots, and satellites. In certain cases or developmental stages of organisms, the walls of the centrioles may contain microtubule doublets (Sulston et al., 1980; Riparbelli et al., 2010; Gottardo et al., 2015), microtubule singlets (Favard et al., 1961; Wolf et al., 1978; Pelletier et al., 2006), or may lack microtubules entirely (Blachon et al., 2009; Uzbekov et al., 2018a, 2018b), while still maintaining nine-beam symmetry. In other instances, the symmetry may also be lost (Fishman et al., 2018).

The structure of centrioles during the formation of cilia or flagella is conserved across the majority of organisms and cell types - 9 MT triplets (aka "9 × 3") with no central MTs (aka "+0"), summarized by the formula "9 × 3 + 0." In rare cases, cilia appear to form a basal body consisting of doublet MTs (Jana et al., 2011; Gottardo et al., 2015; Nechipurenko et al., 2017). The MT triplets consist of three tubules named MT "A," MT "B," and MT "C." Only MT "A" has a closed rounded shape in a transverse section and consists of 13 protofilaments, like most cellular MTs (Ringo, 1967; Tilney et al., 1973). MT "B" takes the shape of an arc adjacent to MT "A" and shares three or four common protofilaments with it. MT "C" also has the shape of an arc adjacent to MT "B" and shares three or four common protofilaments with it. The number of protofilaments in MT "B" and MT "C" can range from 10 to 11 in centrioles from different organisms, meaning the total number of protofilaments in one triplet can be from 33 to 35 (Gibbons and Grimstone, 1960; Tilney et al., 1973; André, 1964; Wheatley, 1982; Ou and Rattner, 2004; Guichard et al., 2013). There are suggestions that 11 protofilaments of MT "B" and MT "C" may have a special biochemical composition (Tilney et al., 1973).

The centriolar cylinders' central nine-fold symmetry is established during the formation of the "cartwheel" structure (Gibbons and Grimstone, 1960; Gall, 1961). This symmetry is generated by the lateral interactions of the N-terminal domains of SAS-6 protein dimers (Kitagawa et al., 2011; van Breugel et al., 2011). The cartwheel structure disassembles in mature centrioles as cells exit mitosis in vertebrates (Gall, 1961; Vorobjev and Chentsov, 1982) but remains in mature centrioles of insects (Friedlander and Wahrman, 1966).

15.2. The Centriole Is a Polar Structure with Two Morphologically and Functionally Different Ends

The centriole's distal end (where the plus ends of the MT are) can serve as the site of cilia formation and associates with distal appendages. A complex of proteins, including CP110 and CEP97, is present at this end and supports cilium formation (Spektor et al., 2007). Subdistal appendages are also more commonly formed closer to the distal end of the centriole. The number of distal appendages (for centrioles of centrosome) and their homologous structures in basal bodies - alar sheets - always equals nine (Szollosi, 1964; Anderson, 1972). The number of subdistal appendages (the original name of these structures was pericentriolar satellites, Bernhard and de Harven, 1960) varies across different cell types and conditions, ranging from 0 to 13 (Stubblefield and Brinkley, 1967; Vorobjev and Chentsov, 1977; Uzbekov and Alieva, 2018).

The foot is a homologous structure of a subdistal appendage in the basal body. One to two feet are found per basal body, and they are positioned closer to the distal end (Anderson, 1972; Doolin and Birge, 1966; Sorokin, 1968). The foot base connects to two or three triplets within a basal body (Wolfe, 1972). In the basal bodies of functionally active cilia, the feet are situated in the plane of the cilia beat (Gibbons, 1961). New centriolar cylinders (procentrioles) typically appear perpendicular to the surface of the mother centriole, closer to its proximal end (Gall, 1961), where the minus ends of the MT triplets are located. The pericentriolar material can also be arranged asymmetrically, surrounding only the proximal part (Vorobjev and Chentsov, 1982; Ou and Rattner, 2000). The proximal end of the MT "A" centriole is covered with a conical structure that has a morphology similar to the γ-tubulin ring complex (γ-TuRC) (Guichard et al., 2010). The polarity of the centriole is evident in the orientation of some of its components. When viewed from the proximal end of the centriole, the MT triplets are always twisted counterclockwise (from the inner MT "A" to the outer MT "C"), while the distal appendages are always twisted clockwise (O'Hara, 1970; Uzbekov and Prigent, 2007).

15.3. The Outer Diameter of the Distal Part of the Centriolar Cylinder Is Smaller Than That of the Proximal End

There are two reasons for the differences in diameter between the centriole ends. (i) The MT triplets transform into doublets of MT as MT "C" is usually shorter than MT "A" and MT "B." (ii) The angle of inclination of the MT triplets (the line passing through the centers of MT "A" and MT "B") to the radius of the centrioles at their distal end is 80–90 degrees (the doublets lie almost in a circle). At the proximal end of the centriole, the triplets' inclination to the radius is 50–55 degrees (the triplets are deployed like turbine blades). The inner radius of the lumen (the distance from the center of the centriole to MT "A") does not change. In contrast, the outer radius (the distance from the center of the centriole to MT "C") increases due to the unfolding of the triplets (Vorobjev and Chentsov, 1982). This twist angle appears to be controlled by the A-C linkers in the proximal segment and an inner scaffold at the distal portion (Le Guennec et al., 2020).

15.4. The Centriolar Cylinder Length Is Highly Regulated

The length of the centriolar cylinder is precisely established and typically ranges from 200 to 700 nm, depending on the cell type and phase of the cell cycle (Avidor-Reiss and Gopalakrishnan, 2013; Basiri et al., 2013). However, in certain cell types, exceptionally long centrioles can be observed (Schreiner and Schreiner, 1905; Gibbons and Grimstone, 1960; Guichard et al., 2012). The elongation of centrioles is regulated by a complex of proteins, including SAS4/CPAP, POC1, and POC5 (Azimzadeh et al., 2009; Blachon et al., 2009; Kohlmaier et al., 2009; Tang et al., 2009). Overexpression of CPAP or its interacting partners, CEP120 and SPICE1 (Spindle and Centriole Associated Protein 1), leads to the formation of excessively long centrioles. The protein CP110 acts as an antagonist to CPAP; CP110 caps the distal end of the growing centriole (Schmidt et al., 2009; Comartin et al., 2013; Lin et al., 2013; Kong et al., 2020).

15.5. In the Centrosome of Proliferating Cells, Two Centrioles Differ Structurally and Functionally

Only the older (mother) centriole has appendages at its distal end. Additionally, the mother centriole has subdistal appendages. Microtubule (MT) nucleating centers are predominantly located on or near the mother centriole. Gamma-tubulin (Oakley and Oakley, 1989; Joshi et al., 1992) forms the basis of two types of protein complexes that nucleate MTs - the large Gamma-TuRC (Zheng et al., 1995) and small Gamma-TuSC (Moritz et al., 1998; Oegema et al., 1999). Other protein complexes on the centrosome anchor the MTs, including CAP350/FOP/EB1 (Yan et al., 2006), PCM/BBS4/ninein/centrin/pericentrin (Kim et al., 2004), and ninein/ODF2/Cep170/centriolin/epsilon-tubulin (Azimzadeh and Bornens, 2007). MT nucleation and attachment activities are found in the heads of subdistal appendages, on the surface of the centrioles, and in the pericentriolar material. Centriolar satellites are dense, rounded structures containing protein complexes of the pericentriolar material located near centrioles and undergo cell cycle-dependent assembly and disassembly (De Harven, 1968; Tilney and Goddard, 1970; Dammermann and Merdes, 2002; Kubo and Tsukita, 2003). They move toward the centrosome along MTs in a dynein-dependent manner, participate in targeting centrin, pericentrin, and ninein to the centrosome, and are implicated in ciliogenesis (Dammermann and Merdes, 2002; Conkar et al., 2019; Prosser and Pelletier, 2020).

15.6. The Proximal Ends of the Mother and Daughter Centrioles Are Connected via a Bundle of Thin Fibers

The proximal ends of the mother and daughter centrioles are connected by a bundle of thin fibers (Bornens et al., 1987; Tournier et al., 1991; Paintrand et al., 1992). This ligament comprises proteins rootletin, beta-catenin, and C-NAP1 (Yang et al., 2002; Bahe et al., 2005; Bahmanyar et al., 2008). Severing this connection can result in centriole separation. The regulation of this separation is mediated by Nek2 kinase (Bahmanyar et al., 2008). Separation of mother and daughter centrioles after mitosis is a necessary prerequisite for the initiation of centriole duplication (Tsou and Stearns, 2006).

The separation of the mother and daughter centrioles during the G1 phase of the cell cycle differs from the separation and divergence of the two centrosomes prior to mitosis. The latter process relies on the integrity of microtubules (MTs) and actin microfilaments (Uzbekov et al., 2002) and is regulated by kinases such as AuroraA (Glover et al., 1995; Roghi et al., 1998; Cowley et al., 2009; Giet et al., 1999b), p34cdc2 (Blangy et al., 1997), and Plk1 (Golsteyn et al., 1996; Lane and Nigg, 1997; Glover et al., 1998). During centrosome separation and movement in the formation of the two poles of the spindle, several types of motor proteins are involved: the cytoplasmic dynein–dynactin–NUMA complex (Gaglio et al., 1996, 1997; Echeverri et al., 1996), Xklp2 (Boleti et al., 1996), kinesin-related motors pEg5 (Sawin et al., 1992; Blangy et al., 1997; Giet et al., 1999a; Uzbekov et al., 1999), and XCTK2 (Walczak et al., 1997).

15.7. The Centriole Lumen Helps in Stabilizing the Centriole

Centriolar MTs exhibit significantly greater stability than spindle or interphase cytosolic MTs. They are resistant to depolymerization by both anti-MT drugs and exposure to cold (De Harven and Bernhard, 1956; Osborn and Weber, 1976; Jensen et al., 1979). However, MT triplets can be disassembled by high salt concentrations (1–2 M NaCl or KCl) once centrioles are isolated (Fais et al., 1986). The stability of MT centrioles is linked to stabilizing proteins like tektins between microtubules (Steffen and Linck, 1988; Stephens, et al., 1998), post-translational polyglutamylation of centriolar tubulin (Plessmann and Weber, 1997; Bobinnec et al., 1998), and the presence of scaffold structures inside the centriole lumen.

Inside the centriolar cylinder, an interconnected system of ligaments connects the MT triplets (Gibbons and Grimstone, 1960; André, 1964; Szollosi, 1964; De Harven, 1968; Guichard et al., 2013). The structure of these connections varies from the proximal end to the distal end. MT "A" has a connection with MT "C" of the neighboring triplet. Additionally, strands of electron-dense material extend inward from each MT "A," linking MT "A" and A–C bundles. At the ultrastructural level, the lumen of the centriolar cylinder appears "empty" near the proximal end and is filled with electron-dense material at the distal end. The protein composition of the A–C linker remains unknown. The distal lumen scaffold consists of POC1B, POC5, FAM161A, CETN, and WDR90 (Le Guennec et al., 2020; Steib et al., 2020). Mutations in some of these proteins lead to destabilization of the centriole cylinder (Pearson et al., 2009).

15.8. Maintaining a Cylindrical Shape of Centrioles Can Be Independent of MT Triplets

Centrioles lack MTs in haploid male larvae trophocytes and hypodermal cells of wasps *Anisopteromalus calandrae* (Uzbekov et al., 2018a). However, their MT-free centrioles' shape and size resemble canonical centrioles with MT triplets of adult (imago) insects or late larvae from various types of wasps (Figure 15.2.). The centriole's symmetry is maintained, and they contain nine prongs of electron-dense material in their walls, forming the structure of the "cogwheel" (Uzbekov et al., 2018a). One possibility is that at this developmental stage, the genes encoding the proteins responsible for constructing MT triplets have not yet been activated. Alternatively, the MTs may have disassembled after the centriole formed. A practically identical cogwheel structure without MT triplets was also found at the base of mature spermatozoa flagella of wasps *Cotesia congregata* (Figure 15.2c). It replaced the "normal" centriole during spermiogenesis, which is present in spermatids (Uzbekov et al., 2018b). The prongs of the cogwheel structure are visible between the triplets or doublets of MTs in centrioles of other insects, particularly in Drosophila. It has been shown that the protein SAS4 (Lattao et al., 2017) concentrates in these regions.

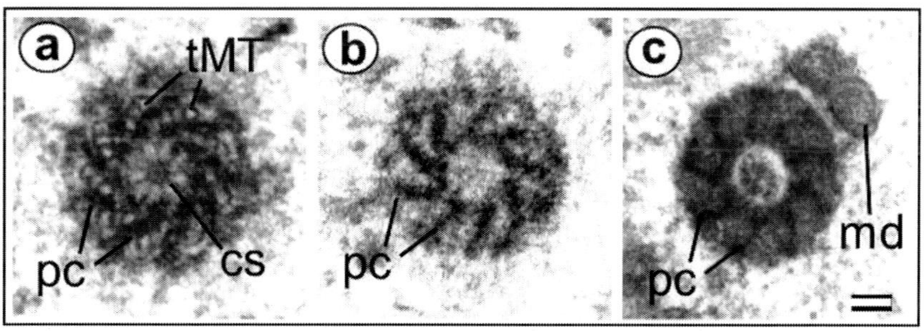

Figure 15.1. Cogwheel structure in the three centrioles of wasps: a) somatic cell of Nasonia vitripennis larvae, where prongs of the cogwheel structure are visible between triplets of MT; b) somatic cell of Anisopteromalus calandrae early larvae, where the centriole has a cogwheel without MT; c) basal part of flagella (basal body) in mature sperm of Cotesia congregata. View from the distal end of the centriole; cs – cartwheel structure, md – mitochondrial derivatives, pc – prongs of cogwheel structure, tMT – triplets of MT. Scale bar: 50 nm. From Uzbekov et al., 2018a, 2018b, with modifications.

After processing the isolated centrioles from bovine spleens with 2 M salts, the triplets' MTs are completely disassembled (Fais et al., 1986). Holes were found at the MT sites; however, the centrioles' cylindrical shape is preserved. The authors called the electron-dense structure surrounding triplets "centriolar rim" (Fais et al., 1986).

Later, it was reported that the PCM is not homogeneous and that some centrosomal proteins are closely associated with the centriole to form a distinct PCM compartment called the "PCM tube." In contrast, other proteins are more peripheral (Bornens, 2002; Ou et al., 2003). These protein complexes surrounding the MT help maintain the cylindrical shape and provide mechanical strength in the absence of MTs.

15.9. The Structure and Activity of Centrosomes Differ in Interphase and Mitotic Cells

The subdistal appendages and primary cilia disappear (Vorobjev and Chentsov, 1982; Rieder et al., 1979), and an amorphous mitotic halo surrounds the centrosome during mitosis. Additionally, the centrosome-associated interphase MTs completely depolymerize before mitosis. Finally, mitotic MT asters form around each of the two centrosomes in prophase. These MTs, along with other MT nucleating activities, organize and form the mitotic spindle in metaphase (Prosser and Pelletier, 2020). All mitotic MTs have their minus ends near the centrosome but have three options for localizing their plus ends: (1) the MT plus end is directed in the opposite direction from the chromosomes and is localized near the cell membrane—astral MTs; (2) the MT plus end is associated with the kinetochore of chromosomes—kinetochore MTs; and (3) the MT plus end interacts with an MT coming from the opposite pole of the spindle—interzonal MTs. The halo also contains many short MTs, which likely later transform into one of the three types. Mitotic MTs have dynamic and biochemical characteristics that differ from interphase MTs; in particular, they are less resistant to anti-microtubule drugs (De Brabander et al., 1986). The ability of centrosomes to nucleate MTs during mitosis increases several times as preparation for cell division (G2 phase of the cell cycle) occurs in a process known as centrosome maturation (Kuriyama and Borisy, 1981b; Vorobjev and Chentsov, 1982; Palazzo et al., 2000).

15.10. New Centrioles Are Usually Formed in Association with Mother Centrioles but Can Be Formed without Preexisting Centriole *De Novo*

Duplication of centrioles occurs only once per cell cycle; on each mature centriole, only one procentriole appears (Figure 3.3.). The regulation accuracy of these processes is controlled by a protein complex that was first identified in *Caenorhabditis elegans*: ZYG-1 (O'Connell et al., 2001; O'Connell 2002), SPD-2 (Pelletier et al., 2004, 2006; Kemp, et al., 2004), SAS-4 (Kirkham et al., 2003; Leidel and Gonczy, 2003), SAS-5 (Delattre et al., 2004), and SAS-6 (Dammermann et al., 2004; Leidel et al., 2005). In human cells, the ZYG-1 homolog is the PLK4 kinase (Habedanck et al., 2005); in Drosophila cells, it is the SAK/PLK4 kinase (Bettencourt-Dias et al., 2005). More than 30 proteins have been described as involved in the duplication of centrioles (Strnad and Gonczy, 2008). In mammals and insects, centriole duplication starts with CEP152/Asterless that recruits PLK4 (Blachon et al., 2008; Guernsey et al., 2010; Hatch and Stearns, 2010). CEP152 forms a ring along with CEP63 and CEP57 around the proximal part of the mother centriole (Brown, et al., 2013; Lukinavicius et al., 2013). Centriole duplication is independent of DNA replication (Rattner and Phillips, 1973; Sluder, 1989; Gorgidze and Vorobjev, 1995) and begins near the end of the G1 phase of the cell cycle (Uzbekov, 2007; Uzbekov and Alieva, 2010). At this time, a complex of biochemical reactions in the cell is initiated, depending on the cell's size, the presence of external growth factors, the growth conditions of the cell, and its interaction with surrounding cells (Hinchcliffe et al., 1999; Lacey, et al., 1999; Meraldi et al., 1999). It has been shown that the Cyclin D/CDK 4/6 complex phosphorylates the pRB protein, which loses its ability to bind the transcription activation factor EF2. The released transcription factor EF2 activates the synthesis of Cyclin E and Cyclin A, which starts the process of centriole duplication. Thus, DNA replication and centriole duplication are regulated by a single cytoplasmic mechanism. However, centriole duplication begins earlier than DNA replication (Uzbekov, 2007), suggesting additional regulatory mechanisms.

In some cases, new centrioles form without preexisting centrioles. In the ciliary epithelial cells, which possess hundreds of cilia, the centrioles form via deuterostomes (Sorokin, 1968; Dirksen, 1971). Deuterosomes can be assembled autonomously from parental centrioles through *de novo* centriole amplification in multiciliated cells (Nanjundappa et al., 2019; Zhao et al.,

2019). In murine early embryonic development, centrioles appear *de novo*, but the mechanism remains unclear (Szollosi et al., 1972). A detailed study of the centrosome ultrastructure in early bovine embryos demonstrated that rodents are no exception to the general rule of centriole formation. In cows, the first centrioles can also form *de novo* (Uzbekov et al., 2023). In this work, the stages of this process were morphologically characterized: at the first stage, "polar corpuscles" localized at the poles of the first cleavage division formed; then they transformed into atypical centrioles with triplets that did not form a cylindrical structure; later, centrioles with triplets of unequal lengths were discovered; and finally, centrioles of the canonical type emerged (Uzbekov et al., 2023). In cells where centrioles were eliminated by micro-irradiation, procentriole formation during the S phase was not prevented. In contrast, in this scenario, many new centrioles are formed during the S phase (Khodjakov et al., 2002). Additionally, centrioles can appear without progenitors during parthenogenetic development in insects (Callaini et al., 1999) and mammals (Uzbekov et al., 2023). These observations suggest that centrioles can assemble independently of preexisting centrioles, and the role of centriole duplication is to limit the number of assembling centrioles to only one new procentriole for each mother centriole.

15.11. The Centrosomes Have Four Types of MT Nucleating Activity

Many of the centrosome's critical functions are mediated by MTs nucleated by it in different cell types and cell cycle phases. This includes four main MTs-nucleating activities:

1. Formation of two types of interphases MTs: (i) radial MTs system around the centriole; and (ii) non-centrosomal (free) cytoplasmic MTs that were polymerized on the centrosome and later released to the cytoplasm.
2. Formation of three types of mitotic spindle microtubules: (i) astral MTs; (ii) kinetochore MTs; and (iii) interzonal MTs.
3. Formation of MT of procentrioles.
4. Formation of cilia or flagellum MTs (Fawcett and Phillips, 1969) or a related structure known as the centriolar adjunct is found in mammalian spermatids (Garanina et al., 2019).

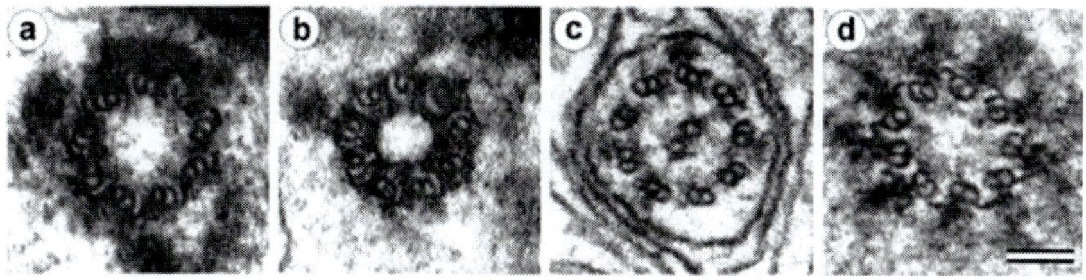

Figure 15.2. Four types of MT-contained structures: (a) proximal centriole in the early pig spermatid—formula "9×3 + 0"; (b) centriolar adjunct in the early pig spermatid—formula "9×2.5 + 0"; (c) flagellum in early pig spermatid—formula "9×2 + 2"; and (d) the cilia-like structure structurally similar to the primary cilium of mammals in the spermatid of the wasp Anisopteromalus calandrae—formula "9×2 + 0." Scale bar: 100 nm. From Garanina et al., 2019 and (d) from (Uzbekov et al., 2018a).

The more mature mother centrioles typically form the primary cilia with the formula 9 × 2 + 0 (Figure 15.2.). The centrioles that arise in the current cell cycle, and are therefore more immature, form the motile cilia. Motile cilia have the formula 9×2 + 2, which consists of nine doublet MTs in the wall and two central MTs. The mature (proximal) centriole creates a centriolar adjunct in mammalian spermatids (Alieva et al., 2018; Garanina et al., 2019). The centriolar adjunct has a formula similar to the primary cilium (9 × 2.5 + 0). Simultaneously, while the centriolar adjunct forms in the spermatid, the daughter (distal) centriole develops a motile flagellum with the formula 9 × 2 + 2 (Figure 15.2.).

15.12. The Complete Process of Centriole Maturation from Procentriole to Mother Centriole Takes More Than One and a Half Cell Cycles in Duration

The entire process of centriole maturation, from procentriole to mother centriole, takes more than one and a half cell cycles (Vorobjev and Chentsov, 1982). The exact timing of procentriole initiation is debated (Sullenberger et al., 2020). Differences may exist among various cell types. The percentage of cells with procentrioles significantly exceeded the S-phase percentage in the cell cycle of synchronized HeLa cells. This difference may indicate that procentriole formation began before DNA replication commenced (Robbins et al., 1968). It was later demonstrated that centriole duplication initiation could occur even in the absence of a nucleus in cytoplasts (Gorgidze and

Vorobjev, 1995) and enucleated sea urchin zygotes (Sluder, 2016). Procentrioles were found near mother centrioles two hours before the onset of DNA replication in the pig kidney cell line, suggesting that procentrioles begin to form prior to DNA replication (Uzbekov, 2007; Uzbekov and Alieva, 2010). However, it remains unknown whether this observation is universal across other cell types.

After its initial appearance, the centriole gradually grows to the size of the mother during the S phase and G2 phase of the cell cycle. Following mitosis in the newly formed cell, it becomes a daughter centriole. The centriole transforms into a mature mother centriole, acquiring a full set of its cellular activities after its second mitosis (Vorobjev and Chentsov, 1982; Sullenberger et al., 2020). The cell (nuclear) cycle and the centriolar (centrosomal) cycle are mutually coordinated at least at two critical points in the cell cycle: the end of the G1 phase and the end of the G2 phase (Nigg and Stearns, 2011; Uzbekov and Alieva, 2013). Additionally, some key events occur during mitosis in a process called "centriole to centrosome conversion" (Izquierdo et al., 2014). Furthermore, many proteins that regulate the cell cycle are concentrated in the centrosome. Disruption of this regulation leads to disturbances in cell cycle progression, ultimately resulting in over-proliferation or degeneration (Nigg and Holland, 2018).

15.13. The Centrosome Is the Center of the Organization of Actin Microfilaments in the Cell

This novel function, "actin organization center," is in addition to the classic centriole function in MT organizing (Inoue et al., 2019). The centrosome occupies a central position in the cytoplasm of many cell types. This position is associated with other interconnected cytoskeletal elements, such as actin microfilaments. The regulation of this position is influenced by the balance of tension forces linked to cytoplasmic dynein (Dujardin and Vallee, 2002) and the repulsive forces generated by the growth of MT (Inoue and Salmon, 1995; Holy et al., 1997). Furthermore, the centrosome's placement depends on the interaction between centrosomal MTs and the actomyosin complex, as well as the activity of the actomyosin complex itself (Salmon et al., 2002; Burakov et al., 2003; Burakov and Nadezhdina, 2020). Additionally, the architecture of intermediate filaments is indirectly reliant on the centrosome, as the intermediate filament system collapses during the depolymerization of MT by colchicine or nocodazole (Goldman, 1971; Alieva et al., 1991).

15.14. Centrioles Can Gain an Atypical Structure and Composition and Become Undetected Using Standard Expectations and Techniques

Centrioles are remodeled and acquire distinct new structures in a species-specific manner that makes them difficult to detect (Avidor-Reiss, 2018; Avidor-Reiss and Fishman, 2019). In some insects, the atypical sperm centrioles lack MT (Gottardo et al., 2015; Khire et al., 2016; Fishman et al., 2017). In contrast, in human and bovine sperm, the MTs are present but arranged around two novel rod structures (Fishman et al., 2017). Like canonical centrioles, these atypical centrioles recruit PCM, form centrosomes and asters, and participate in spindle formation in the zygote. Additionally, canonical centrioles become undetectable during myogenesis, oogenesis, and mouse spermatogenesis (Huettner and Rabinowitz, 1933; Tassin et al., 1985; Manandhar et al., 1998). The process by which centrioles are eliminated in these cells remains unclear. However, it has been suggested that throughout mouse evolution, the structure and function of centrioles were gradually modified, ultimately leading to more robust embryonic development in mice (Avidor-Reiss et al., 2022; Khanal et al., 2024). In amoeboflagellate *Naegleria*, centriolar cylinders are formed *de novo* during the transition from centriole-less amoebae to flagellates, with basal bodies originating from precursor complexes that serve as radiometry centrioles (Fulton and Dingle, 1971; Fritz-Laylin and Fulton, 2016). In early bovine embryos, atypical centrioles consist of irregularly oriented MT triplets, but they can form procentrioles with a canonical structure on their surface (Uzbekov et al., 2023). Therefore, centrioles can acquire novel structures and compositions while still performing many of their classical functions. Although the specialized role of the atypical centrioles underlying their unique structures is unknown, some insight was gained from bovine sperm (Khanal et al., 2021). In this system, the atypical centriole is elastic, moving in coordination with the sperm in the tail and ahead; thus, it has been proposed that the atypical centriole enables the sperm neck to function as a transition system that mechanically links the sperm head and tail.

15.15. The Centrosome Is a Polyfunctional, Multi-Protein, and Cell Regulation Complex; Some of Its Proteins Simultaneously Regulate Several Intracellular Processes

The interactome scheme presented in Figure 15.3. only partially depicts all centrosomal proteins and their interactions. It is designed to be a simplified diagram that helps assess the complexity of the centrosome's biochemical organization. The following publications were used to prepare the presented interactome version: Aprelikova et al., 1999; Arquint et al., 2016; Askham et al., 2002; Astuti et al., 2010; Atorino et al., 2020; Azimzadeh et al., 2009; Bahe et al., 2005; Bahmanyar et al., 2008; Barrera et al., 2010; Basto et al., 2006; Bastos and Barr, 2010; Berrueta et al., 1998; Blachon et al., 2008; Boleti et al., 1996; Bowler et al., 2019; Brown et al., 1996a; Brown et al., 1996b; Brown et al., 2013; Brueckner, 2001; Burakov et al., 2008; Cajanek and Nigg, 2008; Carroll et al., 1999; Casenghi et al., 2003; Casenghi et al., 2005; Chang and Stearns, 2000; Chang et al., 2010; Chartier et al., 2007; Chen et al., 2002; Chen et al., 2003; Chen et al., 2006; Cheng et al., 2007; Choi et al., 2010; Cogswell et al., 2000; Collins et al., 2010; Colombo et al., 2002; Conduit et al., 2010; Dammermann and Merdes, 2002; D'Angiolella et al., 2010; Delattre et al., 2006; Diviani et al., 2000; Doxsey et al., 1994; El Din El Homasany et al., 2005; Ellinger-Ziegelbauer et al., 2000; Eyers et al., 2003; Fabbro et al., 2005; Feng et al., 2006; Fong et al., 2008; Fong et al., 2018; Franck et al., 2011; Frank-Vaillant et al., 2000; Fry et al., 1998a; Fry et al., 1998b; Fukasawa et al., 2008; Giet et al., 1999a; Giet et al., 2002; Glover et al., 1998; Gopalakrishnan et al., 2011; Goto et al., 2010; Graser et al., 2007; Groen et al., 2004; Gromley et al., 2005; Guarguaglini et al., 2005; Gudi et al., 2011; Guillet et al., 2011; Guo et al., 2006; Gurkaslar et al., 2020; Habedanck et al., 2005; Hamel et al., 2017; Hames et al., 2005; Harbour et al., 1999; Haren et al., 2009; Heidi and Nigg, 1997; Hirayama et al., 2008; Hirota et al., 2003; Hosono et al., 2010; Hsu and White, 1998; Huang et al., 2011; Huang et al., 2017; Ibi et al., 2011; Iwamori et al., 2010; Jackman et al., 2003; Jakobsen et al., 2011; Jang et al., 2004; Jin et al., 2009; Jo et al., 2019; Joukov et al., 2006; Kao et al., 2009; Katayama et al., 2004; Kellogg et al., 1994; Keryer et al., 2003; Kim et al., 2004; Kim et al., 2005; Kim et al., 2007; Kleylein-Sohn et al., 2007; Kobayashi et al., 2011; Kobayashi et al., 2014; Kotani and Yamashita, 2005; Krause A. and Hoffmann I., 2010; Kurz et al., 2002; Lacey et al., 1999; Lane et al., 1997; Lee et al., 2001; Lee et al., 2010; Le Guennec et al., 2020; Leidel et al., 2005; Li et al., 2001; Li et al., 2010; Li et al., 2011;

Lin et al., 2006; Lin et al., 2013; Linck et al., 1982; Lou et al., 2004; Lukinavicius et al., 2013; Ma et al., 1999; Ma et al., 2006; Mardin et al., 2010; Mariappan et al., 2019; Matsuo et al., 2010; Maxwell et al., 2005; May-Simera et al.2009; McKendrick et al., 1999; McNally and Vale, 1993; Megraw at al., 1999; Merdes and Cleveland, 1997; Merdes et al., 1996; Mi et al., 2007; Miyoshi et al., 2004; Mogensen et al., 2000; Mogensen, 2004; Molli et al., 2010; Moritz et al., 1998; Morris et al., 2003; Moudjou et al., 1996; Munemitsu et al., 1994; Murphy et al., 2001; Nakamura et al., 1998; Negi and Olson, 2006; Niethammer et al., 2000; Nishimura et al., 2005; Oegema et al., 1999; Okuda et al., 2000; Ou and Rattner, 2000; Ou et al., 2002; Ouchi et al., 2004; Pelletier et al., 2006; Piehl and Cassimeris, 2003; Pietromonaco et al., 1995; Pockwinse et al., 1997; Pugacheva et al., 2007; Purohit et al., 1999; Rapley et al., 2005; Rodrigues-Martins et al., 2007; Rosales et al., 2010; Ruffner et al., 1999; Ruiz et al., 2000; Saavedra et al., 2003; Sasaki et al., 2000; Schafer et al., 1994; Schrøder et al., 2007; Sedjai et al., 2010; Shinmura et al., 2007; Siller et al., 2017; Sillibourne et al., 2002; Sir et al., 2011; Sonn et al., 2011; Soung et al., 2009; Spektor et al., 2007; Sun et al., 2006; Sydor et al., 2018; Szebenyi et al., 2007; Takahashi et al., 1999; Takahashi et al., 2000; Takahashi et al., 2002; Tanos et al., 2013; Terada et al., 2003; Toya et al., 2011; Tsai et al., 2020; Tsang et al., 2006; Tzivion et al., 2001; Uzbekov, 2004; Uzbekov and Alieva, 2013; Vasquez et al., 1999; Vaughan and Vallee, 1995; Wang et al., 1997; Wang et al., 2010; Wang et al., 2011; Wang et al., 2018; Watanabe et al., 2019; Wei et al., 2013; Wei et al., 2020; Wu et al., 2001; Yan et al., 2006; Yao et al., 2004; Yarden and Brody, 1999; Ye et al., 2014; Zach et al., 2012; Zhang and Megraw, 2007; Zhang et al., 1998; Zhang et al., 2005; Zhao et al., 2010; Zimmerman et al., 2004.

Indeed, each year, an increasing number of centrosomal proteins are characterized. These proteins include those that mediate or regulate processes such as signaling pathways, nucleation and anchoring of microtubules, centriole duplication, the separation of the daughter centriole from the mother centriole, conversion of centrioles to basal bodies, and the nucleation of cilia and flagella. It is often impossible to separate the functions of the same protein across different aspects of centrosome functioning.

A single universal classification for centrosomal proteins is simplistic because they can be organized based on several parameters. Many of these proteins have multiple characteristics across different parameters, such as localization, timing, and activity. For localization, some proteins, like tubulins, are found in both the centrioles and pericentriolar material, or in the cilium axoneme.

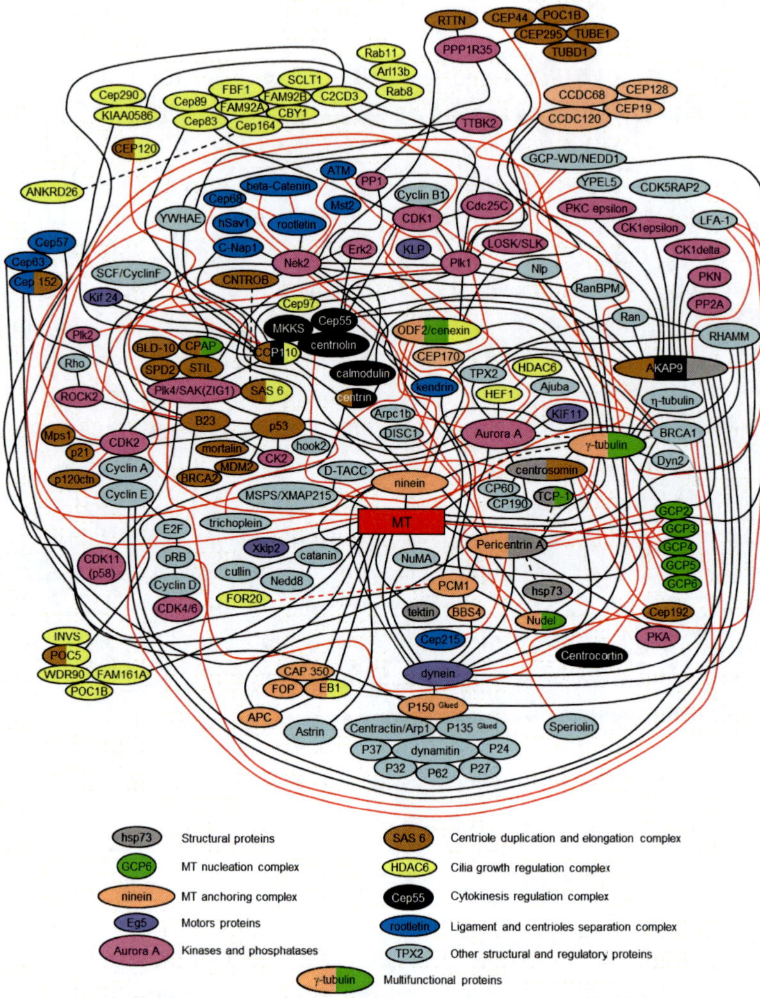

Figure 15.3. Schematic diagram of centrosomal protein interactions and functional complexes (interactome). (Adapted from Uzbekov and Avidor-Reiss, 2020, with modifications). Red and black lines denote centrosomal protein interactions. Sources used to compile this diagram are cited in the text.

Regarding timing, some proteins are consistently present in the centrosome, while others only appear at specific periods in the cell cycle or during particular stages of cell differentiation. In terms of activity, centrosomal proteins can be classified by their biochemical functions - kinases, phosphatases, motors, and structural proteins. Moreover, centrosomal proteins do not exist independently; they form complex, often interconnected functional complexes. Figure 15.3. illustrates how some of these complexes interact with one another.

15.16. The Cellular Genome Contains a Mechanism for Activating Centriole Elimination

Centrioles contain microtubules that are resistant to various influences, achieved through post-translational modifications of the tubulins forming triplets and additional MT-associated proteins. However, in some cell lines, centrioles disassemble. The most well-known examples of centriole elimination occur during oogenesis and, in some organisms, spermatogenesis. Kalbfuss and Gönczy (2023) summarize known examples of centriole elimination across various organisms. In addition to germ cells, this mainly involves terminally differentiated insect cells and Caenorhabditis elegans. Among vertebrate somatic cells, notable centriole loss occurs in muscle cells (Tassin et al., 1985; Connolly et al., 1986; Bugnard et al., 2005) and mouse intestinal cells (Komarova and Vorobjev, 1994, 1995). In both instances, these represent terminally differentiated cells. However, not all terminally differentiated cells lose centrioles. For example, neurons retain centrioles throughout the organism's life, likely due to the presence of sensitive primary cilia in these cells. Thus, the decision to eliminate or retain centrioles is specific to different cell types and is governed by their genetic developmental programs. Unlike DNA, for which mechanisms for repairing various types of damage have been described, the mechanism for centriole repair remains unknown. During successive divisions, "old" centrioles continue to be transmitted to subsequent cell generations, accumulating damage that may act as elimination factors (Kalbfuss and Gönczy, 2023). Since the oocyte cytoplasm likely retains the centriole elimination mechanism before fertilization, we hypothesized that the presence or absence of elimination factors on the sperm centriole (i.e., the relative "old age" of its proximal maternal centriole) could explain the differing fate of this centriole after fertilization in the zygote (Uzbekov and Avidor-Reiss, 2024). Interestingly, the sperm flagellum is formed on a younger daughter distal centriole (Alieva et al., 2018), which is presumably more intact.

The details of the centriole elimination mechanism still need further study, which may have implications for understanding the regulatory processes of interaction between the genome and the centrosome during the functioning of cells from various origins.

Glossary

Alar sheets – Wing-shaped structures at basal bodies, homologous to centriolar distal appendages, mediating membrane anchoring.

Alpha-tubulin – One of two major MT proteins (with beta-tubulin), forming heterodimers with alpha-tubulin positioned toward the MT minus end. Exists in multiple isoforms.

Aneuploidy – Abnormal chromosomal number in a cell (e.g., 45 or 47 chromosomes in human somatic cells vs. the diploid 46).

Apical zone of the cilium – Ciliary region spanning from the transition zone (where MT doublets convert to singlets) to the tip, including the ciliary cap.

Apoptosis – Programmed cell death characterized by membrane blebbing, cell shrinkage, nuclear fragmentation, chromatin condensation, DNA degradation, and mRNA decay. Occurs in multicellular organisms and some unicellular eukaryotes.

Appendages (= distal appendages) – Nine structures at the mother centriole's distal end, anchoring the centriole to the membrane.

Aster of MTs – Radial MT array converging at a single center (e.g., around centrosomes in mitotic prophase). Observed in zygotes and MT repolymerization experiments.

Astral MTs – Mitotic MTs with minus ends anchored at centrosomes (spindle poles) and plus ends extending toward the cell periphery. Position the spindle within the cell.

Astrin – Dual-function protein mediating centrosome cohesion and chromatid attachment, localized to centromeres and centrosomes.

Atypical centriole – Non-canonical centriolar structure (lacking 9 MT triplets) performing centriole-like functions, observed during early embryogenesis.

Aurora A kinase – Serine/threonine kinase localized to centrosomes in late interphase and spindle poles during mitosis. Regulates centrosome maturation and separation via substrate phosphorylation.

Axial grain (= basal plate).

Axis with spokes (= cartwheel).

Axoneme – Core structure of cilia/flagella, typically comprising 9 peripheral MT doublets.

Axosome (= basal plate).

B23 (= Nucleophosmin).

Basal body – Organelle initiating cilium/flagellum growth during ciliogenesis. Homologous to centrioles in centrosomes. Comprises a centriolar cylinder (9×3+0 MT arrangement) with associated structures: stalk, alar sheets, and striated rootlets. Distal end contacts the cell membrane.

Basal plate – Lumenal ciliary structure anchoring central MTs.

BBS-proteins – Protein complex involved in primary cilia biogenesis, homeostasis, and intraflagellar transport. Implicated in Bardet-Biedl Syndrome.

Beta-catenin – Fibrous junction protein linking mother and daughter centrioles in interphase.

Beta-tubulin – Major MT protein (with alpha-tubulin), forming heterodimers oriented with beta-tubulin at the MT plus end. Exists in multiple isoforms.

Blepharoplast (= basal body).

BRCA1 – Tumor suppressor gene product (caretaker gene) encoded by BRCA1, critical for DNA repair.

Bridges (= Y-shaped bundles).

Caltractin (see centrin).

Cap – GTP-bound beta-tubulin region at MT plus ends, hypothesized to prevent depolymerization.

Cartwheel structure (=cartwheel with an indistinct hub and radial spokes) – Proximal centriolar structure (immature centrioles) featuring a 30 nm central hub with radial spokes extending to triplets. Occupies 1/5–3/4 of centriole length; retained in mature insect centrioles.

CDK1 – Centrosome-localized kinase active from late interphase to anaphase onset. Regulates cell/centriolar cycles via cyclin interactions.

Cell center (= centrosome).

Cell cycle – Sequence of cyclical events (G1, S, G2, mitosis) governing cell growth and division.

Cenexin – Maternal centrosome-specific maturation marker protein.

Centractin – Centrosomal actin homolog; dynactin complex component forming 37 nm filaments.

Central bodies (= centrioles, historical term).

Central cap (= cilium cap).

Central capsule – Ring structure encircling axonemal central MTs.
Central case (= central capsule).
Central corpuscles (= centrioles, historical term).
Central hub – 30 nm tubular core of the cartwheel, connecting radial spokes to triplets.
Central MTs – Two singular MTs in ciliary/flagellar axonemes.
Central sheath (= central capsule).
Centrin (= caltractin) – Mother/daughter centriole-localized protein critical for centrosome duplication and ciliogenesis.
Centrinone – Plk4 inhibitor blocking procentriole formation. Induces G1 arrest in non-transformed cells.
Centriolar adjunct – Proximal centriole-derived MT structure (9 × 2.5 + 0) in spermatids, degraded in mature sperm.
Centriolar cycle (= centrosomal cycle) – Centrosome maturation sequence: procentriole formation, growth, disorientation, and separation. Requires >1.5 cell cycles for completion.
Centriolar cylinder (= centriole core) – 9-triplet/doublet/singlet MT cylinder, structurally conserved in centrosomes and basal bodies.
Centriolar rim – Residual centriolar structure post-MT triplet removal.
Centriole – a cylindrical organelle approximately 200 nm in diameter and 300–500 nm in length, characterized by ninefold symmetry. The proximal end typically consists of nine MT triplets, while the distal end transitions to nine MT doublets in most species. However, centriole ultrastructure varies across organisms: some possess MT doublets at both proximal and distal ends, while others exhibit MT singlets throughout their length. Centrioles serve as core structural components of centrosomes and basal bodies.
Centriole disorientation – Loss of orthogonal centriole alignment during telophase/G1, accompanied by centriole separation.
Centriole polarity – Structural/functional asymmetry between distal and proximal ends, marked by triplet counterclockwise twist (proximal view) and appendage clockwise orientation. Reflects MT polarity (minus ends at proximal).
Centriole replication (= erroneous term for centriole duplication).
Centriolin – Subdistal appendage protein (interphase) and centriolar protein (mitosis). Proposed roles in duplication and cell cycle regulation.
Centrobin – Daughter centriole-specific protein regulating duplication and cytokinesis.
Centrosomal cycle (= centriolar cycle).

Centrosome (= cell center) – Centriole complex with pericentriolar material and associated structures.

Centrosome amplification – Aberrant centriole overduplication/structural defects, causing mitotic errors and chromosomal instability. Hallmark of cancer.

Centrosomin – Drosophila centrosomal protein regulating spindle distribution and centrosome organization during syncytial divisions.

Centrosphere (= MT prophase aster, historical term).

CEP 110 – Distal/proximal centriole-localized protein linking centrosomal components.

Chromosomal translocation – Exchange of chromosomal segments between non-homologous chromosomes.

Cilia (singular: cilium) – 200–300 nm × ≤10 μm membrane-bound axonemal projections (9×2+2 MT formula). Shorter than flagella.

Ciliary bud – Initial ciliogenesis structure forming at basal body-membrane vesicle contact sites.

Ciliary cap – Apical cilium region embedding central MT/singlet ends.

Cilium cover (= ciliary sheath).

Ciliary dyskinesia – Motile cilium dysfunction from axonemal dynein mutations.

Ciliogenesis – Multistep process: deuterosome formation, procentriole assembly, centriole maturation/migration, ciliary growth.

Cilium formula – MT arrangement descriptor (e.g., 9×2+2: nine doublets, two central MTs).

Cilium crown – Glycocalyx-linked apical ciliary processes.

Ciliary sheath (= cilium cover).

CLIP-170 – Cytoplasmic linker protein binding MT plus ends.

C-Nap1 – Rootletin tether anchor on mother/daughter centrioles.

Cogwheel structure – Insect centriolar/wasp sperm flagellar base structure with 9 prongs, replacing canonical MT triplets.

Cohesin – Protein mediating sister chromatid cohesion and centriole linkage.

Colcemid – Synthetic colchicine analog arresting cells in metaphase via MT depolymerization.

Colchicine – Colchicum-derived MT-depolymerizing alkaloid.

Condensation forms (= deuterosomes).

CP 110 – Centriole-capping protein regulating length.

CP 60/CP 190 – Nuclear (interphase)/centrosomal (mitosis) proteins potentially involved in γ-tubulin-mediated MT nucleation.

CPAP – Centrosomal tubulin-binding protein regulating procentriole elongation/quantity.

Cytochalasins B/D – Actin microfilament-depolymerizing fungal metabolites.

Cytokinesis – Cytoplasmic division phase of mitosis.

Cytoplast – Enucleated cell retaining centrosomes (cytochalasin B pretreatment).

Cytoskeleton – Filament network (MTs, actin, intermediate filaments) maintaining cell shape, transport, and motility.

Daughter centriole (= immature, inactive) – Younger centriole in a pair, lacking distal/subdistal appendages. In diplosomes, its proximal end faces the mother centriole. Localized at the mitotic halo periphery during mitosis.

De novo centriole formation – the process by which new centrioles are formed without connection to existing centrioles. It occurs during ciliogenesis, early embryonic development, and parthenogenetic development.

Delta-tubulin **(**δ-tubulin**)** – Minor tubulin family protein, initially identified in Chlamydomonas basal bodies. Likely involved in centriolar MT triplet assembly. Absent in yeast.

Deuterosomes (= condensation forms) – Electron-dense precursors (≤0.4 μm) in ciliogenesis, forming generative complexes with up to 14 procentrioles. Feature dense cores and loose peripheral "crowns."

Diplosome – Centrosomal state with two orthogonally oriented centrioles, typical of mitotic and certain terminally differentiated interphase cells.

Distal centriole – Sperm flagellum-originating centriole, partially/fully resorbed during spermiogenesis.

Distal centriole appendages (= appendages) – Counterclockwise-twisted (distal view) wing-like projections from centriolar triplets/doublets, mediating membrane interactions during ciliogenesis.

Distal end – Centriolar region facing the plasma membrane. Contains electron-dense material; triplets exhibit clockwise twist (distal view).

Doublet outer projections (= Y-shaped bundles).

Duplication (erroneously termed replication) – Centriole doubling process, occurring either proximally to mother centrioles or *de novo*.

Dynactin – Centrosomal complex containing centractin and 160–24 kDa proteins. Interacts with dynein/kinesins (e.g., Eg5).

Dynamic MT instability – Theory positing MT length/number depends on assembly/disassembly rates and catastrophe/rescue probabilities.

Dynein arms – MT "A"-associated projections (inner/outer) enabling axonemal motility via MT "B" sliding.

Enatosomes – Hypothetical SAS6 protein dimers forming cartwheel hubs to establish centriolar ninefold symmetry.

Flagellum – Motile cilium-like structure (9×2+2 MT formula) with greater length and specialized components (e.g., sperm tails).

Foot of basal body – Subdistal appendage homolog anchoring basal bodies to 2–3 triplets. Typically 1–2 per basal body.

Free MT convergence foci (= spherical bodies) – Centriole-independent MT nucleation centers resembling subdistal appendage heads.

Free MTs – MTs with unanchored minus/plus ends.

Gamma-tubulin – MT nucleation component forming γ-TuSC/γ-TuRC complexes.

Generative complex – Ciliogenesis structure comprising deuterosomes/mature centrioles and associated procentrioles.

Griseofulvin – Penicillium nigricans-derived MT-depolymerizing antifungal agent.

Hook – Incomplete MT "C" wall at centriolar distal ends.

Intermediate plate – Basal body/cilium luminal partition apical to the terminal plate.

Interzonal MTs – Spindle-stabilizing MTs linking opposite poles. Contribute to anaphase B separation and residual body formation.

Iota-tubulin (ι-tubulin) – Paramecium-specific tubulin isoform with uncharacterized function.

Kappa-tubulin (κ-tubulin) – Paramecium-specific tubulin isoform with uncharacterized function.

Karyoplast – Cytoplasm-depleted cell generated experimentally.

Katanin – Centrosome/spindle pole-localized MT-severing protein.

KIF11 (= pEg5)

Kinetochore – Chromosomal MT attachment site.

Kinetochore MTs – Chromosome-aligning (metaphase) and chromatid-separating (anaphase) MT bundles anchored at centrosomes.

Kinetodesma – Ciliate basal body-associated longitudinal fibrils.

Kinetosome (= basal body).

Knockdown – Experimental reduction of gene expression via genetic modification or antisense oligonucleotides.

Lateral projections (= Y-shaped bundles).

Latrunculin – Actin monomer-binding toxin from marine sponges, inducing microfilament depolymerization.

Main cilium zone – Axonemal region from central MT origins to peripheral doublet-singlet transitions.

Meckelin – TMEM67-encoded primary cilium protein regulating centriole apical migration. Linked to Meckel-Gruber Syndrome.

Membrane vesicle – Ciliary membrane precursor forming near basal body distal ends.

Microtubule "A" – Innermost MT of centriolar triplet/doublet.

Microtubule "B" – Middle (triplet) or outermost (doublet) MT.

Microtubule "C" – Outermost triplet MT.

Microtubules – 25 nm-diameter tubular polymers of α/β-tubulin dimers. Key components of centrioles, spindles, and cilia. Properties depend on tubulin isoforms, post-translational modifications, and associated proteins.

MT nucleation centers (MTNCs) – Ultrastructurally identifiable MT formation/binding sites.

MT organizing centers (MTOCs) – Functional entities (centriole-containing or not) nucleating/anchoring MTs.

MT polarity – α-tubulin (minus end) to β-tubulin (plus end) orientation, governing MT dynamics.

MT associated proteins – Regulatory proteins modulating MT stability/function.

Midbody – Intercellular structure containing interzonal MTs, intermediate filaments, and actin bundles. Maintains daughter cell connectivity post-cytokinesis.

Minus MT end – α-tubulin-terminated end, often γ-tubulin-anchored at centrosomes/spindle poles.

Mitotic halo – ≈1 μm amorphous pericentrosomal matrix housing MT nucleation centers and regulators during mitosis.

MKS1 – Meckel-Gruber Syndrome-associated basal body protein required for ciliogenesis.

Molecular chaperones – Centrosome-localized proteins modulating MT assembly.

Monopolar spindle – Aberrant mitotic structure with non-separated centrosomes.

Mother centriole (= mature, primary, parental, active) – Appendage-bearing centriole occupying diplosome centers. Localized centrally within mitotic halos.

MT nucleation – Initial MT polymerization phase.

MT protofilaments – Tubulin dimer strands (typically 13) composing MT walls.

MT doublets – Laterally linked MT "A" (13 protofilaments) and "B" (14–15 protofilaments).

MT singlets – Individual MTs in early procentrioles or mature centrioles (species-dependent).

MT triplets – Tripartite MT "A"/"B"/"C" complexes (13/14–15/14–15 protofilaments) in centrioles.

Multipolar mitoses – Aberrant divisions with >2 spindle poles.

Necklace – rows of integral protein globules at the junction of the cell membrane with the cilium membrane.

Nek2 – a kinase whose activity leads to the loss of rootletin and β-catenin by centrioles and to the rupture of the connection between the mother and daughter centrioles.

Nexin bundles – structures made of the protein nexin that connect MT "A" of an axonemal MT doublet to MT "B" of the adjacent doublet.

Ninein – a protein localized on the centriolar surface, subdistal appendage heads, and pericentriolar material. Proposed functions include MT stabilization, distribution, and anchoring.

Nlp – a centrosomally localized protein whose stability is dependent on normal BRCA1 function. siRNA-mediated Nlp suppression disrupts chromosomal segregation, causing aberrant spindle formation, aneuploidy, and mitotic failure.

Nocodazole – a reversible MT-depolymerizing agent used in chemotherapy.

*Nodal pit (*ventral nodule) – a ventral indentation in vertebrate embryos, homologous to the dorsal primary nodule. Its primary monocilia exhibit clockwise rotation, generating a leftward extraembryonic fluid flow critical for left-right axis determination.

Nucleophosmin (NPM) – also termed nucleolar phosphoprotein B23 or numatrin, encoded by the NPM1 gene. NPM regulates centrosome duplication and localizes between centrioles in unduplicated centrosomes.

NuMa – a mitotic spindle protein that stabilizes spindle MTs via cross-linking during mitosis and meiosis.

Orthogonal orientation – a right-angle arrangement, e.g., centrioles in a diplosome.

P27 – one of seven cyclin-dependent kinase inhibitor proteins expressed during specific cell cycle phases.

p53 – a regulatory protein (also known as tumor protein p53, cellular tumor antigen p53, or transformation-related protein 53 [TRP53]) frequently mutated in human cancers.

Parabasal fiber (= rhizoplast).

PCM1 (PCM-1) – an acidic centrosomal protein forming complexes with pericentrin-B. Its centrosomal concentration decreases during mitosis. Hypothesized to inhibit MT nucleation.

PCM proteins – centrosomal proteins retained throughout the cell cycle, with reduced concentration during mitosis. They may inhibit MT nucleation.

pEg5 (= KIF11) – a kinesin-like motor protein localized between separating centrosomes at interphase end, spindle poles, and interzonal MTs during mitosis. Disappears post-mitosis; mediates centrosome separation.

Pegs (= Y-shaped bundles).

Pericentrin – a core centrosomal protein forming a structural framework for protein attachment.

Pericentriolar dense bodies (= satellites) – electron-dense pericentriolar structures, distinct from centrioles and subdistal appendages. Potential MT nucleation/binding sites; size-matched to subdistal appendage heads.

Pericentriolar material (= pericentriolar matrix) – fibrillar material linking mother and daughter centrioles in interphase.

Pericentriolar satellites – historical term for subdistal appendages.

Phalloidin – a phallotoxin from Amanita phalloides binding filamentous actin (F-actin) and inhibiting microfilament depolymerization.

*Plk4 (*Polo-like kinase 4) – a kinase essential for centriole duplication initiation. Homologs: Drosophila SAK, C. elegans ZYG-1, Xenopus Plx4.

Plus-end of MT – the MT terminus containing β-tubulin, exhibiting faster growth than the minus-end. Typically oriented toward the cell periphery.

PKHD1 – a cilia-localized gene mutated in autosomal recessive polycystic kidney disease.

PKD1 and *PKD2* – genes encoding ciliary transmembrane proteins (polycystin-1, polycystin-2), responsible for autosomal polycystic kidney disease.

Polar corpuscles – historical term for centrosomes/centrioles at spindle poles. Recently redefined as atypical centrioles in early vertebrate embryogenesis.

Polyglutamylation – a tubulin post-translational modification enhancing MT stability, particularly in centriolar MTs.

RPGR – a cilia-localized gene mutated in X-linked retinitis pigmentosa.

Primary cilium (= rudimentary/abortive cilium) – a non-motile (9×2+0) cilium on mother centrioles, often involved in sensory functions.

Procentriole – an early centriole precursor appearing as an amorphous disk on mother centrioles in late G1. Contains MT singlets progressing to doublets/triplets, reaching maternal size by mitosis.

Proliferative elements (PE) – granular precursors (40-80 nm) of centrioles during ciliogenesis. Deuterosomes form via PE segmentation.

Proximal centriole – in sperm cells, the centriole nearest the nucleus, perpendicular to the flagellum. Nucleates the centriolar adjunct in spermatids.

Proximal end of the centriole – the nucleus-facing end (historically) and procentriole assembly site. Morphologically distinct by cartwheel structures (e.g., in insects).

Radial MT system - a system of MTs whose minus ends are fixed to the centrosome.

Radial spoke (in the axoneme of a cilium/flagellum) – a structure connecting MT "A" of each axonemal doublet to the central sheath (also termed central capsule in some literature).

Radial spoke (in the centriolar structure) – a structure linking the cartwheel's central hub to the MT triplets of the centriole.

Rhizoplast (= parabasal fiber) – a striated rootlet of the basal body in some algae, functioning as a site for spindle MT convergence during cell division.

Rootletin – a protein component of the fibrous ligament connecting mother and daughter centrioles in interphase cells.

Rootlets – structures anchoring the cilium-basal body or flagellum-basal body complex within the cytoplasm. Rootlets may exhibit transverse striations or lack them.

SAS-4 – a protein required for MT assembly around the central hub during early procentriole formation.

SAS-5 – a protein essential for the formation of the central hub at the initial stages of procentriole assembly.

SAS-6 – a protein critical for the formation of the central hub during early procentriole assembly. SAS-6 dimers form the structural basis of the hub's central region, establishing the centriole's ninefold symmetry via lateral interactions.

Separase – a cysteine protease that mediates sister chromatid separation during mitotic anaphase and disengages mother-daughter centrioles.

Shaft of cilium – an elongated structure emerging during cilium bud growth, enclosed by the ciliary sheath until its fusion with the plasma membrane.

Shugoshin – a protein that suppresses sister chromatid separation during mitotic anaphase. A splice variant of Shugoshin localizes to the centrosome, preventing premature disorientation and separation of mother and daughter centrioles.

Situs inversus (situs transversus or oppositus) – a congenital condition characterized by mirror-image reversal of visceral organ positioning, frequently associated with Kartagener syndrome.

SPB (spindle pole body) – the yeast functional equivalent of the centrosome, structurally resembling chromosomal kinetochores.

Sperm aster – a radial array of MTs organized around the sperm basal body during the first zygotic division.

Spherical bodies (= free foci of MT convergence).

Spindle – a dynamic structure composed of MTs and chromosomes, assembled during mitosis and meiosis to ensure equal distribution of genetic material between daughter cells.

Spindle threads – a historical term referring to the kinetochore MT bundles of the mitotic spindle.

Striated feet (= striated rootlets).

Striated rootlets (= kinetodesmata, = striated feet) – bundles of thin (~5 nm) fibrils with periodic transverse striations (12–90 nm periodicity). The proximal (wider) end of the rootlet associates with the centriolar cylinder, while the distal end anchors cilia/flagella and regulates their motility. Striated rootlets act as MT-organizing centers and are occasionally observed in centrosomes as vestigial structures.

Subdistal appendages (= pericentriolar satellites) – centriole-associated structures comprising a globular MT-organizing head and a striated conical stalk. The stalk base attaches to 2–3 centriolar MT triplets. Their number ranges from 0 to 14 per centriole.

Taxol (= paclitaxel) – a plant-derived compound (originally isolated from Taxus brevifolia bark) that promotes tubulin polymerization and stabilizes MTs.

Tektins – a conserved protein family (Tektin A, B, C) integral to cilia and flagella. Tektins stabilize and regulate the length of centriolar MT triplets.

Terminal plate – a septal structure separating the lumen of the basal body's centriolar cylinder from the cilium's lumen.

Tetraploidization – the process of doubling a cell's chromosome complement to tetraploidy (4n). Tetraploid malignant cells exhibit enhanced migratory and invasive capacities *in vitro* and demonstrate higher metastatic potential *in vivo.*

Theta-tubulin (θ-tubulin) – a minor member of the tubulin family identified in the ciliate Paramecium spp., with currently undefined functional roles.

Transition zone (cilium) – the region extending from the distal end of the basal body to the proximal origin of the central MT pair within the cilium.

Treadmilling – a dynamic MT behavior characterized by concurrent tubulin depolymerization at the minus end and polymerization at the plus end under near-equilibrium tubulin concentrations, resulting in directional MT movement.

Tubulins – a protein family comprising ten isoforms: α-, β-, γ-, δ-, ε-, ζ-, η-, θ-, ι-, and κ-tubulin.

Vinblastine – a Vinca rosea L. (Periwinkle)-derived alkaloid that induces MT depolymerization and is utilized in anticancer therapy. Structurally related to vincristine.

Vincristine – an alkaloid isolated from Vinca rosea L. (Periwinkle), which promotes MT disassembly and is employed in chemotherapy. Chemically analogous to vinblastine.

Y-shaped bundles – a system of linkages between peripheral MT doublets in the ciliary transition zone and the ciliary membrane. Alternative designations include pegs, bridges, doublet outer projections, and lateral projections, depending on the organism.

Zeta-tubulin (ζ-tubulin) – a low-abundance tubulin family member found in trypanosomes (Trypanosoma spp.), whose biological function remains uncharacterized.

References

Abdul-Majeed S., Moloney B.C., Nauli S.M. (2012) Mechanisms regulating cilia growth and cilia function in endothelial cells. *Cell Mol. Life Sci.,* 69(1), 165-173.

Abdul-Majeed S., Nauli S.M. (2011) Dopamine receptor type 5 in the primary cilia has dual chemo- and mechano-sensory roles. *Hypertension,* 58(2), 325-331.

Abumuslimov S.S., Nadezhdina E.S., Chentsov Iu.S. (1991) Numerous centers organization of microtubules without centrioles in giant cells of the salivary glands of Chironomus cingulatus. *Tsitologiia,* 33(12), 36-40. In Russian.

Abumuslimov S.S., Nadezhdina E.S., Chentsov Iu.S. (1994) An electron microscopic study of centriole and centrosome morphogenesis in the early development of the mouse. *Tsitologiia,* 36(11), 1054-61. In Russian.

Abumuslimov S.S., Nadezhdina E.S., Chentsov Iu.S. (1996) Centrosome morphogenesis in the early development of mice: an immunofluorescence study using antibodies to the centriole antigen. *Tsitologiia,* 38(1), 5-13. In Russian.

Afzelius B. (1959) Electron microscopy of the sperm tail. Results obtained with a new fixative. *J. Biophysic. Biochem. Cytol.,* 5 (2), 269-278.

Afzelius B.A., Eliasson R., Johnsen O., Lindholmer C. (1975) Lack of dynein arms in immotile human spermatozoa. *J. Cell Biol.,* 66(2), 225-232.

Afzelius, B.A. (2004) Cilia-related diseases. *J. Pathol.,* 204, 470-477.

Aist J.R., Williams P.H. (1972) Ultrastructure and time course of mitosis in the fungus Fusarium oxysporum. *J. Cell Biol.,* 55(2), 368-389.

Albrecht-Buehler G. (1990) The iris diaphragm model of centriole and basal body formation. *Cell Motil. Cytoskeleton,* 17(3), 197-213.

Albrecht-Buehler G. (1992) Rudimentary form of cellular "vision". *Proc. Natl. Acad. Sci. U S A.* 89(17), 8288-8292. doi: 10.1073/pnas.89.17.8288.

Al-Dosari M.S., Shaheen R., Colak D., Alkuraya F.S. (2010) Novel CENPJ Mutation Causes Seckel Syndrome. *J. Med. Genet.,* 47, 411-414.

Alieva I. B., Vorobjev I.A. (1991) Induction of multipolar mitoses in cultured cells: decay and restructuring of the mitotic apparatus; distribution of centrioles. *Chromosoma,* 100, 532-542.

Alieva I., Staub C., Uzbekova S., Uzbekov R. (2018) A Question of flagella origin for spermatids - Mother or daughter centriole? In "Flagella and Cilia: Types, Structure and Functions" (ed. R. Uzbekov). USA, New York: Nova Science Publisher, Inc. Chapter 5; pp. 109-126.

Alieva I.B. and Uzbekov R.E. (2008) Centrosome is a polyfunctional multiprotein cell complex. *Biochemistry (Moscow),* 73(N 6), 626-643. Review.

Alieva I.B., Uzbekov R.E. (2016) Where are the limits of the centrosome? *Bioarchitecture,* 6(3), 47-52. doi: 10.1080/19490992.2016.1168957.

Alieva I.B., Vorob'ev I.A. (1989a) Reaction of the cell center to exposure to the calcium ionophore A-23187. *Tsitologiia,* 31(3), 259-266. In Russian.

Alieva I.B., Vorob'ev I.A. (1989b) Morphofunctional characteristics of the striated centrosome rootlets of cultured embryonic pig kidney cells. *Tsitologiia,* 31(9), 1016-1019. In Russian.

Alieva I.B., Vorob'ev I.A. (1989c) Cellular behavior and centriole distribution during multipolar mitosis induced by nocodazole action. *Tsitologiia,* 31(6), 633-641. In Russian.

Alieva I.B., Vorob'ev I.A. (1989d) Centrosome behavior in the exposure of tissue culture cells to energy metabolism inhibitors. *Nauchnye Dokl. Vyss. Shkoly Biol. Nauki,* (6), 45-49. In Russian.

Alieva I.B., Vorob'ev I.A. (1992a) A stereoscopic analysis of centrosome structure in the cells of a tissue culture under the action of energy metabolism inhibitors. I. 2,4-Dinitrophenol, deoxyglucose, sodium azide and the calcium ionophore A23187. *Tsitologiia,* 34(5), 65-75. In Russian.

Alieva I.B., Vorob'ev I.A. (1992b) A stereoscopic analysis of the centrosome structure in tissue culture cells under the action of carbonyl cyanide p-trifluoromethoxyphenylhydrazone and ouabain. *Tsitologiia,* 34(7), 17-23. In Russian.

Alieva I.B., Vorobjev I.A. (1994) Centrosome behavior under the action of a mitochondrial uncoupler and the effect of disruption of cytoskeleton elements on the uncoupler-induced alterations. *J. Struct. Biol.,* 113(3), 217-224.

Alieva I.B., Vorobjev I.A. (1995) Centrosome behaviour and orientation of centrioles under the action of energy transfer inhibitors. *Cell Biol. Int.,* 19(2), 103-112.

Alieva I.B., Vorobjev I.A. (2004) The vertebrate primary cilium is a sensory part of centrosomal complex in tissue cells but it is "Sleeping Beauty" in cultured cells. *Cell Biol. Int.,* 28(2), 139-150.

Allen R.D. (1969) The morphogenesis of basal bodies and accessory structures of the cortex of the ciliated protozoan Tetrahymena pyriformis. *J. Cell Biol.,* 40(3), 716-733.

Allen R.D. (1971) Fine structure of membranous and microfibrillar systems in the cortex of Paramecium caudatum. *J. Cell Biol.,* 49, 1-20.

Alliegro M.C., Alliegro M.A. (2008) Centrosomal RNA correlates with intron-poor nuclear genes in Spisula oocytes. *Proc. Natl. Acad. Sci. U S A,* 105(19), 6993-6997.

Alliegro, M.C., Alliegro, M.A., and Palazzo, R.E. (2006). Centrosome-associated RNA in surf clam oocytes. *Proc. Natl. Acad. Sci. USA,* 103, 9034-9038.

Amador-Arjona A., Elliott J., Miller A., Ginbey A., Pazour G.J., Enikolopov G., Roberts A.J., Terskikh A.V. (2011) Primary cilia regulate proliferation of amplifying progenitors in adult hippocampus: implications for learning and memory. *J. Neurosci.,* 31(27), 9933-9944.

Amano S. (1957) The structure of the centrioles and spindle body as observed under the electron and phase contrast microscope. S-A new extension-fiber theory concerning mitotic mechanism in animal cells. *Cytologia* (Tokio), 22, 193-212.

Amargant F., Pujol A., Ferrer-Vaquer A., Durban M., Martinez M., Vassena R., and Vernos I. (2021) The human sperm basal body is a complex centrosome important for embryo preimplantation development. *Mol. Hum. Reprod.* 27(11), gaab062. doi: 10.1093/molehr/gaab062.

Andersen R.S. (1991) The cytoskeleton of chromophyte algae. *Protoplasma*, 164, 143-159.

Anderson R.G. (1972) The three-dimensional structure of the basal body from the rhesus monkey oviduct. *J. Cell Biol.,* 54(2), 246-265.

Anderson R.G.W. (1974) Isolation of ciliated or unciliate basal bodies from rabbit oviduct. *J. Cell Biol.,* 60, 393-404.

Anderson W.A. (1967) Cytodifferentiation of spermatozoa in Drosophila melanogaster: the effect of elevated temperature on spermiogenesis. Mol. *Gen. Genet.,* 99(3), 257-273.

André J. (1964) Le centriole et la région centrosomienne. *J. Microscopie*, 3, 1-23. In French.

Andres K.H. (1969) Der olfaktorische Saum der Katze. *Cell Tissue Res.,* [The olfactory epithelium of the cat]. 96, 250-274. In German.

Antonczak A.K., Mullee L.I., Wang Y., Comartin D., Inoue T., Pelletier L., Morrison C.G. (2015) Opposing effects of pericentrin and microcephalin on the pericentriolar material regulate CHK1 activation in the DNA damage response. *Oncogene,* 35(15), 2003-2010. doi:10.1038/onc.2015.257.

Antony D., Brunner H.G., Schmidts M. (2021) Ciliary Dyneins and Dynein Related Ciliopathies. *Cells,* 10(8), 1885.

Aprelikova O.N., Fang B.S., Meissner E.G., Cotter S., Campbell M., Kuthiala A., Bessho M., Jensen R.A., Liu E.T. (1999) BRCA1-associated growth arrest is RB dependent. *Proc. Natl. Acad. Sci. USA,* 96, 11866-11871.

Archer F.L., Wheatley D.N. (1971) Cilia in cell-cultured fibroblasts. II. Incidence in mitotic and post-mitotic BHK 21-C13 fibroblasts. *J. Anat.,* 109(2), 277-292.

Arlot-Bonnemains Y., Klotzbucher A., Giet R., Uzbekov R., Bihan R., and Prigent C. (2001) Identification of a functional destruction box in the Xenopus lavis aurora-A kinase pEg2. *FEBS Letters,* 508 (1), 149-152.

Arnaiz O., Malinowska A., Klotz C., Sperling L., Dadlez M., Koll F., Cohen J. (2009) Cildb: a knowledgebase for centrosomes and cilia. *Database (Oxford)*, bap022.

Arquint C., Nigg E.A. (2016) The plk4-stil-sas-6 module at the core of centriole duplication. *Biochem. Soc. Trans.,* 44, 1253-1263.

Askham J.M., Vaughan K.T., Goodson H.V., Morrison E.E. (2002) Evidence that an interaction between EB1 and p150(Glued) is required for the formation and maintenance of a radial microtubule array anchored at the centrosome. *Mol. Biol. Cell,* 13(10), 3627-3645.

Astuti P., Boutros R., Ducommun B., Gabrielli B. (2010) Mitotic phosphorylation of Cdc25B Ser321 disrupts 14-3-3 binding to the high affinity Ser323 site. *J. Biol. Chem.,* 285(45), 34364-34370.

Atorino E.S., Hata S., Funaya C., Neuner A., Schiebel E. (2020) Cep44 ensures the formation of bona fide centriole wall, a requirement for the centriole-to-centrosome conversion. *Nat. Commun.,* 11, 903.

Avidor-Reiss T. (2018) Rapid evolution of sperm produces diverse centriole structures that reveal the most rudimentary structure needed for function. *Cells,* 7, 67.

Avidor-Reiss T. (2024) Renaissance in sperm cytoplasmic contribution to infertility. *J. Assist. Reprod. Genet.,* 41(2), 293-296. doi: 10.1007/s10815-023-02987-6.

Avidor-Reiss T., Achinger L., Uzbekov R. (2022) The Centriole's Role in Miscarriages. *Front. Cell Dev. Biol.,* 10, 864692. doi: 10.3389/fcell.2022.864692. eCollection 2022.

Avidor-Reiss T., and Gopalakrishnan J. (2013) Building a centriole. *Curr. Opin. Cell Biol.,* 25, 72-77.

Avidor-Reiss T., Fishman E.L. (2019) It takes two (centrioles) to tango. *Reproduction,* 157, R33-R51.

Avidor-Reiss T., Gopalakrishnan J. (2013) Cell cycle regulation of the centrosome and cilium. *Drug Discov. Today. Dis. Mech.,* 10, e119-e124.

Avidor-Reiss T., Khire A., Fishman E.L., Jo K.H. (2015) Atypical centrioles during sexual reproduction. *Front. Cell Dev. Biol.,* 3, 21.

Avidor-Reiss T., Leroux M.R. (2015) Shared and Distinct Mechanisms of Compartmentalized and Cytosolic Ciliogenesis. *Curr. Biol.,* 25(23), R1143-50. doi: 10.1016/j.cub.2015.11.001.

Avidor-Reiss T., Turner K. (2019) The Evolution of Centriole Structure: Heterochrony, Neoteny, and Hypermorphosis. *Results Probl. Cell Differ.* 67, 3-15. doi: 10.1007/978-3-030-23173-6_1.

Azimzadeh J. (2014) Exploring the evolutionary history of centrosomes. Philos. *Trans. R. Soc. Lond. B. Biol. Sci.,* 369(1650), 20130453.

Azimzadeh J., Bornens M. (2007) Structure and duplication of the centrosome. *J. Cell Sci.,* 120, 2139-2142.

Azimzadeh J., Hergert P., Delouvee A., Euteneuer U., Formstecher E., Khodjakov A., Bornens M. (2009) Hpoc5 is a centrin-binding protein required for assembly of full-length centrioles. *J. Cell Biol.,* 185, 101-114.

Azimzadeh J., Wong M.L., Downhour D.M., Sanchez Alvarado A., Marshall W.F. (2012) Centrosome loss in the evolution of planarians. *Science,* 335, 461-463.

Baccetti B, Dallai R. (1977) The spermatozoon of arthropoda. XXIX. The degenerated axoneme and branched acrosome of aleyrodids. *J. Ultrastruct. Res.,* 61, 260-270.

Baccetti B., Dallai R. (1976) The spermatozoon of arthropoda. XXVII. Uncommon axoneme patterns in different species of the cecidomyid flies. *J. Ultrastruct. Res.,* 55, 50-69.

Baccetti B., Dallai R. (1978) The spermatozoon of arthropoda. XXX. The multiflagellate spermatozoon in the termite Mastotermes darwiniensis. *J. Cell Biol.,* 76(3), 569-576.

Baccetti B., Dallai R., Callaini G. (1981) The spermatozoon of Arthropoda: Zootermopsis nevadensis and isopteran sperm phylogeny. *Int. J. Invert. Reprod.,* 3(2), 87-99.

Baccetti B., Dallai R., Giusti F. (1969) The spermatozoon of Arthropoda. VI. Ephemeroptera. J. *Ultrastruct. Res.*, 29, 343-349.

Baccetti B., Dallai R., Rosati F. (1969) The spermatozoon of arthropoda. IV. Corrodentia, Mallophaga and Thysanoptera. *J. Microscopie,* 8, 249-262.

Bahe S., Stierhof Y.D., Wilkinson C.J., Leiss F., Nigg E.A. (2005) Rootletin forms centriole-associated filaments and functions in centrosome cohesion. *J. Cell Biol.,* 171(1), 27-33.

Bahmanyar S., Kaplan D.D., Deluca J.G., Giddings T.H. Jr., O'Toole E.T., Winey M., Salmon E.D., Casey P.J., Nelson W.J., Barth A.I. (2008) beta-Catenin is a Nek2 substrate involved in centrosome separation. *Genes Dev.,* 22(1), 91-105.

Bai R., Wan R., Wang L., Xu K., Zhang Q., Lei J., Shi Y. (2021) Structure of the activated human minor spliceosome. *Science,* 371(6535), eabg0879. doi: 10.1126/science. abg0879.

Bailey A.W, Suri A., Chou P.M., Pundy T., Gadd S., Raimondi S.L., Tomita T., Sredni S.T. (2018). Polo-Like Kinase 4 (PLK4) Is Overexpressed in Central Nervous System Neuroblastoma (CNS-NB). *Bioengineering,* 5 (4), 96. doi:10.3390/bioengineering 5040096.

Bailly E., Pines J., Hunter T., Bornens M. (1992) Cytoplasmic accumulation of cyclin B1 in human cells: assotiation with a detergent-resistent compartment and with the centrosome. *J. Cell Sci.,* 101, 529-545.

Baker J.D., Adhikarakunnathu S., Kernan M.J. (2004) Mechanosensorydefective, male-sterile unc mutants identify a novel basal body protein required for ciliogenesis in Drosophila. *Development,* 131(14), 3411-3422.

Balczon R., Bao L., Zimmer W.E., Brown K., Zinkowski R.P., Brinkley B.R. (1995) Dissociation of centrosome replication events from cycles of DNA synthesis and mitotic division in hydroxyurea-arrested Chinese hamster ovary cells. *J. Cell. Biol.,* 130, 105-115.

Bao J., Yu Y., Chen J., He Y., Chen X., Ren Z., Xue C., Liu L., Hu Q., Li J., Cui G., Sun R. (2018) MiR-126 negatively regulates PLK-4 to impact the development of hepatocellular carcinoma via ATR/CHEK1 pathway. *Cell Death Dis.,* 9(10), 1045 doi: 10.1038/s41419-018-1020-0.

Bào S.N., Kitajima E.W., Callaini G., Lupetti P., Dallai R. (1997) Spermiogenesis in Three Species of *Whitefly* (*Homoptera*, *Aleyrodidae*). *Acta Zool.,* 78, 163-170.

Bareja I., Kučera O., Petitjean II., Sabo J., Braun M., Lansky Z., Koenderink G.H., Dogterom M. (2025) Anillin directly cross-links microtubules with filamentous actin. *bioRxiv preprint.* doi: 10.1101/2025.01.07.631698

Barker A.R., McIntosh K.V., Dawe H.R. (2016) Centrosome positioning in non-dividing cells. *Protoplasma,* 253(4), 1007-1021.

Barnes B.G. (1961) ciliated secretory cells in the pars distalisof the mouse hypophysis. *J. Ultrastruct. Res.,* 5, 453-467.

Baron A.T., Salisbury J.L. (1988) Identification and localization of a novel, cytoskeletal, centrosome associated protein in PtK2 cells. *J. Cell Biol.,* 107, 2669-2678.

Barr M.M., Sternberg P.W. (1999) A polycystic kidney-disease gene homologue required for male mating behaviour in C. elegans. *Nature,* 401, 386-389.

Barrera J.A., Kao L.R., Hammer R.E., Seemann J., Fuchs J.L., Megraw T.L. (2010) Cdk5rap2 regulates centriole engagement and cohesion in mice. *Dev. Cell*, 18, 913-926.

Bartolini F., Gundersen G.G. (2006) Generation of noncentrosomal microtubule arrays. *J. Cell Sci.,* 119 (20), 4155-4163.

Bartoloni L., Blouin J.L., Maiti A.K., Sainsbury A., Rossier C., Gehrig C., She J.X., Marron M.P., Lander E.S., Meeks M., Chung E., Armengot M., Jorissen M., Scott H.S., Delozier-Blanchet C.D., Gardiner R.M., Antonarakis S.E. (2001) Axonemal beta heavy chain dynein DNAH9: cDNA sequence, genomic structure, and investigation of its role in primary ciliary dyskinesia. *Genomics,* 72(1), 21-33.

Basiri M.L., Blachon S., Chim Y.C., Avidor-Reiss T. (2013) Imaging centrosomes in fly testes. *J. Vis. Exp.,* e50938, doi:10.3791/50938.

Basiri M.L., Ha A., Chadha A., Clark N.M., Polyanovsky A., Cook B., Avidor-Reiss T. (2014) A migrating ciliary gate compartmentalizes the site of axoneme assembly in Drosophila spermatids. *Curr. Biol.,* 24(22), 2622-2631.

Basto R., Lau J., Vinogradova T., Gardiol A., Woods C.G., Khodjakov A., Raff J.W. (2006) Flies without centrioles. *Cell,* 125(7), 1375-1386.

Bastos, R.N., Barr, F. A. (2010) Plk1 negatively regulates Cep55 recruitment to the midbody to ensure orderly abscission. *J. Cell Biol.,* 191(4), 751-760.

Bauer M., Cubizolles F., Schmidt A., Nigg E.A. (2016) Quantitative analysis of human centrosome architecture by targeted proteomics and fluorescence imaging. *EMBO J.,* 35 (19), 2152-2166. 10.15252/embj.201694462.

Bernhard W., de Harven E. (1960) L'ultrastructure du centriole et d'autres éléments de l'appareil achromatique. In: Vierter internationaler Kongress fur Elektronenmikroskopie, 1958, [The ultrastructure of the centriole and other elements of the achromatic apparatus. In „*Fourth International Conference on Electron Microscopy*" Vol. 2 (eds. W. Bargmann, G. Mollenstedt, H. Niehrs, D. Peters, E. Ruska and C. Wolpers), pp. 217-227. Germany, Berlin: Springer. In French.

Berns M.W., Aist J., Edwards J., Strahs K., Girton J., McNeill P., Rattner J.B., Kitzes M., Hammer-Wilson M., Liaw L.H., Siemens A., Koonce M., Peterson S., Brenner S., Burt J., Walter R., Bryant P.J., van Dyk D., Colombe J., Chill T., Berns G.S. (1981) Laser microsurgery in cell and developmental biology. *Science,* 213, 505-513.

Berns M.W., Rattner J.B., Brenner S., Meredith S. (1977) The role of the centriolar region in animal cell mitosis. *J. Cell Biol.,* 72, 351-367.

Berrueta L., Kraeft S.K., Tirnauer J.S., Schuyler S.C., Chen L.B., Hill D.E., Pellman D., Bierer BE. (1998) The adenomatous polyposis coli-binding protein EB1 is associated with cytoplasmic and spindle microtubules. *Proc. Natl. Acad. Sci. USA,* 95(18), 10596-10601.

Bessis M., Breton-Gorius J. (1958) Sur une structure inframicroscopique péricentriolaire. Etude au microscope électronique sur des leucocytes Mammifères. [On an pericentral inframicroscopic structure; electron microscopy of mammalian leukocytes]. *Compt. Rend. Acad. Sc.,* 246, 1289-1291. In French.

Bettencourt-Dias M., Glover D.M. (2007) Centrosome biogenesis and function: centrosomics brings new understanding. *Nat. Rev. Mol. Cell Biol.,* 8, 451-463.

Bettencourt-Dias M., Glover D.M. (2009) Snap Shot: centriole biogenesis. *Cell,* 136(1), 188-188.el

Bettencourt-Dias M., Rodrigues-Martins A., Carpenter L., Riparbelli M., Lehmann L., Gatt M.K., Carmo N., Balloux F., Callaini G., Glover D.M. (2005) SAK/PLK4 is required for centriole duplication and flagella development. *Curr. Biol.,* 15(24), 2199-2207. doi: 10.1016/j.cub.2005.11.042.

Beutel R.G., Friedrich F., Ge S.Q., Yang X.K. (2013) *Insect Morphology and Phylogeny: A textbook for students of entomology.* Berlin, Boston: De Gruyter.

Bi E.F., Lutkenhaus J. (1991). FtsZ ring structure associated with division in Escherichia coli". *Nature,* 354 (3-5), 161-164.

Blachon S., Cai X., Roberts K.A., Yang K., Polyanovsky A., Church A., Avidor-Reiss T. (2009) A proximal centriole-like structure is present in Drosophila spermatids and can serve as a model to study centriole duplication. *Genetics,* 182(1), 133-144. doi: 10.1534/genetics.109.101709.

Blachon S., Gopalakrishnan J., Omori Y., Polyanovsky A., Church A., Nicastro D., Malicki J., Avidor-Reiss T. (2008) Drosophila asterless and vertebrate cep152 are orthologs essential for centriole duplication. *Genetics,* 180, 2081-2094.

Blachon S., Khire A., Avidor-Reiss T. (2014) The origin of the second centriole in the zygote of Drosophila melanogaster. *Genetics,* 197(1), 199-205. doi: 10.1534/genetics. 113.160523.

Blachon S., Khire A., Avidor-Reiss T. (2014) The origin of the second centriole in the zygote of Drosophila melanogaster. *Genetics,* 197, 199-205.

Blacque O.E., Leroux M.R. (2006) Bardet-Biedl syndrome: an emerging pathomechanism of intracellular transport. *Cell Mol. Life Sci.,* 63(18), 2145-2161.

Blanchoin L., Boujemaa-Paterski R., Sykes C., Plastino J. (2014) Actin dynamics, architecture, and mechanics in cell motility. *Physiological Reviews,* 94(1), 235-263.

Blangy A., Arnaud L., Nigg E.A. (1997) Phosphorylation by p34cdc2 protein kinase regulates binding of the kinesin-related motor HsEg5 to the dynactin subunit p150. *J. Biol. Chem.,* 272(31), 19418-19424.

Blangy A., Lane H.A., d'Herin P., Harper M., Kress M., Nigg E.A. (1995) Phosphorylation by p34cdc2 regulates spindle association of human Eg5, a kinesin-related motor essential for bipolar spindle formation *in vivo*. *Cell,* 83, 1159-1169.

Blitzer A.L., Panagis L., Gusella G.L., Danias J., Mlodzik M., Iomini C. (2011) Primary cilia dynamics instruct tissue patterning and repair of corneal endothelium. *Proc. Natl. Acad. Sci. U S A,* 108(7), 2819-2824.

Bloodgood R.A. (2009) From central to rudimentary to primary: The history of an underappreciated organelle whose time has come. The primary cilium. *Methods Cell Biol.,* 94, 3-52.

Bobinnec Y., Khodjakov A., Mir L.M., Rieder C.L., Edde B., Bornens M. (1998a) Centriole disassembly *in vivo* and its effect on centrosome structure and function in vertebrate cells. *J. Cell Biol.,* 143, 1575-1589.

Bobinnec Y., Moudjou M., Fouquet J.P., Desbruyeres E., Edde B., Bornens M. (1998b) Glutamylation of centriole and cytoplasmic tubulin in proliferating non-neuronal cells. *Cell Motyl. Cytoskel.,* 8, 238-249.

Boisson C., Mattei X., Mattei C. (1967) Troisieme note sur la spermiogenese de Protopterus annectens (Dipneuste) du Senegale. [Flagella of 9 + 0 type and remarkable extension of the proximal centriole in the spermatids of the Lycodontis after (Bloch, 1795) (Fish Muraenidae)] *Bull. IFAN,* 29, 1097-1127. PMID: 4965026 In French.

Boisvieux-Ulrich E., Sandoz D., Chailley B. (1977) A freeze-fracture and thin section study of the ciliary necklace in quail oviduct. *Biol. Cell,* 30, 245-252.

Bokov K., Steinberg S.V. (2009). A hierarchical model for evolution of 23S ribosomal RNA. *Nature,* 457, 977-980.

Boleti H., Karsenti E., Vernos I. (1996) Xklp2, a novel Xenopus centrosomal kinesin-like protein required for centrosome sepqration during mitosis. *Cell,* 84, 49-59.

Bolli N., Avet-Loiseau H., Wedge D.C., Van Loo P., Alexandrov L.B., Martincorena I., Dawson K.J., Iorio F., Nik-Zainal S., Bignell G.R., Hinton J.W., Li Y., Tubio J.M., McLaren S., O' Meara S., Butler A.P., Teague J W , Mudie L , Anderson E , Rashid N , Tai Y.T., Shammas M.A., Sperling A.S., Fulciniti M., Richardson .PG., Parmigiani G., Magrangeas F., Minvielle S., Moreau P., Attal M., Facon T., Futreal P.A., Anderson K.C., Campbell P.J., Munshi N.C. (2014) Heterogeneity of genomic evolution and mutational profiles in multiple myeloma. *Nat. Commun.*, 5, 2997. doi: 10.1038/ncomms3997.

Bornens M. (2002) Centrosome composition and microtubule anchoring mechanisms. *Curr. Opin. Cell Biol.,* 14, 25-34.

Bornens M. (2021) Centrosome organization and functions. *Curr. Opin. Struct. Biol.,* 66, 199-206.

Bornens M., Paintrand M., Berges J., Marty M.C., Karsenti E. (1987) Structural and chemical characterization of isolated centrosomes. *Cell Motil. Cytoskeleton,* 8(3), 238-249.

Bouck G.B., Brown D.L. (1973) Microtubule biogenesis and cell shape in Ochromonas. I. The distribution of cytoplasmic and mitotic microtubules. *J. Cell Biol.,* 56, 340-359.

Bourke E., Dodson H., Merdes A., Cuffe L., Zachos G., Walker M., Gillespie D., Morrison C.G. (2007) DNA damage induces Chk1-dependent centrosome amplification. *EMBO Rep.,* 8, 603-609.

Boveri T. (1887a) Zellenstudien I. *Jena Zeit. Naturw.,* 21. Citation from Wilson (1900). In German.

Boveri T. (1887b) Ueber den Antheil des Spermatozoon an der Teilung des Eies. Sitzungsber. Ges. Morph. *Physiol. München,* 3, 151-164.

Boveri T. (1888) Zellenstudien II., Die Befruchtung und Teiling des Eies von Ascaris megalocephala. *Jena Zeit. Naturw.*, 22, 685-822. In German. Citation from Wilson (1900).

Boveri T. (1895) Über das Verhalten der Centrosomen bei der Befruchtung des Seeigeleies, etc. *Verhandl. Phys. Med. Ges. Wurzburg, XXIX.* In German. Citation from Wilson (1900).

Boveri T. (1914) Zur Frage der Entstehung maligner Tumoren, 1-64. In German. Concerning the Origin of Malignant Tumours by Theodor Boveri (1914). Translated and annotated by Henry Harris. *J. Cell Sci.,* 2008; 121, 1-84.

Bowler M., Kong D., Sun S., Nanjundappa R., Evans L., Farmer V., Holland A., Mahjoub M.R., Sui H., Loncarek J. (2019) High-resolution characterization of centriole distal appendage morphology and dynamics by correlative storm and electron microscopy. *Nat. Commun.,* 10, 993.

Bowles E.J., Tecirlioglu R.T., French A.J., Holland M.K., St John J.C. (2008) Mitochondrial DNA transmission and transcription after somatic cell fusion to one or more cytoplasts. *Stem Cells,* 26(3), 775-782.

Bragina E., Blanchard E., Uzbekov R. (2018) Ultrastructural and functional abnormalities in various variants of ciliopathy. In "*Flagella and Cilia: Types, Structure and Functions*" (ed. R. Uzbekov). USA, New York: Nova Science Publishers, Inc. Chapter 7, 169-186.

Brailov I., Bancila M., Brisorgueil M., Miquel M., Hamon M., Verge D. (2000) Localization of 5-HT (6) receptors at the plasma membrane of neuronal cilia in the rat brain. *Brain Res.,* 872, 271-275.

Bre M.H., Kreis T.E., Karsenti E. (1987) Control of microtubule nucleation and stability in Madin-Darby canine kidney cells: the occurrence of noncentrosomal, stable detyrosinated microtubules. *J. Cell Biol.,* 105(3), 1283-1296.

Breipohl W., Mendoza A.S., Miragall F. (1980) Freeze-etching studies on the ciliary necklace in the rat and chick. *J. Anat.,* 130, 801-807.

Breland O.P., Barker K.R., Eddleman C.D., Biesele J.J. (1968) Centrioles in the Spermatids of Insects. *Ann. Entomol. SOC. Amer.,* 61(4), 1037-1039.

Brightman M.W., Palay S.L. (1963) The fine structure of ependyma in the brain of the rat. *J. Cell Biol.,* 19, 415-439.

Brinkley B.R., Cox S.M., Pepper D.A., Wible L., Brenner S.L., Pardue R.L. (1981) Tubulin assembly sites and the organization of cytoplasmic microtubules in cultured mammalian cells. *J. Cell Biol.,* 90(3), 554-562.

Brinkley B.R., Stubblefield E. (1970) Ultrastructure and interaction of the kinetochore and centriole in mitosis and meiosis. In "*Advances in Cell biology*" (eds. D.M. Prescott, L. Goldstein, E. Mc Conkey). USA, New York: Appleton-Century Crofts. 1, 119-184.

Brito D.A., Strauss J., Magidson V., Tikhonenko I., Khodjakov A., Koonce M.P. (2005) Pushing forces drive the comet-like motility of microtubule arrays in Dictyostelium. *Mol. Biol. Cell,* 16, 3334-3340.

Brodsky I.B., Burakov A.V., Nadezhdina E.S. (2007) Microtubules' interaction with cell cortex is required for their radial organization, but not for centrosome positioning. *Cell Motil. Cytoskeleton,* 64(6), 407-417.

Brown C.R., Doxsey S.J., Hong-Brown L.Q., Martin R.L., Welch W.J. (1996a) Molecular chaperones and the centrosome. A role for TCP-1 in microtubule nucleation. *J. Biol. Chem.,* 271, 824-832.

Brown C.R., Hong-Brown L.Q., Doxsey S.J., Welch, W.J. (1996b) Molecular Chaperones and the Centrosome A Role for HSP 73 in centrosomal repair followings heat shock treatment. *J. Biol. Chem.,* 271, 833-840.

Brown N.J., Marjanovic M., Luders J., Stracker T.H., Costanzo V. (2013) Cep63 and cep152 cooperate to ensure centriole duplication. *PLoS ONE,* 8, e69986.

Brown R.L., Wible L.J., Brinkley B.R. (1980) Cytoplasmic microtubule assembly-disassembly in enucleated cells and regenerating karyoplasts. *Cell Biol. Int. Rep.,* 4(5), 453-458.

Brownlee C.W., Klebba J.E., Buster D.W., Rogers G.C. (2011) The protein phosphatase 2A regulatory subunit Twins stabilizes Plk4 to induce centriole amplification. *J. Cell Biol.*, 195, 231-243. doi: 10.1083/jcb.201107086.

Brueckner M. (2001) Cilia propel the embryo in the right direction. *Am. J. Med. Genet.,* 101(4), 339-344. doi: 10.1002/1096-8628(20010715)101:4<339::aid-ajmg1442> 3.0.co;2-p.

Buckley P.T., Khaladkar M., Kim J., Eberwine J. (2014) Cytoplasmic intron retention, function, splicing, and the sentinel RNA hypothesis. *Wiley Interdiscip Rev RNA.* 5(2), 223-30. doi: 10.1002/wrna.1203.

Bugnard E., Zaal K.J., Ralston E. (2005). Reorganization of microtubule nucleation during muscle differentiation. *Cell Motil. Cytoskeleton,* 60, 1-13.

Burakov A., Kovalenko O., Semenova I., Zhapparova O., Nadezhdina E., Rodionov V. (2008) Cytoplasmic dynein is involved in the retention of microtubules at the centrosome in interphase cells. *Traffic,* 9, 472-480.

Burakov A., Nadezhdina E., Slepchenko B., Rodionov V. (2003) Centrosome positioning in interphase cells. *J. Cell Biol.,* 162(6), 963-969.

Burakov A.V. Nadezhdina E.S. (2020) Centering and shifting of centrosomes in cells. *Cells,* 9, 1351. doi: 10.3390/cells9061351.

Burakov A.V., Nadezhdina E.S. (2006) Dynein and dynactin as organizers of the system of cell microtubules. *Ontogenez,* 37(5), 323-339. Review. In Russian.

Burakov A.V., Nadezhdina E.S. (2013) Association of nucleus and centrosome: Magnet or velcro? *Cell Biol. Int.,* 37, 95-104.

Burgos M.H., Fawcett D.W. (1955) Studies on the fine structure of the mammalian testis. I. Differenciation of the spermatids in the cat (Felis domestica). *J. Biophys. Biochem. Cytology,* 1(4), 287-300.

Burstein E.A. (1976) Own luminescence of protein. In "*The results of Science and Technology. Ser. "Biophysics"*. USSR, Moscow: VINITI, 6, 214-320. In Russian.

Burute M., Prioux M., Blin G., Truchet S., Letort G., Tseng Q., Bessy T., Lowell S., Young J., Filhol O., Théry M. (2017) Polarity reversal by centrosome repositioning primes cell scattering during epithelial-to-mesenchymal transition. *Dev. Cell,* 40, 168-184.

Busselez J., Uzbekov R.E., Franco B., Pancione M. (2023) New insights into the centrosome-associated spliceosome components as regulators of ciliogenesis and tissue identity. *Wiley Interdiscip Rev RNA,* 14(4), e1776. doi: 10.1002/wrna.1776. Epub 2023 Jan 30. PMID: 36717357.

Bystrevskaya V.B., Lichkun V.V., Krushinsky A.V., Smirnov V.N. (1992) Centriole modification in human aortic endothelial cells. *J. Struct. Biol.,* 109(1), 1-12. doi: 10.1016/1047-8477(92)90061-e.

Cajanek L., Nigg E.A. (2014) Cep164 triggers ciliogenesis by recruiting tau tubulin kinase 2 to the mother centriole. *Proc. Natl. Acad. Sci. USA*, 111, E2841-E2850.

Calarco-Gillam P.D., Siebert M.C., Hubble R., Mitchison T., Kirschner M.W. (1983) Centrosome development in early mouse embryos as defined by an autoantibody against pericentriolar material. *Cell,* 35, 621-629.

Callaini G., Riparbelli M.G., Dallai R. (1999) Centrosome inheritance in insects: Fertilization and parthenogenesis. *Biol. Cell,* 91, 355-366.

Cambray Deakin M.A., Burgoyone R.D. (1987) Acetylated and detyrosinated alpha-tubulins are co-localized in stable microtubules in rat meningeal fibroblasts. *Cell Motil. Cytoskeleton,* 8(3), 284-291.

Cammareri P., Scopelliti A., Todaro M., Eterno V., Francescangeli F., Moyer MP., Agrusa A., Dieli F., Zeuner A., Stassi G. (2010) Aurora-a is essential for the tumorigenic capacity and chemoresistance of colorectal cancer stem cells. *Cancer Res.,* 70(11), 4655-4665. doi: 10.1158/0008-5472.CAN-09-3953.

Caron J.M., Jones A.L., Kirschner M.W. (1985a) Autoregulation of tubulin synthesis in hepatocytes and fibroblasts. *J. Cell Biol.,* 101(5 Pt 1), 1763-1772.

Caron J.M., Jones A.L., Rall L.B., Kirschner M.W. (1985b) Autoregulation of tubulin synthesis in enucleated cells. *Nature,* 317(6038), 648-651.

Carroll P.E., Okuda M., Horn H.F., Biddinger P., Stambrook P.J., Gleich L.L., Li Y., Tarapore P., Fukasawa K. (1999) Centrosome hyperamplification in human cancer: chromosome instability induced by p53 mutation and/or Mdm2 overexpression. *Oncogene*, 18, 1935-1944.

Carson J.L., Collier A.M., Knowles M.R., Boucher R.C., Rose J.G. (1981) Morphometric aspects of ciliary distribution and ciliogenesis in human nasal epithelium. *Proc. Natl. Acad. Sci. USA,* 78, 6996-6999.

Carvajal-Gonzalez J.M., Mulero-Navarro S., Mlodzik M. (2016) Centriole positioning in epithelial cells and its intimate relationship with planar cell polarity. *BioEssays,* 38, 1234-1245.

Casenghi M., Barr F.A., Nigg E.A. (2005) Phosphorylation of Nlp by Plk1 negatively regulates its dynein-dynactin-dependent targeting to the centrosome. *J. Cell Sci.,* 118(21), 5101-5108.

Casenghi M., Meraldi P., Weinhart U., Duncan P.I., Körner R., Nigg E.A. (2003) Polo-like kinase 1 regulates Nlp, a centrosome protein involved in microtubule nucleation. *Dev. Cell,* 5(1), 113-125.

Cassioli C., Baldari C.T. (2019) A ciliary view of the immunological synapse. *Cells,* 8, 789.

Caudron F., Barral, Y. (2009) Septins and the lateral compartmentalization of eukaryotic membranes. *Dev. Cell,* 16, 493-506.

Cavalier-Smith, T. (1974) Basal body and flagellar development during the vegetative cell cycle and the sexual cycle of Chlamydomonas reinhardii. *J. Cell Sci.,* 16, 529-556.

Centouse V.E., Vandre D.D., Borisy G.G. (1986) Growth of microtubules on mitotic centrosomes is blocked by MPM-2. *J. Cell Biol.,* 103, 412a.

Cerulo L., Pezzella N., Caruso F.P., Parente P., Remo A., Giordano G., Forte N., Busselez J., Boschi F., Galiè M., Franco B., Pancione M. (2023) Single-cell proteo-genomic reveals a comprehensive map of centrosome-associated spliceosome components. *iScience,* 26(5), 106602. doi: 10.1016/j.isci.2023.106602.

Chabin-Brion K., Marceiller J., Perez F., Settegrana C., Drechou A., Durand G., Poüs C. (2001) The Golgi is a microtubule-organizing organelle. *Mol. Biol. Cell,* 12, 2047-2060.

Chan J.Y. (2011) A clinical overview of centrosome amplification in human cancers. *Int. J. Biol. Sci.,* 7, 1122-1144. doi: 10.7150/ijbs.7.1122.

Chang B., Khanna H., Hawes N., Jimeno D., He S., Lillo C., Parapuram S.K., Cheng H., Scott A., Hurd R.E., Sayer J.A., Otto E.A., Attanasio M., O'Toole J.F., Jin G., Shou C., Hildebrandt F., Williams D.S., Heckenlively J.R., Swaroop A. (2006) In-frame deletion in a novel centrosomal/ciliary protein CEP290/NPHP6 perturbs its interaction with RPGR and results in early-onset retinal degeneration in the rd16 mouse. *Hum. Mol. Genet.,* 15, 1847-1857.

Chang J., Cizmecioglu O., Hoffmann I., Rhee, K. (2010). PLK2 phosphorylation is critical for CPAP function in procentriole formation during the centrosome cycle. *EMBO J.,* 29(14), 2395-2406.

Chang M.S., Huang C.J., Chen M.L., Chen S.T., Fan C.C., Chu J.M., Lin W.C., Yang Y.C. (2001) Cloning and characterization of hMAP126, a new member of mitotic spindle-associated proteins. *Biochem. Biophys. Res. Commun.,* 287, 116-121.

Chang P. Stearns T. (2000) δ-Tubulin and ε-Tubulin: two new human centrosomal tubulins reveal new aspects of centrosome structure and function. *Nature Cell Biol.,* 2, 30-35.

Chappuis P.O., Rosenblatt J., Foulkes W.D. (1999) The influence of familial and hereditary factors on the prognosis of breast cancer. *Ann. Oncol.,* 10, 1163-1170.

Charlat S., Mercot H. (2001) Cytoplasmic incompatibility and maternal-haploid. *Trends Genet.,* 17, 440-441.

Chartier N.T., Oddou C.I., Lainé M.G., Ducarouge B., Marie C.A., Block M.R., Jacquier-Sarlin M.R. (2007) Cyclin-dependent kinase 2/cyclin E complex is involved in p120 catenin (p120ctn)-dependent cell growth control: a new role for p120ctn in cancer. *Cancer Res.,* 67(20), 9781-9790.

Chen C., Cummings R., Mordovanakis A., Hunt A.J., Mayer M., Sept D., Yamashita Y.M. (2018) Cytokine receptor-Eb1 interaction couples cell polarity and fate during asymmetric cell division. *eLife*, 7, e33685.

Chen C.H., Howng S.L., Cheng T.S., Chou M.H., Huang C.Y., Hong Y.R. (2003) Molecular characterization of human ninein protein: two distinct subdomains required for centrosomal targeting and regulating signals in cell cycle. *Biochem. Biophys. Res. Commun.,* 308(4), 975-983.

Chen C.Y., Olayioye M.A., Lindeman G.J., Tang T.K. (2006) CPAP interacts with 14-3-3 in a cell cycle-dependent manner. *Biochem. Biophys. Res. Commun.,* 342(4), 1203-1210.

Chen H.H., Lu H.Y., Chang C.H., Lin S.H., Huang C.W., Wei P.H., Chen Y.W., Lin Y.R., Huang H.S., Wang P.Y., Tsao Y.P., Chen S.L. (2022) Breast carcinoma-amplified sequence 2 regulates adult neurogenesis via β-catenin. *Stem Cell Res. Ther.,* 13(1), 160. doi: 10.1186/s13287-022-02837-9.

Chen H.Y., Wu C.T., Tang C.C., Lin Y.N., Wang W.J., Tang T.K. (2017) Human microcephaly protein rttn interacts with stil and is required to build full-length centrioles. *Nat. Commun.,* 8, 247.

Chen W.L.J., Fung K.F.K., Chan E.Y. (2024) Nephronophthisis-associated ciliopathy with brachydactyly, medullary cysts, and chronic kidney disease. *Kidney Int.*, 106(4), 759. doi: 10.1016/j.kint.2024.04.022.

Chen Z., Indjeian V.B., McManus M., Wang L., Dynlacht B.D. (2002) CP110, a cell cycle-dependent CDK substrate, regulates centrosome duplication in human cells. *Dev. Cell,* 3(3), 339-350.

Chen Z., Liu X., Kawakami M., Liu X., Baker A., Bhatawadekar A., Tyutyunyk-Massey L., Narayan K., Dmitrovsky E. (2023) CDK2 inhibition disorders centrosome stoichiometry and alters cellular outcomes in aneuploid cancer cells. *Cancer Biol. Ther.,* 24(1), 2279241. doi: 10.1080/15384047.2023.2279241.

Cheng T.S., Hsiao Y.L., Lin C.C., Hsu C.M., Chang M.S., Lee C.I., Yu R.C., Huang C.Y., Howng S.L., Hong Y.R. (2007) hNinein is required for targeting spindle-associated protein Astrin to the centrosome during the S and G2 phases. *Exp. Cell Res.,* 313(8), 1710-1721.

Chernobel'skaia O.A., Alieva I.B., Vorob'ev I.A. (2004) The dynamics of microtubule repolymerization in a cell: rapid growth from the centrosome and slow recovery of free microtubules. *Tsitologiia,* 46 (6), 531-544. In Russian.

Chernobel'skaia O.A., Alieva I.B., Vorob'ev I.A. (2009) Radial-organized microtubules provide maintenance of the cell shape and more effective intercellular transport than in the case of free microtubules. *Tsitologiia,* 51(6), 475-483. In Russian.

Chiplonkar J.M., Vandré D.D., Robinson J.M. (1992) Stimulus-dependent relocation of the microtubule organizing center in human polymorphonuclear leukocytes. *J. Cell Sci.,* 102, 723-728.

Choi Y.K., Liu P., Sze S.K., Dai C., Qi R.Z. (2010) CDK5RAP2 stimulates microtubule nucleation by the gamma-tubulin ring complex. *J. Cell Biol.,* 191(6), 1089-1095.

Chou M.H., Hsieh Y.C., Huang C.W., Chen P.H., Chan S.P., Tsao Y.P., Lee H.H., Wu J.T., Chen S.L. (2015) BCAS2 Regulates Delta-Notch Signaling Activity through Delta Pre-mRNA Splicing in Drosophila Wing Development. *PLoS One,* 10(6), e0130706. doi: 10.1371/journal.pone.0130706.

Chouaib R., Safieddine A., Pichon X., Imbert A., Kwon O.S., Samacoits A., Traboulsi A.M., Robert M.C., Tsanov N., Coleno E., Poser I., Zimmer C., Hyman A., Le Hir H., Zibara K., Peter M., Mueller F., Walter T., Bertrand E. (2020) A Dual Protein-mRNA Localization Screen Reveals Compartmentalized Translation and Widespread Co-translational RNA Targeting. *Dev Cell,* 54(6), 773-791.e5. doi: 10.1016/j.devcel. 2020.07.010.

Chretien D., Buendia B., Fuller S.D., Karsenti E. (1997) Reconstruction of the Centrosome Cycle from Cryoelectron Microgrphs. *J. Struct. Biol.,* 120, 117-133.

Chung C.M., Man C., Jin Y., Jin C., Guan X.Y., Wang Q., Wan T.S., Cheung A.L., Tsao S.W. (2005) Amplification and overexpression of aurora kinase A (AURKA) in immortalized human ovarian epithelial (HOSE) cells. *Mol. Carcinog.* 43, 165-174. doi: 10.1002/mc.20098.

Clark S.W., Meyer D.I. (1992) Centractin is an actin homologue associated with the centrosome. *Nature,* 359, 246-250.

Clark S.W., Meyer D.I. (1994) ACT3: a putative centractin homologue in S. cerevisiae is required for proper orientation of the mitotic spindle. *J. Cell Biol.,* 127, 129-138.

Clement C.A., Kristensen S.G., Mollgard K., Pazour G.J., Yoder B.K., Larsen L.A., Christensen S.T. (2009) The primary cilium coordinates early cardiogenesis and hedgehog signaling in cardiomyocyte differentiation. *J. Cell Sci.*, 122, 3070-3082.

Clevers H. (2006) Wnt/beta-catenin signaling in development and disease. *Cell,* 127(3), 469-80. doi: 10.1016/j.cell.2006.10.018. PMID: 17081971.

Cogswell J.P., Brown C.E., Bisi J.E., Neill S.D. (2000) Dominant-negative polo-like kinase 1 induces mitotic catastrophe independent of cdc25C function. *Cell Growth Differ.,* 11(12), 615-623.

Cole D.G., Saxton W.N., Sheehan K.B., Scholey J.M.A. (1994) "Slow" homotetrameric kinesin-related motor protein purified from Drosophila embryons. *J. Biol. Chem.,* 269, 22913-22916.

Colin A., Singaravelu P., Thery M., Blanchoin L., Gueroui Z. (2018) Actin network architecture regulates microtubule dynamics. *Curr. Biol.,* 28, 2647-2656.e4.

Collins E.S., Hornick J.E., Durcan T.M., Collins N.S., Archer W., Karanjeet K.B., Vaughan K.T., Hinchcliffe E.H. (2010) Centrosome biogenesis continues in the a bsence of microtubules during prolonged s-phase arrest. *J. Cell. Physiol.,* 225, 454-465.

Colombo E., Marine J.C., Danovi D., Falini B., Pelicci P.G. (2002) Nucleophosmin regulates the stability and transcriptional activity of p53. *Nat Cell Biol.,* 4(7), 529-533.

Comartin D., Gupta, G.D., Fussner, E., Coyaud, E., Hasegan, M., Archinti, M., Cheung, S.W., Pinchev, D., Lawo, S., Raught, B., Bazett-Jones D.P., Lüders J., Pelletier L. (2013) Cep120 and spice1 cooperate with cpap in centriole elongation. *Curr. Biol.,* 23, 1360-1366.

Combs J., Kim S.J., Tan S., Ligon L.A., Holzbaur E.L.F., Kuhn J., Poenie M. (2006) Recruitment of dynein to the Jurkat immunological synapse. *Proc. Nat. Acad. Sci. USA,* 103, 14883-14888.

Conduit P.T., Brunk K., Dobbelaere J., Dix C.I., Lucas E.P., Raff J.W. (2010) Centrioles regulate centrosome size by controlling the rate of Cnn incorporation into the PCM. *Curr. Biol.*, 20, 2178-2186. doi: 10.1016/j.cub.2010.11.011.

Conkar D., Bayraktar H., Firat-Karalar E.N. (2019) Centrosomal and ciliary targeting of ccdc66 requires cooperative action of centriolar satellites, microtubules and molecular motors. *Sci. Rep.,* 9, 14250.

Connolly J.A., Kiosses B.W., Kalnins V.I. (1986) Centrioles are lost as embryonic myoblasts fuse into myotubes *in vitro*. *Eur. J. Cell Biol.,* 39 (2), 341-345.

Cordier A.C., Haumont S. (1979) Origin of necklace particles in thymic ciliating cells. *Am. J. Anat.,* 156, 91-97.

Cottee M.A., Johnson S., Raff J.W., Lea S.M. (2017) A key centriole assembly interaction interface between human PLK4 and STIL appears to not be conserved in flies. *Biol Open.,* 6(3), 381-389. doi: 10.1242/bio.024661.

Cowley D.O., Rivera-Perez J.A., Schliekelman M., He Y.J., Oliver T.G., Lu L., O'Quinn R., Salmon E.D., Magnuson T., Van Dyke T. (2009) Aurora-A kinase is essential for bipolar spindle formation and early development. *Mol. Cell Biol.,* 29 (4), 1059-1071.

Craige B., Tsao C.C., Diener D.R., Hou Y., Lechtreck K.F., Rosenbaum, J.L., Witman, G.B. (2010) CEP290 tethers flagellar transition zone microtubules to the membrane and flagellar protein content. *J. Cell Biol.,* 190, 927-940.

Crozet N. (1990) Behavior of the sperm centriole during sheep oocyte fertilization. *Eur. J. Cell Biol.,* 53, 321-332.

Cunha-Ferreira I., Rodrigues-Martins A., Bento I., Riparbelli M., Zhang W., Laue E., Callaini G., Glover D.M., Bettencourt-Dias M. (2009) The SCF/Slimb ubiquitin ligase limits centrosome amplification through degradation of SAK/PLK4. *Curr. Biol.,* 19, 43-49. doi: 10.1016/j.cub.2008.11.037.

D'Angiolella V., Donato V., Vijayakumar S., Saraf A., Florens L., Washburn M.P., Dynlacht B., Pagano M. (2010) Scf(cyclin f) controls centrosome homeostasis and mitotic fidelity through cp110 degradation. *Nature,* 466, 138-142.

Dalen H. (1981) An ultrastructural study of primary cilia, abnormal cilia and ciliary knobs from the ciliated cells of the guinea-pig trachea. *Cell Tissue Res.*, 220(4), 685-697. doi: 10.1007/BF00210455.

Dales S., Hsu K.C., Nagayama A. (1973) The fine structure and immunological labeling of the achromatic mitotic apparatus after disruption of cell membranes. *J. Cell. Biol.,* 59, 643-660.

Dallai R. (2014) Overview on spermatogenesis and sperm structure of Hexapoda. *Arthropod Struct. Dev.,* 43(4), 257-290.

Dallai R., Afzelius B.A. (1995) Sperm structure in Trichoptera. II. The aflagellate spermatozoa of Hydroptila, Orthotrichia and Stactobia (Hydroptilidae). *Int. J. Insect Morphol. Embryol.,* 24(2), 161-170.

Dallai R., Afzelius B.A. (1999) Accessory microtubules in insect spermatozoa: structure, function and phylogenetic significance. In "*The Male Gamete: From Basic Science to Clinical Applications*" (ed. C. Gagnon). Austria, Vienna: Cache River Press, 333-350.

Dallai R., Gottardo M., Beutel R.G. (2016) Structure and Evolution of Insect Sperm: New Interpretations in the Age of Phylogenomics. *Annu. Rev. Entomol.,* 61, 1-23.

Dallai R., Lino-Neto J., Dias G., Nere P.H.A., Mercati D., Lupetti P. (2018) Fine structure of the ladybird spermatozoa (*Insecta, Coleoptera, Coccinellidae*). *Arthropod Struct. Dev.,* 47(3), 286-298.

Dallai R., Mazzini M. (1989) The spermatozoon of the gall-midge oligotrophidi (*Diptera, Cecidomyiidae*). *Boll. Zool.,* 56(1), 13-27.

Dallai R., Mazzini M., Lupetti P. (1993). The spermatozoa of *Contarinia, Allocontarinia, Lestodiplosis and Myricomyia (Diptera, Cecidomyiidae*) with considerations on the systematic relationships within the group. Boll. *Zool.,* 60 , 7-18.

Dallai R., Mercati D., Bu Y., Yin Y.W. (2010) Spermatogenesis and sperm structure of *Acerella muscorum*, (Ionescu, 1930) (*Hexapoda, Protura*). *Tissue Cell,* 42, 97-104.

Dallai R., Mercati D., Bu, Y., Callaini G., Riparbelli M.G. (2010) The spermatogenesis and sperm structure of *Acerentomon microrhinus* (*Protura, Hexapoda*) with considerations on the phylogenetic position of the taxon. *Zoomorphology,* 129, 61-80.

Dallai R., Mercati D., Carapelli A., Nardi F., Machida R., Sekiya K., Frati F. (2011) Sperm accessory microtubules suggest the placement of *Diplura* as the sister-group of *Insecta* s.s. *Arthropod Struct. Dev.,* 40(1), 77-92.

Dallai R., Mercati D., Lino-Neto J., Dias G., Lupetti P. (2017) Evidence of a procentriole during spermiogenesis in the coccinellid insect *Adalia decempunctata* (L): An ultrastructural study. *Arthropod Struct. Dev.,* 46(6), 815-823. https://doi.org/10.1016/j.asd.2017.10.004

Dammermann A., Maddox P.S., Desai A., Oegema K. (2008) SAS-4 is recruited to a dynamic structure in newly forming centrioles that is stabilized by the gamma-tubulin-mediated addition of centriolar microtubules. *J. Cell Biol.,* 180(4), 771-785.

Dammermann A., Merdes A. (2002) Assembly of centrosomal proteins and microtubule organization depends on PCM-1. *J. Cell Biol.,* 159, 255-266.

Dammermann A., Müller-Reichert T., Pelletier L., Habermann B., Desai A., Oegema K. (2004) Centriole assembly requires both centriolar and pericentriolar material proteins. *Dev. Cell,* 7(6), 815-829.

D'Angiolella V., Donato V., Vijayakumar S., Saraf A., Florens L., Washburn M.P., Dynlacht B., Pagano M. (2010) SCF(Cyclin F) controls centrosome homeostasis and mitotic fidelity through CP110 degradation. *Nature,* 466(7302), 138-142.

Daniels M.J., Wang Y., Lee M., Venkitaraman A.R. (2004) Abnormal cytokinesis in cells deficient in the breast cancer susceptibility protein BRCA2. *Science,* 306(5697), 876-879. doi: 10.1126/science.1102574.

Dardenne E. Polay Espinoza M., Fattet L., Germann S., Lambert M.P., Neil H., Zonta E., Mortada H., Gratadou L., Deygas M., Chakrama F.Z., Samaan S., Desmet F.O., Tranchevent L.C., Dutertre M., Rimokh R., Bourgeois C.F., Auboeuf D. (2014) RNA helicases DDX5 and DDX17 dynamically orchestrate transcription, miRNA, and splicing programs in cell differentiation. *Cell Rep.,* 7(6), 1900-13. doi: 10.1016/j.celrep.2014.05.010.

Dauber A., Lafranchi S.H., Maliga Z., Lui J.C., Moon J.E., McDeed C., Henke K.n Zonana J., Kingman G.A., Pers T.H., Baron J., Rosenfeld R.G., Hirschhorn J.N., Harris M.P., Hwa V. (2012) Novel Microcephalic Primordial Dwarfism Disorder Associated with Variants in the Centrosomal Protein Ninein. *J. Clin. Endocrinol. Metab.,* 97, E2140-E2151.

Daungsupawong H., Wiwanitkit V. (2024) RE: 'The psychosocial impact of prostate cancer screening for BRCA1 and BRCA2 carriers'. *BJU Int.,* 134(5), 852. doi: 10.1111/bju.16489. PMID: 39058294.

Davis F.M., Tsao T.Y., Fowler S.K., Rao P.N. (1983) Monoclonal antibodies to mitotic cells. *Proc. Natl. Acad. Sci. USA,* 80, 2926-2930.

Dawe H.R., Smith U.M., Cullinane A.R., Gerrelli D., Cox P., Badano J.L., Blair-Reid S., Sriram N., Katsanis N., Attie-Bitach T., Afford S.C., Copp A.J., Kelly D.A., Gull K., Johnson C.A. (2007) The Meckel-Gruber Syndrome proteins MKS1 and meckelin interact and are required for primary cilium formation. *Hum. Mol. Genet.,* 16(2), 173-186.

De Brabander M., Geuens G., Nuydens R., Willebrords R., Aerts F., De Mey J. (1986) Microtubule dynamics during the cell cycle: the effects of taxol and nocodazole on the microtubule system of Pt K2 cells at different stages of the mitotic cycle. *Int. Rev. Cytol.,* 101, 215-274.

De Harven E. (1968) The centriole and the mitotic spindle. In "*The nucleus*" (eds. A. J. Dalton, F. Haguenau) USA, New-York: Acad. Press, 3, 197-227.

De Harven E., Bernhard W. (1956) Etude au microscope électronique de l'ultrastructure du centriole chez les vertébrés. *Z. Zellforsch. u. Mokr. Anat.,* 45, 1-378. In French.

De Harven E., Dustin P. Jr. (1960) Etude au microscope électronique de la stathmocinése chez le rat. *Colloq. Intern. Centre Natl. Rech. Sci.* (Paris), 88, 189-197. In French.

De Simone A., Spahr A., Busso C., Gönczy P. (2018) Uncovering the balance of forces driving microtubule aster migration in C. elegans zygotes. *Nature Comm.,* 9, 938.

De Thé G. (1964) Cytoplasmic microtubules in different animal cells. *J. Cell Biol.,* 23, 265-275.

Dean M., Boumis J.K., Zhong L., Ingle A., Campbell-Salome G., Davidson L.G., Scherr CL., Brown S.K., Cragun D.L. (2024) Assessing the acceptability and appropriateness of a psychoeducational graphic novel about inherited cancer risk designed for men. *J. Genet. Couns.* 34(2), e1954. doi: 10.1002/jgc4.1954

Debec A. (1978) Haploid cell cultures of *Drosophila melanogaster*. *Nature,* 274, 255-256.

Debec A., Abbadie C. (1989) The acentriolar state of the *Drosophila* cell lines 1182. *Biol. Cell*, 67, 307-311.

Debec A., Detraves C., Montmory C., Geraud G., Wright M. (1995) Polar organization of gamma-tubulin inacentriolar mitotic spindles of Drosophila *melanogaster* cells. *J. Cell. Sci.,* 108, 2645-2653.

Debec A., Loppin B., Zheng C., Liu X., Megraw T.L. (2020) The Enigma of Centriole Loss in the 1182-4 Cell Line. *Cells,* 9(5), 1300. doi: 10.3390/cells9051300.

Debec A., Szollosi A., Szollosi D. (1982). A *Drosophila melanogaster* cell line lacking centriole. *Biol. Cell,* 44, 133-138.

Delabaere L., Orsi G.A., Sapey-Triomphe L., Horard B., Couble P., Loppin B. (2014) The Spartan orthologmaternal haploid is required for paternal chromosome integrity in the Drosophila zygote. *Curr. Biol.,* 24, 2281-2287.

Delattre M., Canard C., Gonczy, P. (2006) Sequential protein recruitment in C. elegans centriole formation. *Curr. Biol.,* 16, 1844-1849.

Delattre M., Leidel S., Wani K., Baumer K., Bamat J., Schnabel H., Feichtinger R., Schnabel R., Gönczy P. (2004) Centriolar SAS-5 is required for centrosome duplication in C. elegans. *Nat. Cell Biol.,* 6(7), 656-664.

Delgehyr N., Rangone H., Fu J., Mao G., Tom B., Riparbelli M.G., Callaini G., Glover D.M. (2012) Klp10A, a microtubule-depolymerizing kinesin-13, cooperates with CP110 to control Drosophila centriole length. *Curr. Biol.,* 22(6), 502-509.

Delgehyr N., Sillibourne J., Bornens M. (2005) Microtubule nucleation and anchoring at the centrosome are independent processes linked by ninein function. *J. Cell Sci.,* 118, 1565-1575.

Deng C.X. (2002) Roles of BRCA1 in centrosome duplication. *Oncogene,* 21(40), 6222-7. doi: 10.1038/sj.onc.1205713.

Denis M.M., Tolley N.D., Bunting M., Schwertz H., Jiang H., Lindemann S., Yost C.C., Rubner F.J., Albertine K.H., Swoboda K.J., Fratto C.M., Tolley E., Kraiss L.W., McIntyre T.M., Zimmerman G.A., Weyrich A.S. (2005) Escaping the nuclear confines: Signal-dependent pre-mRNA splicing in anucleate platelets. *Cell,* 122, 379-391.

Denu R.A., Zasadil L.M., Kanugh C., Laffin J., Weaver B.A., Burkard M.E. (2016) Centrosome amplification induces high grade features and is prognostic of worse outcomes in breast cancer. *BMC Cancer,* 16, 47. doi: 10.1186/s12885-016-2083-x.

Desai A., Mitchison T.J. (1997) Microtubule polymerization dynamics. *Annu. Rev. Cell. Dev. Biol.,* 13, 83-117.

Dias G., Lino-Neto J., Dallai R., Mercati D., Lupetti P. (2020) Sperm ultrastructure of three species of ladybirds (*Coleoptera, Coccinellidae*). *Arthropod Struct. Dev.,* 55, 100919.

Dictenberg J.B., Zimmerman W., Sparks C.A., Young A., Vidair C., Zheng Y., Carrington W., Fay F.S., Doxsey S. J. (1998) Pericentrin and gamma-tubulin form a protein complex and are organized into a novel lattice at the centrosome. *J. Cell Biol.,* 141,163-174.

Dionne L.K., Shim K., Hoshi M., Cheng T., Wang J., Marthiens V., Knoten A., Basto R., Jain S., Mahjoub M.R. (2018) Centrosome amplification disrupts renal development and causes cystogenesis. *J. Cell Biol.,* 217, 2485-2501.

Dippel R.V. (1968) The development of basal bodies in Paramecium. *Proc. Nat. Acad. Sci. (Wash.),* 61, 461-468.

Dirksen E. R. (1991) Centriole and basal body formation during ciliogenesis revisited. *Biol. Cell,* 72, 31-38.

Dirksen E.R. (1961) The presence of centrioles in artificially activated sea urchin eggs. *J. Biophys. Biochem. Cytol.,* 11, 244-247.

Dirksen E.R. (1971) Centriole morphogenesis in developing ciliated epithelium of the mouse oviduct. *J. Cell Biol.,* 51, 286-302.

Dirksen E.R., Croker T.T. (1966) Centriole replication in differentiating ciliated cells of mammalian respiratory epithelium. An electron microscopic study. *J. Microscopie,* 5, 629-644.

Dirksen E.R., Satir P. (1972) Ciliary activity in the mouse oviduct as studied by transmission and scanning electron microscopy. *Tissue Cell,* 4, 389-403.

Distel M., Hocking J.C., Volkmann K., Köster R.W. (2010) The centrosome neither persistently leads migration nor determines the site of axonogenesis in migrating neurons *in vivo*. *J. Cell Biol.,* 191, 875-890.

Diviani D., Langeberg L.K., Doxsey S.J., Scott J.D. (2000) Pericentrin anchors protein kinase A at the centrosome through a newly identified RII-binding domain. *Curr Biol.,* 10(7), 417-420.

Diwanji T.P., Engelman A., Snider J.W., Mohindra P. (2017) Epidemiology, diagnosis, and optimal management of glioma in adolescents and young adults. *Adolesc. Health Med. Ther.*, 8, 99-113. doi: 10.2147/AHMT.S53391.

Dmitrovsky E. (2018) Polo-like kinase 4 inhibition produces polyploidy and apoptotic death of lung cancers. *Proc. Natl. Acad. Sci. USA,* 115, 1913-1918. doi: 10.1073/pnas.1719760115.

Dogterom M., Koenderink G.H. (2019) Actin-microtubule crosstalk in cell biology. *Nature Reviews Mol. Cell Biol.,* 20, 38-54.

Dogterom M., Yurke B. (1997) Measurement of force-velocity relation for growing microtubules. *Science,* 278, 856-860.

Domaratskiĭ K.E., Uvakina E.V., Volkov I.K., Onishchenko G.E. (2005) Specific features of centriole formation and ciliogenesis in ciliary epithelium cells of respiratory tracts in patients with Kartagener syndrome. *Ontogenez,* 36(3), 190-198. In Russian.

Domínguez-Calvo A., Gönczy P., Holland A.J., Balestra F.R. (2021) TRIM37: a critical orchestrator of centrosome function. *Cell Cycle,* 20(23), 2443-2451. doi: 10.1080/15384101.2021.1988289.

Doolin P.F., Birge W.J. (1966) Ultrastructural organization of cilia and basal bodies of the epithelium of the choroid plexus in the chick embryo. *J. Cell Biol.,* 29, 333-345.

Dou X., Li X., Tao L., Hu C., Zhang L., He Q., Yang B., Hu Y. (2013) Synthesis and biological evaluation of novel pyrimido[4,5-b]quinoline-2,4- dione derivatives as MDM2 ubiquitin ligase inhibitors. *Med. Chem.,* 9(4), 581-587.

Doxsey S. (2001) Re-evaluating centrosome function. *Nat. Rev. Mol. Cell Biol.,* 2, 688-698.

Doxsey S., McCollum D., Theurkauf W. (2005a) Centrosomes in cellular regulation. *Annu. Rev. Cell Dev. Biol.,* 21, 411-434. doi:10.1146/annurev.cellbio.21

Doxsey S., Zimmerman W., Mikule K. (2005b) Centrosome control of the cell cycle. *Trends Cell Biol.,* 15(6), 303-311. doi: 10.1016/j.tcb.2005.04.008.

Doxsey S.J., Stein P., Evans L., Calarco P. D., Kirschner M. (1994) Pericentrin, a highly conserved protein involved in microtubule organization. *Cell,* 76, 639-650.

Doyle A.D., Wang F.W., Matsumoto K., Yamada K.M. (2009) One-dimensional topography underlies three-dimensional fibrillar cell migration. *J. Cell Biol.,* 184, 481-490.

Duensing S., Lee B.H., Dal Cin P., Münger K. (2003) Excessive centrosome abnormalities without ongoing numerical chromosome instability in a Burkitt's lymphoma. *Mol. Cancer.,* 2, 30. doi: 10.1186/1476-4598-2-30.

Dujardin D.L., Barnhart L.E., Stehman S.A., Gomes E.R., Gundersen G.G., Vallee R.B. (2003) A role for cytoplasmic dynein and LIS1 in directed cell movement. *J. Cell Biol.,* 163, 1205-1211.

Dujardin D.L., Vallee R.B. (2002) Dynein at the cortex. *Curr. Opin. Cell Biol.,* 14, 44-49.

Dute R., Kung, C. (1978) Ultrastructure of the proximal region of somatic cilia in Paramecium tetraurelia. *J. Cell Biol.,* 78, 451-464.

Dvorak M. (1978) The differentiation of rat ova during cleavage. *Adv. Anat. Embryol. Cell Biol.,* 55 (2), 5-81.

Dye R.B., Flicker P.F., Lien D.Y., Williams R.C. Jr. (1992) End-stabilized microtubules observed *in vitro:* stability, subunit, interchange, and breakage. *Cell Motil. Cytoskeleton,* 21(3), 171-186.

Eakin R.M. (1965) Evolution of photoreceptors. *Cold Spring Harb. Symp. Quant. Biol.,* 30, 363-370.

Echeverri C.J., Paschal B.M., Vaughan K.T., Vallee R.B. (1996) Molecular characterization of the 50-kD subunit of dynactin reveals function for the complex in chromosome alignment and spindle organization during mitosis. *J. Cell Biol.,* 132, 617-633.

Efimov A., Kharitonov A., Efimova N., Loncarek J., Miller P. M., Andreyeva N., Gleeson P., Galjart N., Maia A. R.R., McLeod I.X., Yates J. R., III, Maiato H., Khodjakov A., Akhmanova A., Kaverina I. (2007) Asymmetric CLASP-Dependent Nucleation of Noncentrosomal Microtubules at the trans-Golgi Network. *Dev. Cell,* 12, 917-930.

Eisen J.S., Youssef N.N. (1980) Fine structural aspects of the developing compound eye of the honey bee, Apis mellifera L. *J. Ultrastruct. Res.,* 71(1), 79-94.

El Din El Homasany B.S., Volkov Y., Takahashi M., Ono Y., Keryer G., Delouvée A., Looby E., Long A., Kelleher D. (2005) The scaffolding protein CG-NAP/AKAP450 is a critical integrating component of the LFA-1-induced signaling complex in migratory T cells. *J. Immunol.,* 175(12), 7811-7818.

Ellinger-Ziegelbauer H., Karasuyama H., Yamada E., Tsujikawa K., Todokoro K., Nishida E. (2000) Ste20-like kinase (SLK), a regulatory kinase for polo-like kinase (Plk) during the G2/M transition in somatic cells. *Genes Cells.,* 5(6), 491-498.

Eshel D., Urrestarazu L.A., Vissers S., Jauniaux J.-C., van Vliet-Reedijk J.C., Planta R.J., Gibbons I.R. (1993) Cytoplasmic dynein is required for normal nuclear segregation in yeast. *Proc. Nat. Acad. Sci. USA,* 90, 11172-11176.

Eslava A., Garcia-Puig M., Corripio R. (2021) A 10-Year-Old Boy with Short Stature and Microcephaly, Diagnosed wi th Moyamoya Syndrome and Microcephalic Osteodysplastic Primordial Dwarfism Type II (MOPD II). *Am. J. Case Rep.,* 22, e933919-1.

Essner J.J., Vogan K.J., Wagner M.K., Tabin C.J., Yost H.J., Brueckner M. (2002) Conserved function for embryonic nodal cilia. *Nature,* 418(6893), 37-8. doi: 10.1038/418037a.

Euteneuer U., Schliwa M. (1985) Evidence for an involvement of actin in the positioning and motility of centrosomes. *J. Cell Biol.,* 101, 96-103.

Euteneuer U., Schliwa, M. (1992) Mechanism of centrosome repositioning during the wound response in BSC-1 cells. *J. Cell Biol.,* 116, 1157-1166.

Eyers P.A., Erikson E., Chen L.G., Maller J.L. (2003) A novel mechanism for activation of the protein kinase Aurora A. *Curr Biol.,* 13 (8), 691-697.

Fabbro M., Zhou B., Takahashi M., Sarcevic B., Lal P., Graham M.E., Gabrielli B.G., Robinson P.J., Nigg E.A., Ono Y., Khanna K.K. (2005). Cdk1/Erk2- and Plk1-dependent phosphorylation of a centrosome protein, Cep55, is required for its recruitment to midbody and cytokinesis. *Dev. Cell,* 9(4), 477-488.

Fais D.A., Nadezhdina E.S., Chentsov Y.S. (1986) The centriolar rim. The structure that maintains the configuration of centrioles and basal bodies in the absence of their microtubules. *Exp. Cell Res.,* 164 (1), 27-34.

Faivre-Moskalenko C., Dogterom M. (2002) Dynamics of microtubule asters in microfabricated chambers: The role of catastrophes. *Proc. Nat. Acad. Sci. USA,* 99, 16788-16793.

Fanciulli P.P., Gottardo M., Dallai R. (2013) The spermatogenesis and oogenesis of the springtail *Podura aquatica Linné* 1758 (*Hexapoda: Collembola*). *Tissue Cell,* 45(3), 211-218.

Farhadi L., Fermino Do Rosario C., Debold E.P., Baskaran A., Ross J.L. (2018) Active self-organization of actin-microtubule composite self-propelled rods. *Frontiers in Physics,* 6, 75.

Farina F., Gaillard J., Guérin C., Couté Y., Sillibourne J., Blanchoin L., Théry M. (2016) The centrosome is an actin-organizing centre. *Nature Cell Biol.,* 18, 65-75.

Farruggio D.C., Townsley F.M., Ruderman J.V. (1999) Cdc20 associates with the kinase aurora2/Aik. *Proc. Natl. Acad. Sci. USA,* 96(13), 7306-7311. doi: 10.1073/pnas.96.13.7306.

Favard P. (1961) Evolution des ultrastructures cellulaires au cours de la spermatogenese de l'ascaris (ascaris megalocephala, schrank = parascaris equorum, goerze). *Ann. des Sci. Nat. Zool. et Biol. Anim.,* 12, 52-152.

Favasuli V.K., Ronchetti D., Silvestris I., Puccio N., Fabbiano G., Traini V., Todoerti K., Erratico S., Ciarrocchi A., Fragliasso V., Giannandrea D., Tumiatti F., Chiaramonte R., Torrente Y., Finelli P., Morelli E., Munshi N.C., Bolli N., Neri A., Taìana E. (2024) DIS3 depletion in multiple myeloma causes extensive perturbation in cell cycle progression and centrosome amplification. *Haematologica,* 109(1), 231-244. doi: 10.3324/haematol.2023.283274.

Favord P. (1961) Evolution des ultrastructures cellularies au cours de la spermatogenese de l'Ascaris. *Ann. Sci. Nat. Zool.,* 3, 53-152.

Fawcett D.W. (1958) The structure of the mammalian spermatozoon. *Int. Rev. Cytol.,* 7, 195-235.

Fawcett D.W. (1970) A comparative view of sperm ultrastructure. *Biol. Reprod.,* 2, 90-127.

Fawcett D.W., Phillips D.M. (1969) The fine structure and development of the neck region of the mammalian spermatozoon. *Anat. Rec.,* 165, 153-164.

Fawcett D.W., Porter K.R. (1954) A study of the fine structure of ciliated epithelia. *J. Morphology,* 94, 221-282.

Feng X., Chai Y.H., Jiang K.X., Jiang W.B., Chen W.C., Pan Y. (2024) Bibliometric analysis of olaparib and pancreatic cancer from 2009 to 2022: A global perspective. World. *J. Gastrointest Oncol.,* 16(11), 4489-4505. doi: 10.4251/wjgo.v16.i11.4489.

Feng Y., Yuan J.H., Maloid S.C., Fisher R., Copeland T. D., Longo D.L., Conrads T.P., Veenstra T.D., Ferris A., Hughes S., Dimitrov D.S., Ferris D.K. (2006). Polo-like kinase 1-mediated phosphorylation of the GTP-binding protein Ran is important for bipolar spindle formation. *Biochem. Biophys. Res. Commun.,* 349(1), 144-152.

Field C.M., Lenart P. (2011). Bulk cytoplasmic actin and its functions in meiosis and mitosis. *Curr. Biol.,* 21, R825-R830.

Fink J., Carpi N., Betz T., Bétard A., Chebah M., Azioune A., Bornens M., Sykes C., Fetler L., Cuvelier D., Piel M. (2011). External forces control mitotic spindle positioning. *Nature Cell Biol.,* 13, 771-778.

Fisch C., Dupuis-Williams P. (2011) Ultrastructure of cilia and flagella - back to the future! *Biol. Cell.,* 103(6), 249-270.

Fishman E.L., Jo K., Ha A., Royfman R., Zinn A., Krishnamurthy M., Avidor-Reiss T. (2017) Atypical centrioles are present in Tribolium sperm. *Open Biol.,* 7(3), 160334.

Fishman E.L., Jo K., Nguyen Q.P.H., Kong D., Royfman R., Cekic A.R., Khanal S., Miller A.L., Simerly C., Schatten G., Loncarek J., Mennella V., Avidor-Reis T. (2018) A novel atypical sperm centriole is functional during human fertilization. *Nat. Commun.,* 9, 2210; Erratum in *Nat. Commun.,* 2018, 9, 2800.

Flemming W. (1875) Studien in der Entwicklungsgeschichte der Najaden [Studies in the evolutionary history of the Naiads]. *Sitzungsber Akad. Wissensch Wien.,* 71, 81-147. In German.

Fliegauf M., Horvath J., von Schnakenburg C., Olbrich H., Muller D., Thumfart J., Schermer B., Pazour G.J., Neumann H.P., Zentgraf H., Benzing T., Omran H. (2006) Nephrocystin specifically localizes to the transition zone of renal and respiratory cilia and photoreceptor connecting cilia. *J. Am. Soc. Nephrol.,* 17, 2424-2433.

Flock A., Duvall A.J. 3rd (1965) The ultrastructure of the kinocilium of the sensory cells in the inner ear and lateral line organs. *J. Cell Biol.,* 25, 1-8.

Fong C.S., Ozaki K., Tsou M.B. (2018) Ppp1r35 ensures centriole homeostasis by promoting centriole-to-centrosome conversion. *Mol. Biol. Cell,* 29, 2801-2808.

Fong K.W., Choi Y.K., Rattner J.B., Qi R.Z. (2008) Cdk5rap2 is a pericentriolar protein that functions in centrosomal attachment of the gamma-tubulin ring complex. *Mol. Biol. Cell,* 19, 115-125.

Fonte V.G., Searls R.L., Hilfer S.R. (1971) The relationship of cilia with cell division and differentiation. *J. Cell Biol.,* 49(1), 226-229. doi: 10.1083/jcb.49.1.226.

Forer A. (1965) Local reduction of spindle fiber birefrigence in living (*Nephrotoma surturalis)* (Leow) spermatocytes induced by ultraviolet microbeam irradiation. *J. Cell Biol.,* 25, 95-117.

Forer A., Pickett-Heaps J.D. (1998) Cytochalasin D and latrunculin affect chromosome behaviour during meiosis in crane-fly spermatocytes. *Chromosome Res.,* 6, 533-549.

Forozan F., Karhu R., Kononen J., Kallioniemi A., Kallioniemi O.P. (1997) Genome screening by comparative genomic hybridization. *Trends Genet.,* 13(10), 405-409. doi: 10.1016/s0168-9525(97)01244-4.

Forsyth R., Gunay-Aygun, M. (1993) Bardet-Biedl Syndrome Overview. In " *Gene Reviews (R)*" (eds. M.P. Adam, G.M. Mirzaa, R.A. Pagon, S.E. Wallace, L.J.H. Bean, K.W. Gripp, A. Amemiya) USA, Seattle (WA): University of Washington.

Fraizer G.C., Diaz M.F., Lee I.L., Grossman H.B., Sen S. (2004) Aurora-A/STK15/BTAK enhances chromosomal instability in bladder cancer cells. *J. Oncol.,* 25(6), 1631-1639.

Franck N., Montembault E., Romé P., Pascal A., Cremet J.Y., Giet R. (2011) CDK11(p58) is required for centriole duplication and Plk4 recruitment to mitotic centrosomes. *PLoS One,* 6(1), e14600.

Frank-Vaillant M., Haccard O., Thibier C., Ozon R., Arlot-Bonnemains Y., Prigent C., Jessus C. (2000). Progesterone regulates the accumulation and the activation of Eg2 kinase in Xenopus oocytes. *J. Cell Sci.,* 113, 1127-1138.

Friedländer M. (1980) The sperm centriole persists during early egg cleavage in the insect Chrysopa carnea (Neuroptera, Chrysopidae). *J. Cell Sci.,* 42, 221-226.

Friedländer M., Morse J.C. (1982) The aberrant spermatozoa of Hydropsychidae caddisflies (Trichoptera): An electron microscope analysis on spermiogenesis. *J. Ultrastruct. Res.,* 78(1), 84-94.

Friedländer M., Wahrman J. (1966) Giant centrioles in *Neuropteran* meiosis. *J. Cell Sci.,* 1(1), 129-144.

Friedländer M., Wahrman J. (1970) The spindle as a basal body distributor. A study in the meiosis of the male silkworm moth, Bombyx mori. *J. Cell Sci.,* 7(1), 65-89. doi: 10.1242/jcs.7.1.65.

Friedländer M., Wahrman J. (1971) The number of centrioles in insect sperm. A study in two kinds of differentiating silkworm spermatids. *J. Morph.,* 134(4), 383-397.

Friend K., Kolev N.G., Shu M.D., Steitz J.A. (2008) Minor-class splicing occurs in the nucleus of the Xenopus oocyte. *RNA,* 14(8), 1459-62. doi: 10.1261/rna.1119708.

Fritz-Laylin L.K., Fulton C. (2016) Naegleria: A classic model for *de novo* basal body assembly. *Cilia,* 5, 10. doi: 10.1186/s13630-016-0032-6.

Fritz-Niggli H., Suda T. (1972) Formation and significance of centrioles: A study and new interpretation of the meiosis of Drosophila. *Cytobiologie,* 5, 12-41.

Fry A.M., Mayor T., Meraldi P., Stierhof Y.D., Tanaka K., Nigg E.A. (1998a) C-Nap1, a novel centrosomal coiled-coil protein and candidate substrate of the cell cycle-regulated protein kinase Nek2. *J. Cell Biol.,* 141, 1563-1574.

Fry A.M., Meraldi P., Nigg E.A. (1998b) A centrosomal function for the human Nek2 protein kinase, a member of the NIMA family of cell cycle regulators. *EMBO J.,* 17, 470-481.

Fu J., Hagan I.M., Glover, D.M. (2015) The centrosome and its duplication cycle. *Cold Spring Harb. Perspect. Biol.*, 7, a015800.

Fukasawa K. (2007) Oncogenes and tumour suppressors take on centrosomes. *Nat. Rev. Cancer,* 7, 911-924.

Fukasawa K. (2008) P53, cyclin-dependent kinase and abnormal amplification of centrosomes. *Biochim. Biophys. Acta,* 1786(1), 15-23. doi: 10.1016/j.bbcan.2008.04.002.

Fulka J. Jr, Fulka H. (2007) Somatic cell nuclear transfer (SCNT) in mammals: the cytoplast and its reprogramming activities. *Adv. Exp. Med. Biol.,* 591, 93-102.

Fulton C. (1971) Centrioles. In "*Origin and continuity of cell organelles*". USA, New York: Springer-Verlag New-York Inc., 170-221.

Fulton C., Dingle A.D. (1971) Basal bodies, but not centrioles, in *Naegleria. J. Cell Biol.,* 51, 826-836.

Gachet J., Thiéry J.P. (1964) Application de la méthode de tirage photographique avec rotation ou translation a l'étude de macromolécules et de structures biologiques [Application of the method of photographic printing with rotation or translation to the study of macromolecules and biological structures]. *J. Microscopie,* 3, 253-368. In French.

Gaglio T., Dionne M.A., Compton D.A. (1997) Mitotic spindle poles are organized by structural and motor proteins in addition to centrosomes. *J. Cell Biol.,* 138, 1055-1066.

Gaglio T., Saredi A., Bingham J.B., Hasbani M.J., Gill S.R., Schroer T.A., Compton D.A. (1996). Opposing motor activities are required for the organization of the mammalian mitotic spindle pole. *J. Cell Biol.,* 135, 399-414.

Gaglio T., Saredi A., Compton D.A. (1995). NuMA is required for the organization of microtubules into aster-like mitotic arrays. *J. Cell Biol.,* 131, 693-708.

Gaino E., Mazzin M. (1991) Aflagellate sperm in three species of *Leptophlebiidae* (*Ephemeroptera*). *Int. J. Insect Morphol. Embryol.,* 20, 119-125.

Gall J. G. (1961) Centriole Replication. A study of Spermatogenesis in the Snail Viviparus. *J. Biophys.Biochem. Cytology,* 10, 163-193.

Gans, M.; Audit, C.; Masson, M. (1975) Isolation and characterization of sex-linked female-sterile mutants in Drosophila melanogaster. *Genetics,* 81, 683-704.

Garanina A.S., Alieva I.B., Bragina E.E., Blanchard E., Arbeille B., Guerif F., Uzbekova S., Uzbekov R.E. (2019) The Centriolar Adjunct - Appearance and Disassembly in Spermiogenesis and the Potential Impact on Fertility. *Cells*, 8, E180.

Garrido-Jimenez S., Roman A.C., Alvarez-Barrientos A., Carvajal-Gonzalez J.M. (2018) Centriole planar polarity assessment in Drosophila wings. *Development,* 145(23), dev169326. doi: 10.1242/dev.169326.

Gavin R.H. (1997) Microtubule - microfilament synergy in the cytoskeleton. *Int. Rev. Cytol.,* 173, 207-242.

Gazzin A., Pala F., Bosticardo M., Niemela J., Stoddard J., Biasin E., Quarello P., Carli D., Ferroni F., Delmonte O.M., Montin D., Rosenzweig S.D., Licciardi F., Notarangelo L.D. (2023) Mulibrey nanism and immunological complications: a comprehensive case report and literature review. *Front. Immunol.,* 14, 1303251. doi: 10.3389/fimmu. 2023.1303251.

Geiger B., Rosen D., Berke G. (1982) Spatial relationships of microtubule-organizing centers and the contact area of cytotoxic T lymphocytes and target cells. *J. Cell Biol.,* 95, 137-143.

Geimer S., Melkonian M. (2004) The ultrastructure of the Chlamydomonas reinhardtii basal apparatus: identification of an early marker of radial asymmetry inherent in the basal body. *J. Cell Sci.,* 117, 2663-2674.

Ghadimi B.M., Sackett D.L., Difilippantonio M.J., Schröck E., Neumann T., Jauho A., Auer G., Rie T. (2000) Centrosome amplification and instability occurs exclusively in aneuploid, but not in diploid colorectal cancer cell lines, and correlates with numerical chromosomal aberrations. *Genes, chromosomes and cancer,* 27, 183-190. PMID: 10612807.

Giangaspero F., Burger P.C. (1983) Correlations between cytologic composition and biologic behavior in the glioblastoma multiforme. A postmortem study of 50 cases. *Cancer,* 52, 2320-2333.

Gibbons I. R., (1961) The relationship between the fine structure and direction of beat in gill cilia of a lamellibranch mollusc. *J. Biophys. Biochem Cytology,* 11, 179-205.

Gibbons I.R., Grimstone A.V. (1960) On flagellar structure in certain flagellates. *J. Biophys. Biochem. Cytology,* 7, 697-716.

Giehl M., Fabarius A., Frank O., Hochhaus A., Hafner M., Hehlmann R., Seifarth W. (2005) Centrosome aberrations in chronic myeloid leukemia correlate with stage of disease and chromosomal instability. *Leukemia,* 19, 1192-1197. doi: 10.1038/sj.leu. 2403779.

Giet R., McLean D., Descamps S., Lee M.J., Raff J.W., Prigent C., Glover D.M. (2002) Drosophila Aurora A kinase is required to localize D-TACC to centrosomes and to regulate astral microtubules. *J. Cell Biol.,* 156(3), 437-451.

Giet R., Uzbekov R., Cubizolles F., Le Guellec K., Prigent C. (1999a) The Xenopus laevis aurora-related protein kinase pEg2 associates with and phosphorylates the kinesin-related protein XlEg5. *J. Biol. Chem.,* 274, 15005-15013.

Giet R., Uzbekov R., Kireev I., and Prigent C. (1999b) The Xenopus laevis centrosome aurora-related kinase. *Biol. Cell,* 91, 461-470.

Gill S.R., Schroer T.A., Szilak I., Steuer E.R., Sheetz M.P., Cleveland D.W. (1991) Dynactin, a conserved, ubiquitously expressed component of an activator of vesicle motility mediated by cytoplasmic dynein. *J. Cell Biol.,* 115(6), 1639-1650.

Gilula N.B., Satir P. (1972) The ciliary necklace. A ciliary membrane specialization. *J. Cell Biol.,* 53, 494-509.

Giménez-Abián J.F., Sumara I., Hirota T., Hauf S., Gerlich D., de la Torre C., Ellenberg J., Peters J.M. (2004) Regulation of sister chromatid cohesion between chromosome arms. *Curr Biol.,* 14(13), 1187-1193. doi: 10.1016/j.cub.2004.06.052.

Glanzer J., Miyashiro K.Y., Sul J-Y., Barrett L., Belt B., Haydon P., Eberwine J. (2005) RNA splicing capability of live neuronal dendrites. *Proc. Natl. Acad. Sci. USA,* 102, 16859-16864.

Global Burden of Disease Cancer Collaboration. Fitzmaurice C., Allen C., Barber R.M., Barregard L., Bhutta Z.A., et al. (2017) Global, regional, and national cancer incidence, mortality, years of life lost, years lived with disability, and disability-adjusted life-years for 32 cancer groups, 1990 to 2015: a systematic analysis for the global burden of disease study. *JAMA Oncol.,* 3, 524-548. doi: 10.1001/ jamaoncol.2016.5688.

Glover D.M., Hagan I.M., Tavares A.A.M. (1998) Polo-like-kinases: a team that play throughout mitosis. *Genes Dev.,* 12, 3777-3787.

Glover D.M., Leibowitz M.H., McLean D.A., Parry H. (1995) Mutations in aurora prevent centrosome separation leading to the formation of monopolar spindles. *Cell,* 81(1), 95-105.

Gluenz E., Hoog J.L., Smith A.E., Dawe H.R., Shaw M.K., Gull K. (2010) Beyond 9+0: noncanonical axoneme structures characterize sensory cilia from protists to humans. *FASEB J.,* 24, 3117-3121.

Godinho S.A., Kwon M., Pellman D. (2009) Centrosomes and cancer: how cancer cells divide with too many centrosomes. *Cancer Metastasis Rev.*, 28, 85-98.

Godinho S.A., Pellman D. (2014) Causes and consequences of centrosome abnormalities in cancer. *Philos. Trans. R. Soc. Lond. B Biol. Sci.,* 369(1650):20130467. doi: 10.1098/rstb.2013.0467.

Gohram G.W., Merselis J.G. (1959) Kartagener's triad: a family study. *Case reports,* 104(1), 11-6. PMID: 13618690

Goldman R.D. (1971) The role of three cytoplasmic fibers in bhk-21 cell motility. I. Microtubules and the effects of colchicine. *J. Cell Biol.,* 51, 752-762.

Goldman R.D., Pollack R., Chang C.M., Bushnell A. (1975) Properties of enucleated cells. III. Changes in cytoplasmic architecture of enucleated BHK21 cells following trypsinization and replating. *Exp. Cell Res.,* 93(1), 175-183.

Goldman R.D., Pollack R., Hopkins N.H. (1973) Preservation of normal behavior by enucleated cells in culture. *Proc. Natl. Acad. Sci. USA,* 70(3), 750-4. doi: 10.1073/pnas.70.3.750.

Goldspink D.A., Matthews Z.J., Lund E.K., Wileman T., Mogensen M.M. (2017) Immunofluorescent labeling of microtubules and centrosomal proteins in *ex vivo* intestinal tissue and 3D *in vitro* intestinal organoids. *J. of Visualized Experiments,* 130, 56662.

Goldspink D.A., Rookyard C., Tyrrell B.J., Gadsby J., Perkins J., Lund E.K., Galjart N., Thomas P., Wileman T., Mogensen M.M. (2017) Ninein is essential for apico-basal microtubule formation and CLIP-170 facilitates its redeployment to non-centrosomal microtubule organizing centres. *Open Biology,* 7, 160274.

Golsteyn R.M., Lane H.A., Mundt K.E., Arnaud L., Nigg E.A. (1996) The family of polo-like kinases. *Prog. Cell Cycle Res.,* 2, 107-114.

Gonçalves J., Nolasco S., Nascimento R., Lopez Fanarraga M., Zabala J.C., Soares H. (2010) TBCCD1, a new centrosomal protein, is required for centrosome and Golgi apparatus positioning. *EMBO Reports,* 11, 194-200.

Gonçalves J., Pelletier L. (2017) The ciliary transition zone: Finding the pieces and assembling the gate. *Mol. Cells,* 40(4), 243-253.

Gonczy P., Pichler S., Kirkham M., Hyman A.A. (1999). Cytoplasmic dynein is required for distinct aspects of MTOC positioning, including centrosome separation, in the one cell stage Caenorhabditis elegans embryos. *J. Cell Biol.* 147, 135-150.

Goodrich L.V., Strutt D. (2011) Principles of planar polarity in animal development. *Development,* 138, 1877-1892.

Gopalakrishnan J., Mennella V., Blachon S., Zhai B., Smith A.H., Megraw T.L., Nicastro, D., Gygi S.P., Agard D.A., Avidor-Reiss T. (2011) Sas-4 providesa scaffold for cytoplasmic complexes and tethers them in a centrosome. *Nat. Commun.,* 2, 359.

Gorgidze L.A., Oshemkova S.A., Vorob'ev I.A. (1991) The inhibition of mitosis in tissue culture cells by blue light. *Tsitologiia*, 33(1), 50-56. In Russian.

Gorgidze L.A., Oshemkova S.A., Vorobjev I.A. (1998) Blue light inhibits mitosis in tissue culture cells. *Biosci. Rep.,* 18(4), 215-224. doi: 10.1023/a:1020104914726.

Gorgidze L.A., Vorob'ev I.A. (1992) The ultrastructure of the centrosome in cytoplasts and its alteration under the action of ouabain. *Tsitologiia*, 34(10), 45-51. In Russian.

Gorgidze L.A., Vorob'ev I.A. (1994) Centrioles can replicate in cultured cells in the absence of a nucleus. *Tsitologiia*, 36(8), 837-843. In Russian.

Gorgidze L.A., Vorobjev I.A. (1995) Centrosome and microtubules behavior in the cytoplasts. *J. Submicrosc. Cytol. Pathol.,* 27(3), 381-389.

Gosti-Testu F., Marty M-C., Berges J., Maunoury R., Bornens M. (1986) Identification of centrosomal proteins in a human lymphoblastic cell line. *EMBO J.,* 5 (10), 2545-2550.

Goto M., O'Brien D.A., Eddy E.M. (2010) Speriolin is a novel human and mouse sperm centrosome protein. *Hum. Reprod.,* 25(8), 1884-1894.

Gottardo M., Callaini G., Riparbelli M.G. (2013) The cilium-like region of the Drosophila spermatocyte: an emerging flagellum? *J. Cell Sci.,* 126(Pt 23), 5441-5452.

Gottardo M., Callaini G., Riparbelli M.G. (2014) Procentriole assembly without centriole disengagement - a paradox of male gametogenesis. *J. Cell Sci.,* 127(Pt 16), 3434-3439.

Gottardo M., Callaini G., Riparbelli M.G. (2015a) Structural characterization of procentrioles in Drosophila spermatids. *Cytoskeleton (Hoboken),* 72(11), 576-584.

Gottardo M., Callaini G., Riparbelli M.G. (2015b) The Drosophila centriole - conversion of doublets into triplets within the stem cell niche. *J. Cell Sci.,* 128(14), 2437-2442.

Gottardo M., Callaini G., Riparbelli M.G. (2016a) Klp10A modulates the localization of centriole-associated proteins during Drosophila male gametogenesis. *Cell Cycle,* 15(24), 3432-3441.

Gottardo M., Callaini G., Riparbelli M.G. (2016b) Does Unc-GFP uncover ciliary structures in the rhabdomeric eye of Drosophila? *J. Cell Sci.,* 129(14), 2726-2731.

Gottardo M., Pollarolo G., Llamazares S., Reina J., Riparbelli M.G., Callaini G., Gonzalez C. (2015) Loss of centrobin enables daughter centrioles to form sensory cilia in drosophila. *Curr. Biol.,* 25, 2319-2324.

Gottardo M., Riparbelli M.G., Callaini G., Megraw T.L. (2023) Evidence for intraflagellar transport in butterfly spermatocyte cilia. *Cytoskeleton (Hoboken),* 80(5-6), 112-122.

Gould R.R., Borisy G. (1977) The pericentriolar material in Chinese hamster ovary cells nucleates microtubule formation. *J. Cell Biol.,* 73, 601-615.Graham C.F., Morgan R.W. (1966) Changes in the cell cycle during early amphibian development. *Dev. Biol.,* 14, 439-460.

Graham C.F., Morgan R.W. (1966) Changes of the cell cycle during early amphibian development. *Dev. Biol.,* 14, 439-460.

Graser S., Stierhof Y.D., Nigg E.A. (2007) Cep68 and Cep215 (Cdk5rap2) are required for centrosome cohesion. *J. Cell Sci.,* 120(24), 4321-4331.

Graser S., Stierhof Y-D, Lavoie S.B., Gassner O.S., Lamla S., Le Clech M., Nigg E.A. (2007) Cep164, a novel centriole appendage protein required for primary cilium formation. *J. Cell Biol.,* 179(2), 321-330.

Graveley B.R. (2005) Coordinated control of splicing and translation. *Nat. Struct. Mol. Biol.,* 12(12), 1022-1023.

Green J.S., Parfrey P.S., Harnett J.D., Farid N.R., Cramer B.C., Johnson G., Heath O., McManamon P.J., O'Leary E., Pryse-Phillips W. (1989) The cardinal manifestations of Bardet-Biedl syndrome, a form of Laurence-Moon-Biedl syndrome. *N. Engl. J. Med.,* 321, 1002-1009, doi: 10.1056/NEJM198910123211503.

Gregson H.C., Schmiesing J.A., Kim J.S., Kobayashi T., Zhou S., Yokomori K. (2001) A potential role for human cohesin in mitotic spindle aster assembly. *J. Biol. Chem.*, 276(50), 47575-82. doi: 10.1074/jbc.M103364200. PMID: 11590136.

Groen A.C., Cameron L.A., Coughlin M., Miyamoto D.T., Mitchison T.J., Ohi R. (2004) XRHAMM functions in ran-dependent microtubule nucleation and pole formation during anastral spindle assembly. *Curr. Biol.*, 14(20), 1801-1811.

Gromley A., Yeaman C., Rosa J., Redic, S., Chen C., Mirabelle S., Guha M., Sillibourne J., Doxsey S. J. (2005). Centriolin anchoring of exocyst and SNARE complexes at the midbody is required for secretory-vesicle-mediated abscission. *Cell,* 123(1), 75-87.

Gros O.J., Damstra H.G.J., Kapitein L.C., Akhmanova A., Berger F. (2021). Dynein self-organizes while translocating the centrosome in T-cells. *Mol. Biol. Cell,* 32, 855-868.

Gruber J., Harborth J., Schnabel J., Weber K., Hatzfeld M. (2002) The mitotic-spindle-associated protein astrin is essential for progression through mitosis. *J. Cell Sci.,* 115, 4053-4059.

Guarguaglini G., Duncan P.I., Stierhof Y.D., Holmstrom T., Duensing S., Nigg E.A. (2005). The forkhead-associated domain protein Cep170 interacts with Polo-like kinase 1 and serves as a marker for mature centrioles. *Mol. Biol. Cell,* 16(3), 1095-1107.

Guderian G., Westendorf J., Uldschmid A., Nigg E.A. (2010) Plk4 trans-autophosphorylation regulates centriole number by controlling βTrCP-mediated degradation. *J. Cell Sci.,* 123 (13), 2163. 10.1242/jcs.068502.

Gudi R., Zou C., Li J., Gao Q. (2011) Centrobin-tubulin interaction is required for centriole elongation and stability. *J. Cell Biol.,* 193(4), 711-725.

Gudima G.O., Vorob'ev I.A., Chentsov Iu.S. (1984) Cell center of macrophages, granulocytes and lymphocytes during *in vitro* cell spreading, polarization and movement. *Tsitologiia,* 26(9), 1002-1007. In Russian.

Guernsey D.L., Jiang H., Hussin J., Arnold M., Bouyakdan K., Perry S., Babineau-Sturk T., Beis J., Dumas N., Evans S.C., Ferguson M., Matsuoka M., Macgillivray C., Nightingale M., Patry L, Rideout A.L., Thomas A., Orr A., Hoffmann I., Michaud J.L., Awadalla P., Meek D.C., Ludman M., Samuels M.E. (2010) Mutations in centrosomal protein cep152 in primary microcephaly families linked to mcph4. *Am. J. Hum. Genet.,* 87, 40-51.

Gueth-Hallonet C., Antony C., Aghion J., Santa Maria A., Lajoie Mazenc I., Wright M., Maro B. (1993) gamma-Tubulin is present in acentriolar MTOCs during early mouse development. *J. Cell Sci.,* 105 (1), 157-166.

Guichard P., Chretien D., Marco S., Tassin A.M. (2010) Procentriole assembly revealed by cryo-electron tomography. *EMBO J.,* 29, 1565-1572.

Guichard P., Desfosses A., Maheshwari A., Hachet V., Dietrich C., Brune A., Ishikawa T., Sachse C., Gonczy P. (2012) Cartwheel architecture of trichonympha basal body. *Science,* 337, 553.

Guichard P., Hachet V., Majubu N., Neves A., Demurtas D., Olieric N., Fluckiger I., Yamada A., Kihara K., Nishida Y., Moriya S., Steinmetz M.O., Hongoh Y., Gönczy P. (2013) Native architecture of the centriole proximal region reveals features underlying its 9-fold radial symmetry. *Curr. Biol.,* 23, 1620-1628.

Guillet V., Knibiehler M., Gregory-Pauron L., Remy M.H., Chemin C., Raynaud-Messina B., Bon C., Kollman J.M., Agard D.A., Merdes A., Mourey L. (2011) Crystal structure of γ-tubulin complex protein GCP4 provides insight into microtubule nucleation. *Nat. Struct. Mol Biol.,* 18(8), 915-919.

Gundersen G.G. (2002) Evolutionary conservation of microtubule-capture mechanisms. *Nature Reviews Mol. Cell Biol.,* 3, 296-304.

Gunsalus K.C., Bonaccorsi S., Williams E., Verni F., Gatt, M., Goldberg M.L. (1995) Mutations in twinstar, a Drosophila gene encoding a cofilin/ADF homologue, result in defects in centrosome migration and cytokinesis. *J. Cell Biol.,* 131, 1243-1259.

Guo J., Yang Z., Song W., Chen Q., Wang F., Zhang Q., Zhu X. (2006) Nudel contributes to microtubule anchoring at the mother centriole and is involved in both dynein-dependent and -independent centrosomal protein assembly. *Mol. Biol. Cell,* 17(2), 680-689.

Gupta A., Kitagawa D. (2018) Ultrastructural diversity between centrioles of eukaryotes. *J. Biochem.,* 164(1), 1-8. doi: 10.1093/jb/mvy031.

Gurkaslar H.K., Culfa E., Arslanhan M.D., Lince-Faria M., Firat-Karalar E.N. (2020) Ccdc57 cooperates with microtubules and microcephaly protein cep63 and regulates centriole duplication and mitotic progression. *Cell Rep.,* 31, 107630.

Haag N., Schüler S., Nietzsche S., Hübner C.A., Strenzke N., Qualmann B., Kessels M.M. (2018) The actin nucleator Cobl is critical for centriolar positioning, postnatal planar cell polarity refinement, and function of the cochlea. *Cell Reports,* 24, 2418-2431.

Haass W., Kleiner H., Weiss C., Haferlach C., Schlegelberger B., Muller M.C., Hehlmann R., Hofmann W.K., Fabarius A., Seifarth W. (2015) Schweizerische Arbeitsgemeinschaft für Klinische Krebsforschung; German CML Study Group. Clonal Evolution and Blast Crisis Correlate with Enhanced Proteolytic Activity of Separase in BCR-ABL b3a2 Fusion Type CML under Imatinib Therapy. *PLoS One,* 10(6), e0129648. doi: 10.1371/journal.pone.0129648.

Habedanck R., Stierhof Y.D., Wilkinson C.J., Nigg E.A. (2005) The Polo kinase Plk4 functions in centriole duplication. *Nat. Cell Biol.,* 7(11), 1140-1146. doi: 10.1038/ncb1320.

Hale C.M., Chen W.C., Khatau S.B., Daniels B.R., Lee J.S., Wirtz D. (2011) SMRT analysis of MTOC and nuclear positioning reveals the role of EB1 and LIC1 in single-cell polarization. *J. Cell Sci.,* 124, 4267-4285.

Hall J.G., Flora C., Scott C.I., Jr., Pauli R.M., Tanaka K.I. (2004) Majewski Osteodysplastic Primordial Dwarfism Type II (MOPD II): Natural History and Clinical Findings. *Am. J. Med. Genet. A,* 130A, 55-72.

Hamaguchi M.S., Hiramoto Y. (1986) Analysis of the role of astral rays in pronuclear migration in sand dollar eggs by the colcemid-UV method. *Development, Growth and Differentiation,* 28, 143-156.

Hamel V., Steib E., Hamelin R., Armand F., Borgers S., Fluckiger I., Busso C., Olieric N., Sorzano C.O.S., Steinmetz M.O., Guichard P., Gönczy P. (2017) Identification of chlamydomonas central core centriolar proteins reveals a role for human wdr90 in ciliogenesis. *Curr. Biol.,* 27, 2486-2498.e6.

Hames R.S., Crookes R.E., Straatman K.R., Merdes A., Hayes M.J., Faragher A.J., Fry, A. M. (2005) Dynamic recruitment of Nek2 kinase to the centrosome involves microtubules, PCM-1, and localized proteasomal degradation. *Mol. Biol. Cell,* 16(4), 1711-1724.

Han H., Braunschweig U., Gonatopoulos-Pournatzis T., Weatheritt R.J., Hirsch C.L., Ha K.C.H., Radovani E., Nabeel-Shah S., Sterne-Weiler T., Wang J., O'Hanlon D., Pan Q., Ray D., Zheng H., Vizeacoumar F., Datti A., Magomedova L., Cummins C.L., Hughes T.R., Greenblatt J.F., Wrana J.L., Moffat J., Blencowe B.J. (2017) Multilayered Control of Alternative Splicing Regulatory Networks by Transcription Factors. *Mol. Cell,* 65(3), 539-553.e7. doi: 10.1016/j.molcel.2017.01.011.

Han Y.G., Kim H.J., Dlugosz A.A., Ellison D.W., Gilbertson R.J., Alvarez-Buylla A. (2009) Dual and opposing roles of primary cilia in medulloblastoma development. *Nat. Med.,* 15, 1062-1065.

Handel M., Schulz A., Stanarius M., Schreff M., Erdtmann-Vourliotis H., Schmidt G., Wolf G., Hollt V. (1999) Selective targeting of somatostatin receptor 3 to neuronal cilia. *Neuroscience,* 89, 909-926.

Hannaford M.R., Liu R., Billington N., Swider Z.T., Galletta B.J., Fagerstrom C.J., Combs C., Sellers J.R., Rusan N.M. (2022). Pericentrin interacts with kinesin-1 to drive centriole motility. *J. Cell Biol.,* 221, e202112097.

Hannaford M.R., Rusan N.M. (2024) Positioning centrioles and centrosomes. *J. Cell Biol.,* 223(4), e202311140.

Harbour J.W., Luo R.X., Dei Santi A., Postigo A.A., Dean D.C. (1999). Cdk phosphorylation triggers sequential intramolecular interactions that progressively block Rb functions as cells move through G1. *Cell,* 98, 859-869.

Haren L., Stearns T., Lüders J. (2009) Plk1-dependent recruitment of gamma-tubulin complexes to mitotic centrosomes involves multiple PCM components. *PLoS One,* 4(6), e5976.

Hartill V., Szymanska K., Sharif S.M., Wheway G., Johnson C.A. (2017) Meckel-Gruber Syndrome: an update on diagnosis, clinical management, and research advances. *Front. Pediatr.,* 5, 244.

Hatch E., Stearns T. (2010) The life cycle of centrioles. *Cold Spring Harb. Symp. Quant. Biol.,* 75, 425-431.

Hauck L., Harms C., An J., Rohne J., Gertz K., Dietz R., Endres M., von Harsdorf R. (2008) Protein kinase CK2 links extracellular growth factor signaling with the control of p27(Kip1) stability in the heart. *Nat. Med.,* 14, 315-324. doi: 10.1038/nm1729.

Haward F., Maslon M.M., Yeyati P.L., Bellora N., Hansen J.N., Aitken S., Lawson J., von Kriegsheim A., Wachten D., Mill P., Adams I.R., Caceres J.F. (2021) Nucleo-cytoplasmic shuttling of splicing factor SRSF1 is required for development and cilia function. *Elife,*10:e65104. doi:10.7554/eLife.65104.

Hayden J.H., Bowser S.S., Rieder C.L. (1990) Kinetochores capture astral MT during chromosome attachment to the mitotic spindle: direct visualization in live newt lung cells. *J. Cell Biol.,* 111, 1039-1045.

Hebert A.M., Dubo B., Casaletto J.B., Gladden A.B., McClatchey A.I. (2012). Merlin/ERM proteins establish cortical asymmetry and centrosome position. *Genes and Development,* 26, 2709-2723.

Heck M.M., Pereira A., Pesavento P., Yannoni Y., Spradling A.C., Goldstein L.S. (1993) The kinesin-like protein KLP61F is essential for mitosis in Drosophila. *J. Cell Biol.,* 123, 665-679.

Heidemann S.R., Sander G. (1977) Evidence for a functional role of RNA in centrioles. *Cell,* 10(3), 337-350.

Heidi A.L., Nigg E.A. (1997) Cell-cycle control: POLO-like kinases join the outer circle. *Trends Cell Biol.,* 7, 63-68.

Heindenhain M. (1907) Plasma und Zelle, Pt. I, Fischher, Jena. In German.

Helps N.R., Luo X., Barker H.M., Cohen P.T. (2000) NIMA-related kinase 2 (Nek2), a cell-cycle-regulated protein kinase localized to centrosomes, is complexed to protein phosphatase 1. *Biochem J.*, 349(Pt 2), 509-18. doi: 10.1042/0264-6021:3490509.

Henneguy L.F. (1898) Sur les rapports des cils vibratiles avec les centrosomes [On the relationship between cilia and centrosomes]. *Arch. Anat. Microsc.,* 1, 481-496. In French.

Hepler P.K., Pickett-Heaps J.D., Gunning B.E.S. (2013) Some retrospectives on early studies of plant microtubules. *Plant J.,* 75(2), 189-201.

Herai R.H., Negraes P.D., Muotri A.R. (2017) Evidence of nuclei-encoded spliceosome mediating splicing of mitochondrial RNA. *Hum Mol. Genet.,* 26(13), 2472-2479. doi: 10.1093/hmg/ddx142. Erratum in: *Hum. Mol. Genet.* 26(13), 2590. doi: 10.1093/hmg/ddx220.

Hertwig O. (1875) Beitrage zur Kenntniss der Bildung, Befruchtung und Theilung des thierischen Eies [Contributions to the knowledge of the formation, fertilization and division of the animal egg]. *Morphol. Jb.,* 1, 347-434. In German.

Heuser, T., Raytchev, M., Krell, J., Porter, M.E., Nicastro, D. (2009) The dynein regulatory complex is the nexin link and a major regulatory node in cilia and flagella. *J. Cell Biol.,* 187, 921-933.

Hildebrandt F., Benzing T., Katsaniset N. (2011) Ciliopathies. *N. Engl. J. Med.,* 364, 1533-1543.

Hildebrandt F., Zhou W. (2007) Nephronophthisis-associated ciliopathies. *J. Am. Soc. Nephrol.,* 18, 1855-1871.

Hinchcliffe E.H., Li C., Thompson E. A., Maller J.L., Sluder G. (1999) Requirement of Cdk2-cyclin E activity for repeated centrosome reproduction in Xenopus egg extracts. *Science,* 283, 851-854.

Hinchcliffe E.H., Miller F.J., Cham M., Khodjakov A., Sluder G. (2001) Requirement of a centrosomal activity for cell cycle progression through G1 into S phase. *Science,* 291(5508), 1547-1550. doi: 10.1126/science.1056866.

Hirayama S., Yamazaki Y., Kitamura A., Oda Y., Morito D., Okawa K., Kimura H., Cyr D.M., Kubota H., Nagata K. (2008) MKKS is a centrosome-shuttling protein degraded by disease-causing mutations via CHIP-mediated ubiquitination. *Mol. Biol. Cell,* 19(3), 899-911.

Hirota T., Kunitoku N., Sasayama T., Marumoto T., Zhang D., Nitta M., Hatakeyama K., Saya H. (2003) Aurora-A and an interacting activator, the LIM protein Ajuba, are required for mitotic commitment in human cells. *Cell,* 114, 585-598.

Hofmann J.C., Tegha-Dunghu J., Dräger S., Will C.L., Lührmann R., Gruss O.J. (2013) The Prp19 complex directly functions in mitotic spindle assembly. *PLoS One,* 8(9), e74851. doi: 10.1371/journal.pone.0074851.

Höing S., Yeh T.Y., Baumann M., Martinez N.E., Habenberger P., Kremer L., Drexler H.C.A., Küchler P., Reinhardt P., Choidas A., Zischinsky M.L., Zischinsky G., Nandini S., Ledray A.P., Ketcham S.A., Reinhardt L., Abo-Rady M., Glatza M., King S.J., Nussbaumer P., Ziegler S., Klebl B., Schroer T.A., Schöler H.R., Waldmann H., Sterneckert J. (2018) Dynarrestin, a novel inhibitor of cytoplasmic dynein. *Cell Chem. Biol.,* 25, 357-369.e6.

Holland A.J., Lan W., Niessen S., Hoover H., Cleveland D.W. (2010) Polo-like kinase 4 kinase activity limits centrosome overduplication by autoregulating its own stability. *J. Cell Biol.,* 188, 191-198. doi: 10.1083/jcb.200911102.

Holleran E.A., Karki S., Holzbaur E.L.F. (1998) The role of the dynactin complex in intracellular motility. *Int. Review Cytology,* 182, 69-109.

Holleran E.A., Tokito M.K., Karki S., Holzbaur E.L.F. (1996) centractin (ARP1) associated with spectrin revealing a potential mechanism to link dynactin to intracellular organelles. *J. Cell Biol.,* 135, 1815-1830.

Holstein A.F., Roosen-Runge E.S. (1981) *Atlas of human spermatogenesis.* Germany, Berlin: Grosse Verlag.

Holy T.E., Dogterom M., Yurke B., Leibler S. (1997) Assembly and positioning of microtubule asters in microfabricated chambers. *Proc. Natl. Acad. Sci. USA,* 94, 6228-6231.

Home E.M. (1972) Centrioles and associated structures in the retinula cells of insect eyes. *Tissue Cell,* 4(2), 227-234.

Home E.M. (1976). The fine structure of some carabid beetle eyes, with particular reference to ciliary structures in the retinula cells. *Tissue Cell,* 8(2), 311-333.

Hoogerbrugge N., Jongmans M.C. (2016) Finding all BRCA pathogenic mutation carriers: best practice models. *Eur. J. Hum. Genet.*, 24 (Suppl 1), S19-26.

Horst C.J., Forestner D.M., Besharse J.C. (1987) Cytoskeletal-membrane interactions: a stable interaction between cell surface glycoconjugates and doublet microtubules of the photoreceptor connecting cilium. *J. Cell Biol.,* 105, 2973-2987.

Hosono K., Noda S., Shimizu A., Nakanishi N., Ohtsubo M., Shimizu N., Minoshima S. (2010) YPEL5 protein of the YPEL gene family is involved in the cell cycle progression by interacting with two distinct proteins RanBPM and RanBP10. *Genomics,* 96(2), 102-111.

Houlahan K.E., Bihie M., Contreras J.G., Fulop D.J., Lopez G., Huang H.H., Van Loo P., Curtis C., Boutros P.C., Huang K.L. (2024) Deletions Rate-Limit Breast and Ovarian Cancer Initiation. *bioRxiv*, doi:10.1101/2024.10.17.618945.

Houliston E., Pickering S.J., Maro B. (1987) Redistribution of microtubules and pericentriolar material during the development of polarity in mouse blastomeres. *J. Cell Biol.,* 104(5), 1299-308.

Howard A., Pelc S.R. (1953) Synthesis of Desoxyribonucleic acid in normal and irradiated cells and its relation to chromosome breakage. *Heredity Suppl.,* 6, 261-273.

Howard J., Hyman A.A. (2003) Dynamics and mechanics of the microtubule plus end. *Nature,* 422, 753-758.

Hsu L.C., White R.L. (1998) BRCA1 is associated with the centrosome during mitosis. *Proc. Natl. Acad. Sci. USA,* 95(22), 12983-12988.

Hu Q., Milenkovic L., Jin H., Scott M.P., Nachury M.V., Spiliotis E.T., Nelson W.J. (2010) A septin diffusion barrier at the base of the primary cilium maintains ciliary membrane protein distribution. *Science,* 329, 436-439.

Hu W., Kavanagh J.J., Deaver M., Johnston D.A., Freedman R.S., Verschraegen C.F., Sen S. (2005) Frequent overexpression of STK15/Aurora-A/BTAK and chromosomal instability in tumorigenic cell cultures derived from human ovarian cancer. *Oncol. Res.*, 15, 49-57. doi: 10.3727/096504005775082101.

Huang H., Regan K.M., Lou Z., Chen J., Tindall D.J. (2006) CDK2-dependent phosphorylation of FOXO1 as an apoptotic response to DNA damage. *Science,* 314, 294-297. doi: 10.1126/science.1130512.

Huang H., Tindall D.J. (2007) Dynamic FoxO transcription factors. *J. Cell Sci.,* 120, 2479-2487. doi: 10.1242/jcs.001222.

Huang N., Xia Y., Zhang D., Wang S., Bao Y., He R., Teng J., Chen J. (2017) Hierarchical assembly of centriole subdistal appendages via centrosome binding proteins ccdc120 and ccdc68. *Nat. Commun.,* 8, 15057.

Huang X.F., Xiang L., Fang X.L., Liu W.Q., Zhuang Y.Y., Chen Z.J., Shen R.J., Cheng W., Han R.Y., Zheng S.S., Chen X.J., Liu X., Jin Z.B. (2019) Functional characterization of CEP250 variant identified in nonsyndromic retinitis pigmentosa. *Hum. Mutat.,* 40(8), 1039-1045. doi: 10.1002/humu.23759.

Huang Y-H, Wu C-C, Chou C-K, Huang C-YF. (2011) A Translational Regulator, PUM2, Promotes Both Protein Stability and Kinase Activity of Aurora-A. *PLoS One*, 6(5), e19718.

Huettner A.F., Rabinowitz M. (1933) Demonstration of the central body in the living cell. *Science,* 78, 367-368.

Hughes J., Ward C.J., Peral B., Aspinwall R., Clark K., San Millán J.L., Gamble V., Harris P.C. (1995) The polycystic kidney disease 1 (PKD1) gene encodes a novel protein with multiple cell recognition domains. *Nat. Genet.,* 10(2), 151-160. doi: 10.1038/ng0695-151.

Hughes K., Forer A., Wilson P., Leggiadro C. (1988) Ultraviolet microbeam irradiation of microtubules *in vitro*. The action spectrum for local depolymerization of marginal band microtubules *in vitro* matches that for reducing birefringence of chromosomal spindle fibres *in vivo*. *J. Cell Sci.,* 91(4), 469-478.

Hyman A.A., White J.G. (1987) Determination of cell division axes in the early embryogenesis of Caenorhabditis eleguns. *J. Cell Biol.,* 100, 1607-1616.

Iaconis, D., Monti, M., Renda, M., van Koppen A., Tammaro R., Chiaravalli M., Cozzolino F., Pignata P., Crina C., Pucci P., Boletta A., Belcastro V., Giles R.H., Surace E.M., Gallo S., Pende M., Franco B. (2017) The centrosomal OFD1 protein interacts with the translation machinery and regulates the synthesis of specific targets. *Sci. Rep.,* **7**, 1224.

Ibi M., Zou P., Inoko A., Shiromizu T., Matsuyama M., Hayashi Y., Enomoto M., Mori D., Hirotsune S., Kiyono T., Tsukita S., Goto H., Inagaki M. (2011) Trichoplein controls microtubule anchoring at the centrosome by binding to Odf2 and ninein. *J. Cell Sci.,* 124(6), 857-864.

Icard P., Fournel L., Wu Z., Alifano M., Lincet H. (2019) Interconnection between metabolism and cell cycle in cancer. *Trends Biochem. Sci.,* 44, 490-501. doi: 10.1016/j.tibs.2018.12.007.

Ilık İ.A., Aktaş T. (2021) Nuclear speckles: dynamic hubs of gene expression regulation. *FEBS J.,* 289(22), 7234-7245 doi: 10.1111/febs.16117.

Inoue D., Obino D., Pineau J., Farina F., Gaillard J., Guerin C., Blanchoin L., Lennon-Dumenil A.M., Thery M. (2019) Actin filaments regulate microtubule growth at the centrosome. *EMBO J.,* 38, e99630.

Inoue S., Salmon E.D. (1995) Force generation by microtubule assembly / disassembly in mitosis and related movements. *Mol. Biol. Cell.,* 6, 1619-1640.

Insinna C., Besharse J.C. (2008) Intraflagellar transport and the sensory outer segment of vertebrate photoreceptors. *Dev. Dyn.,* 237, 1982-1992.

Ishikawa H., Kubo A., Tsukita S., Tsukita S. (2005) Odf2-deficient mother centrioles lack distal/subdistal appendages and the ability to generate primary cilia. *Nature Cell Biol.,* 7(5), 517- 524.

Ito K.K., Watanabe K., Kitagawa D. (2020) The Emerging Role of ncRNAs and RNA-Binding Proteins in Mitotic Apparatus Formation. *Noncoding RNA*, 6(1), 13. doi: 10.3390/ncrna6010013.

Iwamori T., Iwamori N., Ma L., Edson M.A., Greenbaum M.P., Matzuk M.M. (2010) TEX14 interacts with CEP55 to block cell abscission. *Mol. Cell Biol.,* 30(9), 2280-2292.

Izquierdo D., Wang W.J., Uryu K., Tsou M.F. (2014) Stabilization of cartwheel-less centrioles for duplication requires cep295-mediated centriole-to-centrosome conversion. *Cell Rep.,* 8, 957-965.

Jackman M., Lindon C., Nigg E.A., Pines J. (2003) Active cyclin B1-Cdk1 first appears on centrosomes in prophase. *Nat. Cell Biol.,* 5(2), 143-148.

Jakobsen L., Vanselow K., Skogs M., Toyoda Y., Lundberg E., Poser I., Falkenby L.G., Bennetzen M., Westendorf J., Nigg E.A., Uhlen M., Hyman A.A., Andersen J.S. (2011) Novel asymmetrically localizing components of human centrosomes identified by complementary proteomics methods. *EMBO J.,* 30, 1520-1535.

Jana S.C., Girotra M., Ray K. (2011) Heterotrimeric kinesin-ii is necessary and sufficient to promote different stepwise assembly of morphologically distinct bipartite cilia in drosophila antenna. *Mol. Biol. Cell,* 22, 769-781.

Jana S.C., Mendonça S., Machado P., Werner S., Rocha J., Pereira A., Maiato H., Bettencourt-Dias M. (2018) Differential regulation of transition zone and centriole proteins contributes to ciliary base diversity. *Nat. Cell Biol.,* 20(8), 928-941.

Jang Y.J., Ji J.H., Ahn J.H., Hoe K.L., Won M. Im D.S., Chae S.K., Song S., Yoo, H. S. (2004). Polo-box motif targets a centrosome regulator, RanGTPase. *Biochem. Biophys. Res. Commun.,* 325(1), 257-264.

Januschke J., Reina J., Llamazares S., Bertran T., Rossi F., Roig J., Gonzalez C. (2013) Centrobin controls mother-daughter centriole asymmetry in Drosophila neuroblasts. *Nat. Cell Biol.,* 15(3), 241-248.

Jarvik J.W., Suhan, J.P. (1991) The role of the flagellar transition region: inferences from the analysis of a Chlamydomonas mutant with defective transition region structures. *J. Cell Sci.,* 99, 731-740.

Jauregui A.R., Nguyen K.C., Hall D.H., Barr M.M. (2008) The Caenorhabditis elegans nephrocystins act as global modifiers of cilium structure. *J. Cell Biol.,* 180, 973-988.

Jean C., Tollon Y., Raynaud-Messina B., Wright M. (1999) The mammalian interphase centrosome: two independent units maintained together by the dynamics of the microtubule cytoskeleton. *Eur. J. Cell Biol.,* 78, 549-560.

Jensen C.G., Jensen L.C., Rieder C.L. (1979) The occurrence and structure of primary cilia in a subline of Potorous tridactylus. *Exp. Cell Res.,* 123(2), 444-449.

Jiang J., Tian S., Yu C., Chen M., Sun C. (2016) TRIM37 promoted the growth and migration of the pancreatic cancer cells. *Tumour Biol*., 37(2), 2629-2634. doi: 10.1007/s13277-015-4078-7.

Jiang K., Toedt G., Montenegro Gouveia S., Davey N.E., Hua S., Van Der Vaart B., Grigoriev I., Larsen J., Pedersen L.B., Bezstarosti K., Lince-Faria M., Demmers J., Steinmetz M.O., Gibson T.J., Akhmanova A. (2012). A proteome-wide screen for mammalian SxIP motif-containing microtubule plus-end tracking proteins. *Curr. Biology*, 22, 1800-1807.

Jiang S.T., Chiou Y.Y., Wan, E., Chien Y.L., He H.H., Tsai F.J., Lin C.Y., Tsai S.P., Li H. (2009) Essential role of nephrocystin in photoreceptor intraflagellar transport in mouse. *Hum. Mol. Genet.,* 18, 1566-1577.

Jimenez A.J., Schaeffer A., De Pascalis C., Letort G., Vianay B., Bornens M., Piel M., Blanchoin L., Théry M. (2021) Acto-myosin network geometry defines centrosome position. *Curr. Biology,* 31(6), 1206-1220.e5.

Jin H., White S.R., Shida T., Schulz S., Aguiar M., Gygi S.P., Bazan J.F., Nachury M.V. (2010) The conserved Bardet-Biedl syndrome proteins assemble a coat that traffics membrane proteins to cilia. *Cell,* 141, 1208-1219.

Jin L., Chen Y., Crossman D.K., Datta A., Vu T., Mobley J.A., Basu M.K., Scarduzio M., Wang H., Chang C., Datta P.K. (2020) STRAP regulates alternative splicing fidelity during lineage commitment of mouse embryonic stem cells. *Nat. Commun.,* 11, 5941.

Jin S., Gao H., Mazzacurati L., Wang Y., Fan W., Chen Q., Yu W., Wang M., Zhu X., Zhang C., Zhan Q. (2009) BRCA1 interaction of centrosomal protein Nlp is required for successful mitotic progression. *J. Biol. Chem.,* 284(34), 22970-22977.

Jo K.H., Jaiswal A., Khanal S., Fishman E.L., Curry A.N., Avidor-Reiss T. (2019) Poc1B and Sas-6 Function Together during the Atypical Centriole Formation in Drosophila melanogaster. *Cells,* 8(8), 841.

Johnson C.A., Malicki J.J. (2019) The Nuclear Arsenal of Cilia. *Dev. Cell,* 49(2), 161-170. doi: 10.1016/j.devcel.2019.03.009.

Johnson E., Berry S.J. (1977) Cellular Differentiation in a Highly Specialized Insect Secretory Organ. *Differentiation*, 8(1-3), 39-52.

Johnson U.G., Porter K.R. (1968) Fine structure of cell division in Chlamydomonas reinhardi. Basal bodies and microtubules. *J. Cell Biol.,* 38, 403-425.

Jones C., Chen P. (2007) Planar cell polarity signaling in vertebrates. *BioEssays,* 29, 120-132.

Jorda R., Buckova Z., Reznickova E., Bouchal J., Krystof V. (2018) Selective inhibition reveals cyclin-dependent kinase 2 as another kinase that phosphorylates the androgen receptor at serine 81. *Biochimica et biophysica acta. Molecular cell research,* 1865, 354-363. doi: 10.1016/j.bbamcr.2017.11.011.

Joseph B., Lai E.C. (2021) The Exon Junction Complex and intron removal prevent re-splicing of mRNA. *PLoS Genet.,* 17(5), e1009563. doi: 10.1371/journal.pgen.1009563.

Joshi H.C., Palacios M.J., McNamara L., Cleveland D.W. (1992) Gamma-tubulin is a centrosomal protein required for cell cycle-dependent microtubule nucleation. *Nature,* 356(6364), 80-83.

Joukov V., Groen A.C., Prokhorova T., Gerson R., White E., Rodriguez A., Walter J.C., Livingston D.M. (2006) The BRCA1/BARD1 heterodimer modulates ran-dependent mitotic spindle assembly. *Cell,* 127(3), 539-552.

Jurand A. (1974) Some aspects of the development of the notochord in mouse embryos. *J. Embryol. Exp. Morphol.,* 32(1), 1-33.

Jurca A.D., Petchesi C.D., Jurca S., Severin E., Jurca A.A., Jurca C.M. (2024) Clinical Challenges in Diagnosing Primordial Dwarfism: Insights from a MOPD II Case Study. *Medicina (Kaunas),* 60(11), 1906. doi: 10.3390/medicina60111906.

Justine J.-L., Le Brun N., Mattei X. (1985) The aflagellate spermatozoon of Diplozoon (Platyhelminthes: Monogenea: Polyopisthocotylea): a demonstrative case of relationship between sperm ultrastructure and biology of reproduction. *J. Ultrastruct. Res.,* 92, 47-54.

Kalay E., Yigit G., Aslan Y., Brown K.E., Pohl E., Bicknell L.S., Kayserili H., Li Y., Tüysüz B., Nürnberg G., Kiess W., Koegl M., Baessmann I., Buruk K., Toraman B., Kayipmaz S., Kul S., Ikbal M., Turner D.J., Taylor M.S., Aerts J., Scott C., Milstein K., Dollfus H., Wieczorek D., Brunner H.G., Hurles M., Jackson A.P., Rauch A., Nürnberg P., Karagüzel A., Wollnik B. (2011) CEP152 Is a Genome Maintenance Protein Disrupted in Seckel Syndrome. *Nat. Genet.,* 43, 23-26.

Kallenbach R. J. (1982) *'De novo'* centrioles originate at sites associated with annulate lamellae in sea-urchin eggs. *Biosci. Rep.,* 2(11), 959-966.

Kalnins V.I., Porter K.R. (1969) Centriole replication During Ciliogenesis in the Chick Tracheal Epithelium. *Z. Zellforsch.,* 100, 1-30.

Kang E., Wu G., Ma H., Li Y., Tippner-Hedges R., Tachibana M., Sparman M., Wolf D.P., Schöler H.R., Mitalipov S. (2014) Nuclear reprogramming by interphase cytoplasm of two-cell mouse embryos. *Nature,* 509(7498), 101-4. doi: 10.1038/nature13134. Erratum in: *Nature,* 516(7530), 276. PMID: 24670652; PMCID: PMC4124901.

Kao L.R., Megraw T.L. (2009) Centrocortin cooperates with centrosomin to organize Drosophila embryonic cleavage furrows. *Curr Biol.,* 19(11), 937-942.

Kaplan D.D., Meigs T.E., Kelly P., Casey P.J. (2004) Identification of a role for beta-catenin in the establishment of a bipolar mitotic spindle. *J. Biol. Chem.,* 279(12), 10829-32. doi: 10.1074/jbc.C400035200.

Karki S., Holzbaur E.L. (1995) Affinity chromatography demonstrates a direct binding between cytoplasmic dynein and the dynactin complex. *J. Biol. Chem.,* 270(48), 28806-28811.

Karlberg S., Lipsanen-Nyman M., Lassus H., Kallijärvi J., Lehesjoki A.E., Butzow R. (2009) Gynecological tumors in Mulibrey nanism and role for RING finger protein TRIM37 in the pathogenesis of ovarian fibrothecomas. *Mod. Pathol.,* 22(4), 570-578. doi: 10.1038/modpathol.2009.13.

Karsenti E., Kobayashi S., Mitchison T., Kirschner M. (1984a) Role of the centrosome in organizing the interphase microtubule array: Properties of cytoplasts containing or lacking centrosomes. *J. Cell Biol.,* 98, 1763-1776.

Karsenti E., Newport J., Hubble R., Kirschner M. (1984b) Interconversion of metaphase and interphase microtubule arrays, as studied by the injection of centrosomes and nuclei into Xenopus eggs. *J. Cell Biol.,* 98(5), 1730-1745.

Katayama H., Sasai K., Kawai H., Yuan Z., Bondaruk J., Suzuki F., Fujii S., Arlinghaus R.B., Czerniak B.A., Sen S. (2004) Phosphorylation by aurora kinase A induces Mdm2-mediated destabilization and inhibition of p53. *Nature genetics,* 36(1), 55-62.

Kawakami M., Mustachio L.M., Zheng L., Chen Y., Rodriguez-Canales J., Mino B., Kurie J.M., Roszik J., Villalobos P.A., Thu K.L., Silvester J., Cescon D.W., Wistuba I.I., Mak T.W., Liu X., Dmitrovsky E. (2018) Polo-like kinase 4 inhibition produces polyploidy and apoptotic death of lung cancers. *Proc. Natl. Acad. Sci. USA,* 115, 1913-1918. doi: 10.1073/pnas.1719760115.

Kazazian K., Go C., Wu H., Brashavitskaya O., Xu R., Dennis J.W., Gingras A.C., Swallow C.J. (2017) Plk4 Promotes Cancer Invasion and Metastasis through Arp2/3 Complex Regulation of the Actin Cytoskeleton. *Cancer Res.,* 77, 434-447. doi: 10.1158/0008-5472.CAN-16-2060.

Keating T. J., Borisy G. G. (1999) Centrosomal and non-centrosomal microtubules. *Biol. Cell,* 91, 321-329.

Keating T.J., Peloquin J.G., Rodionov V.I., Momcilovic D., Borisy G.G. (1997) Microtubule release from the centrosome. *Proc. Natl. Acad. Sci. USA*, 94 (10), 5078-5083.

Keil T.A. (2012) Sensory cilia in arthropods. *Arthropod Struct. Dev.,* 41(6), 515-534.

Kellogg D.R., Moritz M., Alberts B.M. (1994) The centrosome and cellular organization. *Annu. Rev. Biochem.,* 63, 639-674.

Kemp C.A., Kopish K.R., Zipperlen P., Ahringer J., O'Connell K.F. (2004) Centrosome maturation and duplication in C. elegans require the coiled-coil protein SPD-2. *Dev. Cell,* 6(4), 511-523.

Kennerly D.A. (1987) Preparation and characterization of mast cell cytoplasts. *J. Immunol. Methods,* 97(2), 173-183.

Kerketta L.S., Ghosh K., Nadkarni A., Madkaikar M., Vundinti B.R. (2017) Centrosome Aberration Frequency and Disease Association in B-Acute Lymphoblastic Leukemia. *In Vivo*, 31(2), 215-220. doi: 10.21873/invivo.11048.

Kerr C.L., Bol G.M., Vesuna F., Raman V. (2019) Targeting RNA helicase DDX3 in stem cell maintenance and teratoma formation. *Genes Cancer,* 10(1-2), 11-20. doi: 10.18632/genesandcancer.187.

Keryer G., Di Fiore B., Celati C., Lechtreck K.F., Mogensen M., Delouvee A., Lavia P., Bornens M., Tassin A.M. (2003) Part of Ran is associated with AKAP450 at the centrosome: involvement in microtubule-organizing activity. *Mol. Biol. Cell.,* 14(10), 4260-4271.

Keryer G., Ris H., Borisy G.G. (1984) Centriolar distribution during tripolar mitosis in Chinese hamster ovary cells. *J. Cell Biol.,* 98, 2222-2229.

Khan Z.S., Saini S.K., Chua W.J., Jacky Liao H.T., Manikkam S. (2024) Kartagener syndrome with pectus excavatum and upper lobar bronchiectasis. *Radiol. Case Rep.,* 19(9), 3952-3958.

Khanal S., Jaiswal A., Chowdanayaka R., Puente N., Turner K., Assefa K.Y., Nawras M., Back E.D., Royfman A., Burkett J.P., Cheong S.H., Fisher H.S., Sindhwani P., Gray J., Ramachandra N.B., Avidor-Reiss T. (2024) The evolution of centriole degradation in mouse sperm. *Nat. Commun.,* 15(1), 117. doi: 10.1038/s41467-023-44411-8.

Khanal S., Leung M.R., Royfman A., Fishman E.L., Saltzman B., Bloomfield-Gadêlha H., Zeev-Ben-Mordehai T., Avidor-Reiss T. (2021) A dynamic basal complex modulates mammalian sperm movement. *Nat. Commun.,* 12(1), 3808. doi: 10.1038/s41467-021-24011-0.

Khetarpal P., Das S., Panigrahi I., Munshi A. (2016) Primordial Dwarfism: Overview of Clinical and Genetic Aspects. *Mol. Genet. Genom.,* 291, 1-15.

Khire A., Jo K.H., Kong D., Akhshi T., Blachon S., Cekic A.R., Hynek S., Ha A., Loncarek J., Mennella V., Avidor-Reiss T. (2016) Centriole Remodeling during Spermiogenesis in Drosophila. *Curr. Biol.,* 26(23), 3183-3189.

Khire A., Jo K.H., Kong D., Akhshi T., Blachon S., Cekic A.R., Hynek S., Ha A., Loncarek J., Mennella V., Avidor-Reiss T. (2016) Centriole Remodeling during Spermiogenesis in Drosophila. *Curr. Biol.,* 26(23), 3183-3189. doi: 10.1016/j.cub.2016.07.006.

Khire A., Vizuet A.A., Davila E., Avidor-Reiss T. (2015) Asterless reduction during spermiogenesis is regulated by Plk4 and is essential for zygote development in Drosophila. *Curr. Biol.,* 25(22), 2956-2963. doi: 10.1016/j.cub.2015.09.045.

Khodiakov A.L., Onishchenko G.E., Chentsov Iu.S. (1986) Distribution of chromosomes and centrioles in three-polar metaphases induced by 2-mercaptoethanol in cultured Chinese hamster cells. *Dokl. Akad. Nauk SSSR,* 290(1), 225-227. In Russian.

Khodjakov A., Cole R.W., Rieder C.L. (1997) A synergy of technologies: combining laser microsurgery with green fluorescrnt protein tagging. *Cell Motil. Cytoskeleton.,* 38, 311-317.

Khodjakov A., Rieder C.L. (2001) Centrosomes enhance the fidelity of cytokinesis in vertebrates and are required for cell cycle progression. *J. Cell Biol.,* 153, 237-242.

Khodjakov A., Rieder C.L., Sluder G., Cassels G., Sibon O., Wang C-L. (2002) *De novo* formation of centrosomes in vertebrate cells arrested during S phase. *J. Cell Biol.,* 158, 1171-1181.

Kim E., Walz G. (2007) Sensitive cilia set up the kidney. *Nat. Med.,* 13, 1409-1411.

Kim H., Kim Y., Jeoung D. (2017) DDX53 Promotes Cancer Stem Cell-Like Properties and Autophagy. *Mol. Cells,* 40(1), 54-65. doi: 10.14348/molcells.2017.2258.

Kim H.S., Takahashi M., Matsuo K., Ono Y. (2007) Recruitment of CG-NAP to the Golgi apparatus through interaction with dynein-dynactin complex. *Genes Cells,* 12(3), 421-434.

Kim I., Fu Y., Hui K., Moeckel G., Mai W., Li C., Liang D., Zhao P., Ma J., Chen X-Z., George A.L.Jr., Coffey R.J., Feng Z-P., Wu G. (2008) Fibrocystin/polyductin modulates renal tubular formation by regulating polycystin-2 expression and function. *J. Am. Soc. Nephrol.,* 19, 455-468.

Kim J.C., Badano J.L., Sibold S., Esmail M.A., Hill J., Hoskins B.E., Leitch C.C., Venner K., Ansley S. J., Ross A.J., Leroux M.R, Katsanis N., Beales P.L. (2004). The Bardet-

Biedl protein BBS4 targets cargo to the pericentriolar region and is required for microtubule anchoring and cell cycle progression. *Nat. Genet.,* 36, 462-470.

Kim J.C., Ou Y.Y., Badano J.L., Esmail M.A., Leitch C.C., Fiedrich E., Beales P.L., Archibald J.M., Katsanis N., Rattner J. B., Leroux M.R. (2005) MKKS/BBS6, a divergent chaperonin-like protein linked to the obesity disorder Bardet-Biedl syndrome, is a novel centrosomal component required for cytokinesis. *J. Cell Sci.,* 118 (5), 1007-1020.

Kim J.H., Shinde D.N., Reijnders M.R.F., et al. (2016) *De Novo* Mutations in SON Disrupt RNA Splicing of Genes Essential for Brain Development and Metabolism, Causing an Intellectual-Disability Syndrome. *Am. J. Hum. Genet.*, 99(3), 711-719. doi:10.1016/j.ajhg.2016.06.029.

Kimura K., Kimura A. (2011a) A novel mechanism of microtubule length-dependent force to pull centrosomes toward the cell center. *BioArchitecture,* 1, 74-79.

Kimura K., Kimura A. (2011b) Intracellular organelles mediate cytoplasmic pulling force for centrosome centration in the Caenorhabditis elegans early embryo. *Proc. Nat. Acad. Sci. USA,* 108, 137-142.

King P.E., Ahmed K.S. (1989) Sperm structure in Psocoptera. *Acta Zool.,* 70(1), 57-61.

King S.M. (2010) Axonemal dyneins winch the cilium. *Nat. Struct. Mol. Biol.,* 17, 673-674.

Kinoshita N., Goebl M., Yanagida M. (1991) The fission yeast dis3+ gene encodes a 110-kDa essential protein implicated in mitotic control. *Mol. Cell Biol.,* 11(12), 5839-5847. doi: 10.1128/mcb.11.12.5839-5847.1991.

Kirkham M., Müller-Reichert T., Oegema K., Grill S., Hyman A.A. (2003) SAS-4 is a C. elegans centriolar protein that controls centrosome size. *Cell,* 112(4), 575-587.

Kirsch-Volders M., Mišík M., de Gerlache J. (2024) Tetraploidy as a metastable state towards malignant cell transformation within a systemic approach of cancer development. *Mutat. Res. Genet. Toxicol. Environ. Mutagen*, 896, 503764. doi: 10.1016/j.mrgentox.2024.503764.

Kiss A., Horvath P., Rothballer A., Kutay U., Csucs G. (2014) Nuclear motility in glioma cells reveals a cell-line dependent role of various cytoskeletal components. *PLoS ONE*, 9, e93431.

Kitagawa D., Vakonakis I., Olieric N., Hilbert M., Keller D., Olieric V., Bortfeld M., Erat M.C., Flückiger I., Gönczy P., Steinmetz M.O. (2011) Structural basis of the 9-fold symmetry of centrioles. *Cell,* 144(3), 364-375.

Kitajima T.S., Sakuno T., Ishiguro K., Iemura S., Natsume T., Kawashima S.A., Watanabe Y. (2006) Shugoshin collaborates with protein phosphatase 2A to protect cohesin. *Nature,* 441(7089), 46-52. doi: 10.1038/nature04663.

Klein A., Reichardt W., Jung V., Zang K.D., Meese E., Urbschat S. (2004) Overexpression and amplification of STK15 in human gliomas. *Int. J. Oncol.*, 25(6), 1789-1794. PMID: 15547718.

Klein C., Wolf K.W. (1997) Spermatocytes of the caddisfly Potamophylax rotundipennis (Trichoptera, Insecta): a fine structure study with emphasis on synaptonemal complex plates associated with chromatin. *Tissue Cell,* 29(3), 283-291.

Klein G. (1983) Specific chromosomal translocations and the genesis of B-cell-derived tumors in mice and men. *Cell,* 32, 311-315. doi: 10.1016/0092-8674(83)90449-x.

Kleylein-Sohn J., Pöllinger B., Ohmer M., Hofmann F., Nigg E.A., Hemmings B.A., Wartmanne M. (2012) Acentrosomal spindle organization renders cancer cells dependent on the kinesin HSET. *J. Cell Sci.,* 125(Pt 22), 5391-5402. doi: 10.1242/jcs. 107474.

Kleylein-Sohn J., Westendorf J., Le Clech M., Habedanck R., Stierhof Y.D., Nigg E.A. (2007) Plk4-induced centriole biogenesis in human cells. *Dev. Cell,* 13(2), 190-202.

Kloc M., Kubiak J.Z., Ghobrial R.M. (2014) The newly found functions of MTOC in immunological response. *J. of Leukocyte Biol.,* 95, 417-430.

Kobayashi H., Ohno S., Sasaki Y., Matsuura M. (2013) Hereditary Breast and Ovarian Cancer Susceptibility Genes (Review). *Oncol. Rep.,* 30, 1019-1029.

Kobayashi T., Kim S., Lin Y.C., Inoue T., Dynlacht B.D. (2014) The cp110-interacting proteins talpid3 and cep290 play overlapping and distinct roles in cilia assembly. *J. Cell Biol.,* 204, 215-229.

Kobayashi T., Tsang W.Y., Li J., Lane W., Dynlacht B.D. (2011) Centriolar kinesin Kif24 interacts with CP110 to remodel microtubules and regulate ciliogenesis. *Cell,* 145(6), 914-925.

Kochanski R.S., Borisy G. G. (1990) Mode of Centriole Duplication and Distribution. *J. Cell Biol.,* 110, 1599-1605.

Koh H.M., Jang B.G., Hyun C.L., Kim Y.S., Hyun J.W., Chang W.Y., Maeng Y.H. (2017) Aurora Kinase A Is a Prognostic Marker in Colorectal Adenocarcinoma. *J. Pathol Transl. Med.,* 51(1), 32-39. doi: 10.4132/jptm.2016.10.17.

Kohlmaier G., Loncarek J., Meng X., McEwen B.F., Mogensen M.M., Spektor A., Dynlacht B.D., Khodjakov A., Gönczy P. (2009) Overly long centrioles and defective cell division upon excess of the SAS-4-related protein CPAP. *Curr Biol.,* 19(12), 1012-1018.

Komarova Iu.A., Vorob'ev I A (1994) The ultrastructure of the cell center in the enterocytes of mouse embryos and newborn mice. *Ontogenez,* 25(2),76-88. PMID: 8190453 In Russian.

Komarova Iu.A., Vorob'ev I A (1995) The centrosome structure in enterocytes in the histogenesis of the mouse intestine. *Ontogenez,* 26(5), 390-399. PMID: 8524547. In Russian.

Komarova Iu.A., Vorob'ev I.A. (1993) Ultrastructural changes in the cell center during enterocyte differentiation in the mouse. *Tsitologiya,* 35, 36-43. In Russian.

Komarova Y.A., Vorobjev I.A., Borisy G.G. (2002a) Life cycle of MTs: persistent growth in the cell interior; asymmetric transition fraquencies and effects of cell boundary. *J. Cell Sci.,* 115 (17), 3517-3539.

Komarova Yu.A., Ryabov E.A., Alieva I.B., Uzbekov R.E., Uzbekova S.V., Vorobjev I.A. (1996) Polyclonal antibodies against human gamma-tubulin stain centrioles in mammalian cells from different tissues. *Biol. Membranes,* 13(5), 468-475. In Russian.

Komesli S., Tournier F., Paintrand M., Margolis R.L., Job D., Bornens M. (1989) Mass Isolation of Calf Thymus Centrosomes: Identification of a Specific Configuration. *J. Cell Biol.,* 109(6), 2869-2878.

Kong D., Sahabandu N., Sullenberger C., Vasquez-Limeta A., Luvsanjav D., Lukasik K., Loncarek J. (2020) Prolonged mitosis results in structurally aberrant and over-elongated centrioles. *J. Cell Biol.*, 219, e201910019.

König H., Matter N., Bader R., Thiele W., Müller F. (2007) Splicing Segregation: The Minor Spliceosome Acts outside the Nucleus and Controls Cell Proliferation. *Cell,* 131, 718-729.

Koonce M.P., Kohler J., Neujahr R., Schwartz J.M., Tikhonenko I., Gerish G. (1999) Dynein motor stabilizes interphase microtubule arrays and determines centrosome position. *EMBO J.,* 18, 6786-6792.

Korzeniewski N., Hohenfellner M., Duensing S. (2012) CAND1 promotes PLK4-mediated centriole overduplication and is frequently disrupted in prostate cancer. *Neoplasia,* 14, 799-806. doi: 10.1593/neo.12580.

Kotani T., Yamashita, M. (2005) Overexpression of truncated gamma-tubulins disrupts mitotic aster formation in Xenopus oocyte extracts. *Biochem. J.,* 389(3), 611-617.

Krämer A., Neben K., Ho A.D. (2005) Centrosome aberrations in hematological malignancies. *Cell Biol Int.,* 29(5), 375-83. doi: 10.1016/j.cellbi.2005.03.004.

Krämer H., Cagan R.L., Zipursky S.L. (1991) Interaction of bride of sevenless membrane-bound ligand and the sevenless tyrosine-kinase receptor. *Nature,* 352(6332), 207-212.

Kratz A.-S., Bärenz F., Richter K.T., Hoffmann I. (2015) Plk4-dependent phosphorylation of STIL is required for centriole duplication. *Biol. Open,* 4 (3), 370. 10.1242/bio.201411023.

Krause A., Hoffmann I. (2010) Polo-like kinase 2-dependent phosphorylation of NPM/B23 on serine 4 triggers centriole duplication. *PLoS One,* 5(3), e9849.

Krem M.M., Horwitz M.S. (2013) Mitotic errors, aneuploidy and micronuclei in Hodgkin lymphoma pathogenesis. *Commun. Integr. Biol.,* 6(3), e23544. doi: 10.4161/cib. 23544.

Krem M.M., Luo P., Ing B.I., Horwitz M.S. (2012) The kelch protein KLHDC8B guards against mitotic errors, centrosomal amplification, and chromosomal instability. *J. Biol. Chem.,* 287(46), 39083-93. doi: 10.1074/jbc.M112.390088.

Krem M.M., Press O.W., Horwitz M.S., Tidwell T. (2015) Mechanisms and clinical applications of chromosomal instability in lymphoid malignancy. *Br. J. Haematol.,* 171(1), 13-28. doi: 10.1111/bjh.13507.

Krendel M., Zenke F.T., Bokoch G.M. (2002) Nucleotide exchange factor GEF-H1 mediates cross-talk between microtubules and the actin cytoskeleton. *Nature Cell Biol.,* 4, 294-301.

Krioutchkova M., Onishchenko G.E., Chentsov Yu.S. (1994b) An ultrastructural study of the centrosome and centrioles in gametogenesis and early embryogenesis of Limnaea stagnalis L. II. Centrosomes and centrioles in oogenesis and early embryogenesis. *J. Struct. Biol.,* 112, 59-69.

Krioutchkova M., Onishchenko G.E., Chentsov Yu.S. (1994a) An ultrastructural study of the centrosome and centrioles in gametogenesis and early embryogenesis of Limnaea stagnalis L. I. Centrosome and centrioles in spermatogenesis. *J. Struct. Biol.,* 112, 49-58.

Krioutchkova M.M, Onishchenko G.E., Chentsov Iu.S. (1989) The ultrastructure of the spindle poles of the first cleavage divisions in Lymnaea stagnalis. *Ontogenez,* 20 (N5), 525-531. In Russian.

Krioutchkova M.M., Onishchenko G.E. (1999) Structural and functional characteristics of the centrosome in gametogenesis and early embryogenesis of animals. *Int. Rev. Cytol.,* 185, 107-156.

Krstic K. (1976) *Ultrastructure of the Mammalian Cell.* USA, New-York: Springer-Verlag. 114-115.

Kryukova E., Kryukov F., Hajek R. (2016) Centrosome amplification and clonal evolution in multiple myeloma: Short review. *Crit. Rev. Oncol. Hematol.,* 98, 116-21. doi: 10.1016/j.critrevonc.2015.10.019.

Kubo A., Tsukita S. (2003) Non-membranous granular organelle consisting of pcm-1: Subcellular distribution and cell-cycle-dependent assembly/disassembly. *J. Cell Sci.,* 116, 919-928.

Kuhn C. 3rd., Engleman W. (1978) The structure of the tips of mammalian respiratory cilia. *Cell Tissue Res.,* 186, 491-498.

Kumar R. (2016) Separase: Function Beyond Cohesion Cleavage and an Emerging Oncogene. *J. Cell Biochem.,* 118(6), 1283-1299. doi: 10.1002/jcb.25835.

Kupfer A., Swain S.L., Janeway C.A., Jr., Singer S.J. (1986) The specific direct interaction of helper T cells and antigen-presenting B cells. *Proc. Nat. Acad. Sci. USA,* 83, 6080-6083.

Kuriyama R. (2009) Centriole Assembly in CHO cells expressing Plk4/SAS6/SAS4 is similar to centriogenesis in ciliated epithelial cells. *Cell Motil. Cytoskeleton,* 66(8), 588-596.

Kuriyama R., Borisy G.G. (1981a) Centriole cycle in chinese hamster ovary cells as determined by whole-mount electron microscopy. *J. Cell Biol.,* 91, 814-821.

Kuriyama R., Borisy G.G. (1981b) Microtubule-nucleating activity of centrosomes in Chinese hamster ovary cells is independent of the centriole cycle but coupled to the mitotic cycle. *J. Cell Biol.,* 91(3), 822-826.

Kurz T., Pintard L., Willis J.H., Hamill D.R., Gonczy P., Peter M., Bowerman B. (2002) Cytoskeletal regulation by the Nedd8 ubiquitin-like protein modification pathway. *Science,* 295 (5558), 1294-1298.

Kwon O.S., Mishra R., Safieddine A., Coleno E., Alasseur Q., Faucourt M., Barbosa I., Bertrand E., Spassky N., Le Hir H. (2021) Exon junction complex dependent mRNA localization is linked to centrosome organization during ciliogenesis. *Nat. Commun.,* 12 (1), 1351. doi: 10.1038/s41467-021-21590-w

La Terra S., English C.N., Hergert P., McEwen B.F., Sluder G., Khodjakov A. (2005). The *de novo* centriole assembly pathway in HeLa cells: cell cycle progression and centriole assembly/maturation. *J. Cell Biol.,* 168(5), 713-722.

Laan L., Pavin N., Husson J., Romet-Lemonne G., van Duijn M., López M.P., Vale R.D., Jülicher F., Reck-Peterson S.L., Dogterom M. (2012) Cortical dynein controls microtubule dynamics to generate pulling forces that position microtubule asters. *Cell,* 148, 502-514.

Laan L., Roth S., Dogterom M. (2012) End-on microtubule-dynein interactions and pulling-based positioning of microtubule organizing centers. *Cell Cycle,* 11, 3750-3757.

Lacey K.R., Jackson P.K., Stearns T. (1999) Cyclin-dependent kinase control of centrosome duplication. *Proc. Natl. Acad. Sci. USA,* 96 (6), 2817-2822.

LaFountain J.R. Jr. (1976) Analysis of birefringence and ultrastructure of spindles in primary spermatocytes of Nephrotoma suturalis during anaphase. *J. Ultrastruct. Res.,* 54(3), 333-346.

Lambert A.M. (1993) Microtubule-organizing centers in higher plants. *Curr. Opin. Cell Biol.,* 5, 116-122.

Lane H.A., Nigg E.A. (1997) Cell-cycle control: Polo-like kinases join the outer circle. *Trends Cell Biol.,* 7, 63-68.

Laporte M.H., Gambarotto D., Bertiaux E., Bournonville L., Louvel V., Nunes J.M., Borgers S., Hamel V., and Guichard P. (2024) Time-series reconstruction of the molecular architecture of human centriole assembly. *Cell,* 187, 2158-2174 e2119.

Lash J.T., Holtzer H., Swift H. (1957) Regeneration of mature skeletal muscle. *Anat. Rec.,* 128(4), 679-697.

Latta H., Maunsbach A.B., Madden S.C. (1961) Cilia in different segments of the rat nephron. *J. Biophys. Biochem. Cytol.,* 11(1), 248-252. doi: 10.1083/jcb.11.1.248.

Lattao R., Kovacs L., Glover D.M. (2017) The centrioles, centrosomes, basal bodies, and cilia of drosophila melanogaster. *Genetics,* 206, 33-53.

Lauweryns J.M., Boussauw L. (1973) Striated filamentous bundles associated with centrioles in pulmonary lymphatic endothelial cells. *J. Ultrastruct. Research,* 42, 25-28.

Le Guen P., Crozet N. (1989) Microtubule and centrosome distribution during sheep fertilization. *Eur. J. Cell Biol.,* 48, 239-249.

Le Guennec M., Klena N., Gambarotto D., Laporte M.H., Tassin A.M., van den Hoek H., Erdmann P.S., Schaffer M., Kovacik L., Borgers S., Goldie K.N., Stahlberg H., Bornens M., Azimzadeh J., Engel B.D., Hamel V., Guichard P. (2020) A helical inner scaffold provides a structural basis for centriole cohesion. *Sci. Adv.,* 6, eaaz4137.

Le Hir H., Saulière J., Wang Z. (2016) The exon junction complex as a node of post-transcriptional networks. *Nat. Rev. Mol. Cell Biol.,* 17(1), 41-54.

Leal-Esteban L.C., Fajas L. (2020) Cell cycle regulators in cancer cell metabolism. *Biochim. Biophys. Acta, Mol. Basis Dis.,* 1866, 165715. doi: 10.1016/j.bbadis.2020.165715.

Lechtreck K.F., Teltenkotter A., Grunow, A. (1999) A 210 kDa protein is located in a membrane-microtubule linker at the distal end of mature and nascent basal bodies. *J. Cell Sci.,* 112, 1633-1644.

Lee M.J., Gergely F., Peak-Chew S.Y., Raff J.W. (2001) MSPS/XMAP215 interacts with the centrosomal protein D-TACC to regulate microtubule behaviour. *Nat. Cell Biol.,* 3, 643-649.

Lee S., Rhee K. (2010) CEP215 is involved in the dynein-dependent accumulation of pericentriolar matrix proteins for spindle pole formation. *Cell Cycle,* 9(4), 774-783.

Lehmann-Che J., Renault N., Giron M.L., Roingeard P., Clave E., Tobaly-Tapiero J., Bittoun P., Toubert A., de Thé H., Saïb A. (2007) Centrosomal latency of incoming foamy viruses in resting cells. *PLoS Pathog.,* 3(5), e74. doi: 10.1371/journal.ppat.0030074.

Leidel S., Delattre M., Cerutti L., Baumer K., Gonczy P. (2005) SAS-6 defines aprotein family required for centrosome duplication in C. elegans and in human cells. *Nat. Cell Biol.,* 7, 115-125.

Leidel S., Gönczy P. (2003) SAS-4 is essential for centrosome duplication in C. elegans and is recruited to daughter centrioles once per cell cycle. *Dev. Cell,* 4(3), 431-439.

Leightner A.C., Hommerding C.J., Peng Y., Salisbury J.L., Gainullin V.G., Czarnecki P.G., Sussman C.R., Harris P.C. (2013) The Meckel syndrome protein meckelin (TMEM67) is a key regulator of cilia function but is not required for tissue planar polarity. *Hum. Mol. Genet.*, 22(10), 2024-2040.

Leslie R.J. (1989) A detrgent insoloble component of the interphase cytoskeleton forms aggregates at the spindle poles during prophase. J. *Cell Biol.,* 109, 88a.

Leslie R.J. (1990) Recruitment: The ins and outs of spindle pole formation. *Cell Motil. Cytoskeleton,* 16(4), 225-228.

Leslie R.J., Wilson L. (1987) A human autoimmune serum recognized an 80 kD protein in sea urchun mitotic spindle poles. *J. Cell Biol.,* 105, 284a.

Letort G., Nedelec F., Blanchoin L., Théry M. (2016) Centrosome centering and decentering by microtubule network rearrangement. *Mol. Biol. Cell,* 27, 2833-2843.

Leung M.R., Roelofs M.C., Ravi R.T., Maitan P., Henning H., Zhang M., Bromfield E.G., Howes S.C., Gadella B.M., Bloomfield-Gadelha H., and Zeev-Ben-Mordehai T. (2021) The multi-scale architecture of mammalian sperm flagella and implications for ciliary motility. *EMBO J.,* 40, e107410.

Levine M.S., Bakker B., Boeckx B., Moyett J., Lu J., Vitre B., Spierings D.C., Lansdorp P.M., Cleveland D.W., Lambrechts D., Foijer F., Holland A.J. (2017) Centrosome amplification is sufficient to promote spontaneous tumorigenesis in mammals. *Dev. Cell,* 40, 313-322.e5. doi: 10.1016/j.devcel.2016.12.022.

Li J., Wang J., Jiao H., Liao J., Xu, X. (2010) Cytokinesis and cancer: Polo loves ROCK'n' Rho(A). *J. Genet. Genomics,* 37(3), 159-172.

Li J., Zhan Q. (2011) The role of centrosomal Nlp in the control of mitotic progression and tumourigenesis. *Br. J. Cancer.,* 104(10), 1523-1528.

Li Q., Hansen D., Killilea A., Joshi H.C., Palazzo R.E., Balczon R. (2001) Kendrin/pericentrin-B, a centrosome protein with homology to pericentrin that complexes with PCM-1. *J. Cell Sci.,* 114(4), 797-809.

Li W., Yi P., Zhu Z., Zhang X., Li W., Ou G. (2017) Centriole translocation and degeneration during ciliogenesis in Caenorhabditis elegans neurons. *EMBO J.,* 36, 2553-2566.

Li Y.-Y., Yeh E., Hays T., Bloom K. (1993) Disruption of mitotic spindle orientation in a yeast dynein mutant. *Proc. Nat. Acad. Sci. USA,* 90, 10096-10100.

Li Z., Dai K., Wang C., Song Y., Gu F., Liu F., Fu L. (2016) Expression of Polo-Like Kinase 4(PLK4) in Breast Cancer and Its Response to Taxane-Based Neoadjuvant Chemotherapy. *J. Cancer,* 7, 1125-1132. doi: 10.7150/jca.14307.

Lima J.T., Pereira A.J., Ferreira J.G. (2024) The LINC complex ensures accurate centrosome positioning during prophase. *Life Science Alliance,* 7, e202302404.

Lin C.C., Cheng T.S., Hsu C.M., Wu C.H., Chang L.S., Shen Z.S., Yeh H.M., Chang L.K., Howng S.L., Hong Y.R. (2006) Characterization and functional aspects of human ninein isoforms that regulated by centrosomal targeting signals and evidence for docking sites to direct gamma-tubulin. *Cell Cycle,* 5(21), 2517-2527.

Lin Y.N., Wu C.T., Lin Y.C., Hsu W.B., Tang C.J., Chang C.W., Tang T.K. (2013) Cep120 interacts with cpap and positively regulates centriole elongation. *J. Cell Biol.,* 202, 211-219.

Linck R.W., Albertini D.F., Kenney D.M., Langevin G.L. (1982) Tektin filaments: chemically unique filaments of sperm flagellar microtubules. *Prog. Clin. Biol. Res.,* 80, 127-132.

Lindemann C.B., Lesich K.A. (2010) Flagellar and ciliary beating: the proven and the possible. *J. Cell Sci.,* 123, 519-528.

Ling H., Hanashiro K., Luong T.H., Benavides L., Fukasawa K. (2015) Functional relationship among PLK2, PLK4 and ROCK2 to induce centrosome amplification. *Cell Cycle,* 14(4), 544-553.

Lingle W.L., Lutz W.H., Ingle J.N., Maihle N.J., Salisbury J.L. (1998) Centrosome hypertrophy in human breast tumors: Implications for genomic stability and cell polarity. *Proc. Natl. Acad. Sci.,* 95, 2950-2955. doi: 10.1073/pnas.95.6.2950.

Liu X., Kapoor T.M., Chen J.K., Huse M. (2013) Diacylglycerol promotes centrosome polarization in T cells via reciprocal localization of dynein and myosin II. *Proc. Nat. Acad. Sci. USA,* 110, 11976-11981.

Liu Y., Pelletier L. (2019) A magic bullet for targeting cancers with supernumerary centrosomes. *EMBO J.,* 38, e101134. doi: 10.15252/embj.2018101134.

Lohret T.A., Zhao L., Quarmby L.M. (1999) Cloning of Chlamydomonas p60 katanin and localization to the site of outer doublet severing during deflagellation. *Cell Motil. Cytoskeleton,* 43, 221-231.

Loktev A.V., Zhang Q., Beck J.S., Searby C.C., Scheetz T.E., Bazan J.F., Slusarski D.C., Sheffield V.C., Jackson P.K., Nachury M.V. (2008) A BBSome subunit links ciliogenesis, microtubule stability, and acetylation. *Dev. Cell,* 15, 854-865.

Lomakin A.J., Semenova I., Zaliapin I., Kraikivski P., Nadezhdina E., Slepchenko B.M., Akhmanova A., Rodionov V. (2009). CLIP-170-dependent capture of membrane organelles by microtubules initiates minus-end directed transport. *Dev. Cell,* 17, 323-333.

Lombardi M.L., Jaalouk D.E., Shanahan C.M., Burke B., Roux K.J., Lammerding J. (2011) The interaction between nesprins and sun proteins at the nuclear envelope is critical for force transmission between the nucleus and cytoskeleton. *J. Biol. Chem.,* 286, 1-20.

Loncarek J., Hergert P, Khodjakov A. (2010) Centriole reduplication during prolonged interphase requires procentriole maturation governed by Plk1. *Curr. Biol.,* 20, 1277-1282.

Longo F.J., Anderson T. (1969) Cytological aspects of fertilization in the lamellibranch, Mytilus edulis. II. Development of the male pronucleus and the association of the maternally and paternally derived chromosomes. *J. Exp. Zool.,* 172, 69-96.

Longoria R.A., Shubeita G.T. (2013) Cargo transport by cytoplasmic Dynein can center embryonic centrosomes. *PLoS ONE,* 8, e67710.

Lopez M.P., Huber F., Grigoriev I., Steinmetz M.O., Akhmanova A., Koenderink G.H., Dogterom M. (2014). Actin-microtubule coordination at growing microtubule ends. *Nature Comm.,* 5, 4778.

Loppin B., Berger F., Couble P. (2001) Paternal chromosome incorporation into the zygote nucleus is controlled bymaternal haploid in Drosophila. *Dev. Biol.,* 231, 383-396.

Lou Y., Xie W., Zhang D.F., Yao J.H., Luo Z.F., Wang Y.Z., Shi Y.Y., Yao X.B. (2004). Nek2A specifies the centrosomal localization of Erk2. *Biochem. Biophys. Res. Commun.,* 321(2), 495-501.

Lucas J.J., Szekely E., Kates J.R. (1976) The regeneration and division of mouse L-cell karyoplasts. *Cell,* 7(1), 115-122. doi: 10.1016/0092-8674(76)90261-0.

Lucas J.J., Zorn G.A., Brings A., Szekely E., Kates J.R. (1978) Recent developments with karyoplast regeneration and nuclear transplantation. *Natl. Cancer Inst. Monogr.,* 48, 37-44. PMID: 748759

Lukinavicius G., Lavogina D., Orpinell M., Umezawa K., Reymond L., Garin N., Gonczy P., Johnsson K. (2013) Selective chemical crosslinking reveals a cep57-cep63-cep152 centrosomal complex. *Curr. Biol.,* 23, 265-270.

Luo L., Roy S., Li L., Ma M. (2023) Polycystic kidney disease: novel insights into polycystin function. *Trends Mol. Med.,* 29(4), 268-281. doi: 10.1016/j.molmed.2023.01.005.

Luxton G.W., Gundersen G.G. (2011) Orientation and function of the nuclear-centrosomal axis during cell migration. *Curr. Opin. Cell Biol.,* 23, 579-588.

Lvov K.M., Aslanov R.B., Gasymov O.K., Mamedov Sh.V. (1975) Photoproduction of radicals N – CH, HCO in proteins after UV irradiation. *Biophizika,* 20, 763-767. In Russian.

Ma S., Triviños-Lagos L., Gräf R., Chisholm R.L. (1999) Dynein intermediate chain mediated dynein-dynactin interaction is required for interphase microtubule organization and centrosome replication and separation in Dictyostelium. *J. Cell Biol.,* 147(6), 1261-1274.

Ma Z., Kanai M., Kawamura K., Kaibuchi K., Ye K., Fukasawa K. (2006) Interaction between ROCK II and nucleophosmin/B23 in the regulation of centrosome duplication. *Mol. Cell Biol.,* 26(23), 9016-9034.

Mack G.J., Compton D.A. (2001) Analysis of mitotic microtubule-associated proteins using mass spectrometry identifi es astrin, a spindle-associated protein. *Proc. Natl. Acad. Sci. USA.,* 98, 14434-14439.

Maekawa T., Leslie R. J., Kuriyama R. (1989) Aggregation of centrosomal components in mitotic cells is blocked by both taxol and nocodazole. *J. Cell Biol.,* 109, 346a.

Mahowald A.P., Strassheim J.M. (1970) Intercellular migration of centrioles in the germarium of Drosophila melanogaster. An electron microscopic study. *J. Cell Biol.,* 45(2), 306-320.

Malik S., Saito H., Takaoka M., Miki Y., Nakanishi A. (2016) BRCA2 mediates centrosome cohesion via an interaction with cytoplasmic dynein. *Cell Cycle,* 15(16), 2145-2156.

Malikov V., Kashina A., Rodionov V. (2004). Cytoplasmic dynein nucleates microtubules to organize them into radial arrays *in vivo. Mol. Biol. Cell.,* 15, 2742-2749.

Maller J., Poccia D., Nishioka D., Kidd P., Gerhart J., Hartman H. (1976) Spindle formation and cleavage in Xenopus eggs injected with centriole-containing fractions from sperm. *Exp. Cell Res.,* 99(2), 285-294.

Manandhar G., Simerly C., Schatten G. (2000) Highly degenerated distal centrioles in rhesus and human spermatozoa. *Hum. Reprod.,* 15, 256-263.

Manandhar G., Sutovsky P., Joshi H.C., Stearns T., Schatten G. (1998) Centrosome reduction during mouse spermiogenesis. *Dev. Biol.,* 203(2), 424-434. doi: 10.1006/dbio.1998.8947.

Mandelkow E.M., Schultheiss R., Rapp R., Muller M., Mandelkow E. (1986) On the surface lattice of microtubules: helix starts, protofilament number, seam, and handedness. *J. Cell Biol.,* 102(3), 1067-1073.

Maniotis A., Schliwa M. (1991) Microsurgical removal of centrosomes blocks cell reproduction and centriole generation in BSC-1 cells. *Cell,* 67, 495-504.

Manton I. (1957) Observations with the electron microscope on the cell structure of the antheridium and spermatozoid of Sphagnum. *J. Exp. Bot.,* 8, 382-400.

Manton I., Clarke B. (1952) An electron microscope study of the spermatozoid of Sphagnum. *J. Exp. Bot.,* 3, 265-275.

Mardin B.R., Lange C., Baxter J.E., Hardy T., Scholz S.R., Fry A.M., Schiebel E. (2010) Components of the Hippo pathway cooperate with Nek2 kinase to regulate centrosome disjunction. *Nat. Cell Biol.,* 12(12), 1166-1176.

Mariappan A., Soni K., Schorpp K., Zhao F., Minakar A., Zheng X., Mandad S., Macheleidt I., Ramani A., Kubelka T., Dawidowski M., Golfmann K., Wason A., Yang C., Simons J., Schmalz H-G., Hyman A.A., Aneja R., Ullrich R., Urlaub H., Odenthal M., Büttner R., Li H., Sattler M., Hadian K., Gopalakrishnan J. (2019) Inhibition of cpap-tubulin interaction prevents proliferation of centrosome-amplified cancer cells. *EMBO J.,* 38, e99876.

Maro B., Bornens M. (1980) The centriole-nucleus association: effects of cytochalasin B and nocodazole. *Biol. Cell,* 39, 287-290.

Maro B., Gueth-Hallonet C., Aghion J., Antony C. (1991). Cell polarity and microtubule organisation during mouse early embryogenesis. *Dev. Suppl.,* 1, 17-25.

Maro B., Howlett S. K., Webb M. (1985). Non-spindle microtubule organizing centers in metaphase II-arrested mouse oocytes. *J. Cell Biol.,* 101, 1665-1672.

Martin C.-A., Ahmad I., Klingseisen A., Hussain M.S., Bicknell L.S., Leitch A., Nürnberg G., Toliat M.R., Murray J.E., Hunt D., Khan F., Ali Z., Tinschert S., Ding J., Keith C., Harley M.E., Heyn P., Müller R., Hoffmann I., Cormier-Daire V., Dollfus H., Dupuis L., Bashamboo A., McElreavey K., Kariminejad A., Mendoza-Londono R., Moore A.T., Saggar A., Schlechter C., Weleber R., Thiele H., Altmüller J., Höhne W., Hurles M.E., Noegel A.A., Baig S.M., Nürnberg P., Jackson A.P. (2014) Mutations in PLK4, Encoding a Master Regulator of Centriole Biogenesis, Cause Microcephaly, Growth Failure and Retinopathy. *Nat. Genet.,* 46, 1283-1292.

Martín-Cófreces N.B., Alarcón B., Sánchez-Madrid F. (2011) Tubulin and actin interplay at the T cell and antigen-presenting cell interface. *Front. Imm.,* 2, 24.

Martín-Cófreces N.B., Robles-Valero J., Cabrero J.R., Mittelbrunn M., Gordón-Alonso M., Sung C.H., Alarcón B., Vázquez J., Sánchez-Madrid F. (2008) MTOC translocation modulates IS formation and controls sustained T cell signaling. *J. Cell Biol.,* 182, 951-962.

Martín-Subero J.I, Knippschild U., Harder L., Barth T.F, Riemke J., Grohmann S., Gesk S., Höppner J., Möller P., Parwaresch R.M., Siebert R. (2003) Segmental chromosomal aberrations and centrosome amplifications: pathogenetic mechanisms in Hodgkin and Reed-Sternberg cells of classical Hodgkin's lymphoma? *Leukemia,* (11), 2214-2219. doi: 10.1038/sj.leu.2403129.

Mathur R.S. (1960) Studies on the normal and X-irradiated spermatids of Stenobothrus viridulus and Schistoccrca gregaria (Orthoptera). *La Cellule,* 61, 171-190.

Matsumoto Y., Maller J.L. (2004) A centrosomal localization signal in cyclin E required for Cdk2-independent S phase entry. *Science,* 306(5697), 885-888.

Matsuo K., Nishimura T., Hayakawa A., Ono Y., Takahashi M. (2010) Involvement of a centrosomal protein kendrin in the maintenance of centrosome cohesion by modulating Nek2A kinase activity. *Biochem. Biophys. Res. Commun.,* 398(2), 217-223.

Matsuura K., Lefebvr P. A., Kamiya R., Hirono M. (2004) Bld10p, a novel proteinessential for basal bodyassembly in Chlamydomonas: localization to thecartwheel, the first ninefold symmetrical structure appearing during assembly. *J. Cell Biol.,* 165, 663-671.

Matsuzaki O., Bakin R.E., Cai X., Menco B.P., Ronnett G.V. (1999) Localization of the olfactory cyclic nucleotide-gated channel subunit 1 in normal, embryonic and regenerating olfactory epithelium. *Neuroscience,* 94, 131-140.

Matthies H.J.G., McDonald H.B., Goldstein L.S.B., Theurkauf W.E. (1996) Anastral meiotic spindle morphogenesis: role of the non-claret disjunctional kinesin-like protein. *J. Cell Biol.,* 134, 455-464.

Maxwell C.A., Keats J.J., Belch A.R., Pilarski L.M., Reiman T. (2005) Receptor for hyaluronan-mediated motility correlates with centrosome abnormalities in multiple myeloma and maintains mitotic integrity. *Cancer Res.,* 65(3), 850-860. PMID: 15705883.

Mayer G. (2006). Structure and development of onychophoran eyes: what is the ancestral visual organ in arthropods? *Arthropod Struct. Dev.,* 35(4), 231-245.

Mayor T, Hacker U, Stierhof YD, Nigg EA. (2002) The mechanism regulating the dissociation of the centrosomal protein C-Nap1 from mitotic spindle poles. *J. Cell Sci.,* 115(Pt 16), 3275-84. doi: 10.1242/jcs.115.16.3275.

Mayor T., Stierhof Y.D., Tanaka K., Fry A.M., Nigg E.A. (2000) The centrosomal protein C-Nap1 is required for cell cycle-regulated centrosome cohesion. *J. Cell Biol.,* 151(4), 837-46. doi: 10.1083/jcb.151.4.837.

May-Simera H.L., Ross A., Rix S., Forge A., Beales P.L., Jagger D.J. (2009) Patterns of expression of Bardet-Biedl syndrome proteins in the mammalian cochlea suggest noncentrosomal functions. *J. Comp. Neurol.,* 514(2), 174-188.

Mazel T., Biesemann A., Krejczy M., Nowald J., Müller O., Dehmelt L. (2014) Direct observation of microtubule pushing by cortical dynein in living cells. *Mol. Biol. Cell,* 25, 95-106.

Mazia D. (1961) *The Cell* (eds. J. Brachet and A. Mirsky) USA and United Kingdom, New York and London: Academic Press. Vol. III,

Mazia D. (1963) *Mitosis and physiology of cell division.* USSR, Moscow: Mir. In Russian.

Mazia D. (1984) Centrosomes and mitotic poles. *Exp. Cell Res.,* 153(1), 1-15.

McIntosh J.R., Euteneuer U. (1984) Tubulin hooks as probes for microtubule polarity: an analysis of the method and an evaluation of data on microtubule polarity in the mitotic spindle. *J. Cell Biol.,* 98(2), 525-533.

McKendrick L., Milne D., Meek D. (1999) Protein kinase CK2-dependent regulation of p53 function: evidence that the phosphorylation status of the serine 386 (CK2) site of p53 is constitutive and stable. *Mol. Cell Biochem.,* 191(1-2), 187-199.

McNally F., Vale R. (1993) Identification of katanin, an ATPase that severs and disassembles stable microtubules. *Cell,* 75, 419-429.

McNally K.P., Bazirgan O.A., McNally F.J. (2000) Two domains of p80 katanin regulate microtubule severing and spindle pole targeting by p60 katanin. *J. Cell Sci.,* 113 (9), 1623-1633.

Megraw T.L., Kao L.R., Kaufman T.C. (2001) Zygotic development without functional mitotic centrosomes. *Curr. Biol.,* 11, 116-120.

Megraw T.L., Li K., Kao L.R., Kaufman T.C. (1999) The centrosomin protein is required for centrosome assembly and function during cleavage in Drosophila. *Development,* 126(13), 2829-2839.

Meitinger F., Ohta M., Lee K.Y., Watanabe S., Davis R.L., Anzola J.V., Kabeche R., Jenkins D.A., Shiau A.K., Desai A., Oegema K. (2020) TRIM37 controls cancer-specific vulnerability to PLK4 inhibition. *Nature,* 585(7825), 440-446. doi: 10.1038/s41586-020-2710-1.

Menco B.P., Bruch R.C., Dau B., Danho W. (1992) Ultrastructural localization of olfactory transduction components: the G protein subunit Golf alpha and type III adenylyl cyclase. *Neuron,* 8, 441-453.

Menco M. (1980) Qualitative and quantitative freeze-fracture studies on olfactory and respiratory epithelial surfaces of frog, ox, rat, and dog. IV. Ciliogenesis and ciliary necklaces (including high-voltage observations). *Cell Tissue Res.,* 212, 1-16.

Mendiola M., Barriuso J., Marino-Enriquez A., Redondo A., Dominguez-Caceres A., Hernandez-Cortes G., Perez-Fernandez E., Sanchez-Navarro I., Vara JA, Suarez A., Espinosa E., Gonzalez-Baron M., Palacios J., Hardisson D. (2009) Aurora kinases as prognostic biomarkers in ovarian carcinoma. *Hum. Pathol.*, 40, 631-638. doi: 10.1016/j.humpath.2008.10.011.

Meraldi P., Lukas J., Fry A. M., Bartek J., Nigg E. A. (1999) Centrosome duplication in mammalian somatic cells requires E2F and Cdk2-cyclin A. *Nat. Cell Biol.,* 1, 88-93.

Merdes A., Cleveland D.W. (1997) Pathways of spindle pole formation: different mechanisms; conserved components. *J. Cell Biol.,* 138, 953-956.

Merdes A., Ramyar K., Vechio J. D., Cleveland D. W. (1996) A complex of NuMA and cytoplasmic dynein is essential for mitotic spindle assembly. *Cell,* 87, 447-458.

Meselson M., Stahl F. W. (1958) The replication of DNA in Escherichia coli. *Proc. Nat. Acad. Sci. USA,* 44, 671-682.

Meves F. (1903) On oligophrenic and pyrenic spermatozoa and their origin, based on observations on Paludina and Pygaera [On oligophrenic and pyrenic spermatozoa and their origin, based on observations on Paludina and Pygaera]. *Arch. Mikrosk. Anat.,* 61, 1-84. In German.

Mi J., Guo C., Brautigan D.L., Larner J.M. (2007) Protein phosphatase-1alpha regulates centrosome splitting through Nek2. *Cancer Res.,* 67(3), 1082-1089.

Mignot J.P. (1996) The centrosomal big bang: from a unique central organelle towards a constellation of MTOCs. *Biol. Cell,* 86, 81-91.

Mikhailov A.V., Gundersen G.G. (1995) Centripetal transport of microtubules in motile cells. *Cell Motil. Cytoskeleton,* 32, 173-181.

Mikina W., Hałakuc P., Milanowski R. (2024) Transposon-derived introns as an element shaping the structure of eukaryotic genomes. *Mob. DNA,* 15(1), 15. doi: 10.1186/s13100-024-00325-w. Erratum in: *Mob. DNA,* 2024 15(1), 21. doi: 10.1186/s13100-024-00329-6.

Milner J., Ponder B., Hughes-Davies L., Seltmann, M. Kouzarides T. (1997) Transcriptional activation functions in BRCA2. *Nature,* 386(6627), 772-773. doi: 10.1038/386772a0.

Mitchison T., Kirschner M.W. (1984) Microtubule assembly nucleated by isolated centrosomes. *Nature,* 312, 232-237.

Mitchison T., Kirschner M.W. (1984a) Microtubule assembly nucleated by isolated centrosomes. *Nature Lond.,* 312(5991), 237-242.

Mitchison T., Kirschner M.W. (1984b) Dynamic instability of microtubule growth. *Nature Lond.,* 312(5991), 237-242.

Mittal K., Ogden A., Reid M.D., Rida P.C., Varambally S., Aneja R. (2015) Amplified centrosomes may underlie aggressive disease course in pancreatic ductal adenocarcinoma. *Cell Cycle,* 14(17), 2798-2809. doi: 10.1080/15384101.2015.1068478.

Miyoshi K., Asanuma M., Miyazaki I., Diaz Corrales F.J., Katayama T., Tohyama M., Ogawa N. (2004) DISC1 localizes to the centrosome by binding to kendrin. *Biochem. Biophys. Res. Commun.,* 317(4), 1195-1199.

Mochizuki T., Wu G., Hayashi T., Xenophontos S.L., Veldhuisen B., Saris J.J., Reynolds D.M., Cai Y., Gabow P.A., Pierides A., Kimberling W.J., Breuning M.H., Deltas C.C., Peters D.J., Somlo S. (1996) PKD2, a gene for polycystic kidney disease that encodes an integral membrane protein. *Science,* 272(5266), 1339-1342. doi: 10.1126/science.272.5266.1339.

Mogensen M.M. (2004) Microtubule organizing centers in polarized epithelial cells. In "*Centrosomes in development and disease*" (ed. E.A. Nigg). Germany, Weinheim: Wiley-VCH Verlag GmbH and Co., KGaA.

Mogensen M.M., Mackie J.B., Doxsey S.J., Stearns T., Tucker J.B. (1997) Centrosomal deployment of gamma-tubulin and pericentrin: evidence for a microtubule-nucleating domain and a minus-end docking domain in certain mouse epithelial cells. *Cell Motil. Cytoskeleton,* 36, 276-290.

Mogensen M.M., Malik A., Piel M., Bouckson Castang V., Bornens M. (2000) Microtubule minus-end anchorage at centrosomal and non-centrosomal sites: the role of ninein. *J. Cell Sci.,* 113 (17), 3013-3023.

Mogensen M.M., Tucker J.B., Mackie J.B., Prescott A.R., Nathke I.S. (2002) The adenomatous polyposis coli protein unambiguously localizes to microtubule plus ends and is involved in establishing parallel arrays of microtubule bundles in highly polarized epithelial cells. *J. Cell Biol.,* 157, 1041-1048.

Mohammed A.A., Karasiewicz J., Modliński J.A. (2008) Developmental potential of selectively enucleated immature mouse oocytes upon nuclear transfer. *Mol. Reprod. Dev.,* 75(8), 1269-1280.

Molli P.R., Li D.Q., Bagheri-Yarmand R., Pakala S.B., Katayama H., Sen S., Iyer J., Chernoff J., Tsai M.Y., Nair S.S., Kumar R. (2010) Arpc1b, a centrosomal protein, is both an activator and substrate of Aurora A. *J. Cell Biol.,* 190(1), 101-114.

Morante J., Desplan C., Celik, A. (2007) Generating patterned arrays of photoreceptors. *Curr. Opin. Genet. Dev.,* 17(4), 314-319.

Morden C.R., Farrell A.C., Sliwowski M., Lichtensztejn Z., Altman A.D., Nachtigal M.W., McManus K.J. (2021) Chromosome instability is prevalent and dynamic in high-grade serous ovarian cancer patient samples. *Gynecol. Oncol.,* 161(3), 769-778. doi: 10.1016/j.ygyno.2021.02.038.

Moreno S, Nurse P. (1994) Regulation of progression through the G1 phase of the cell cycle by the rum1+ gene. *Nature,* 367(6460), 236-242.

Morgan G.J., Walker B.A., Davies F.E. (2012) The genetic architecture of multiple myeloma. *Nat. Rev. Cancer,* 12(5), 335-348. doi: 10.1038/nrc3257.

Morita Y., Kinoshita K., Wakisaka S., Makihara, S. (1991) Claws of cilia: further observation of ciliated epithelium in neurenteric cyst. Virchows *Arch. A Pathol. Anat. Histopathol.,* 418 (3), 263-265.

Moritz M., Agard D. 2001. g-tubulin complexes and microtubule nucleation. *Curr. Opin. Struct. Biol.,* 11,174-181.

Moritz M., Braunfeld M.B., Guenebaut V., Heuser J., Agard D.A. (2000) Structure of the gamma-tubulin ring complex: a template for microtubule nucleation. *Nat. Cell Biol.,* 2(6), 365-370.

Moritz M., Zheng Y., Alberts B.M., Oegema K. (1998) Recruitment of the gamma-tubulin ring complex to Drosophila salt-stripped centrosome scaffolds. *J. Cell Biol.,* 142(3), 775-786.

Morris J.A., Kandpal G., Ma L., Austin C.P. (2003) DISC1 (Disrupted-In-Schizophrenia 1) is a centrosome-associated protein that interacts with MAP1A, MIPT3, ATF4/5 and NUDEL: regulation and loss of interaction with mutation. *Hum. Mol. Genet.,* 12(13), 1591-1608.

Moskvin-Tarkhanov M.I., Onishchenko G.E. (1978) Centrioles in mouse bone marrow megakaryocytes. *Tsitologiia,* 20(12), 1436-1438. In Russian.

Moss D.K., Bellett G., Carter J.M., Liovic M., Keynton J., Prescott A.R., Lane E.B., Mogensen M.M. (2007) Ninein is released from the centrosome and moves bi-directionally along microtubules. *J. Cell Sci.,* 120, 3064-3074.

Mottier-Pavie V., Megraw T.L. (2009) Drosophila bld10 is a centriolar protein that regulates centriole, basal body, and motile cilium assembly. *Mol. Biol. Cell,* 20(10), 2605-2614.

Moudjou M., Bordes N., Paintrand M., Bornens M. (1996) gamma-Tubulin in mammalian cells: the centrosomal and the cytosolic forms. *J. Cell Sci.,* 109 (4), 875-887.

Mountain V., Simerly C., Howard L., Ando A., Schatten G., Compton D.A. (1999) The kinesin-related protein, HSET, opposes the activity of Eg5 and cross-links microtubules in the mammalian mitotic spindle. *J. Cell Biol.,* 147, 351-366.

Moutinho-Pereira S., Debec A., Maiato H. (2009) Microtubule cytoskeleton remodeling by acentriolar microtubule-organizing centers at the entry and exit from mitosis in Drosophila somatic cells. *Mol. Biol. Cell,* 20(11), 2796-2808.

Moyer J.H., Lee-Tischler M.J., Kwon H.Y., Schrick J.J., Avner E.D., Sweeney W.E., Godfrey V.L., Cacheiro N.L., Wilkinson J.E., Woychik R.P. (1994) Candidate gene associated with a mutation causing recessive polycystic kidney disease in mice. *Science,* 264(5163), 1329-1333. doi: 10.1126/science.8191288.

Moyer T.C., Clutario K.M., Lambrus B.G., Daggubati V., Holland A.J. (2015) Binding of STIL to Plk4 activates kinase activity to promote centriole assembly. *J. Cell Biol.*, 209 (6), 863. 10.1083/jcb.201502088.

Munemitsu S., Souza B., Müller O., Albert I., Rubinfeld B., Polakis P. (1994) The APC gene product associates with microtubules in vivo and promotes their assembly *in vitro*. *Cancer Res.*, 54(14), 3676-3681.

Munger B.L. (1958) A light and electron microscope study of cellular differentiation in the pancreatic islets of the mouse. *Am. J. Anat.*, 103, 275-311.

Muňoz-Cuevas M.A. (1975). Modèle ciliaire du développement du photorécepteur chez *Ischyropsalis luteipes* (*Opiliones, Arachnida*). [Ciliary model of the development of the photoreceptor in *Ischyropsalis luteipes* (*Opiliones, Arachnida*)]. *C. R. Séances Acad. Sci. D.,* 280(6), 725-728. In French.

Munshi N.C., Anderson K.C. (2013) New strategies in the treatment of multiple myeloma. *Clin Cancer Res.,* 19(13), 3337-3344.

Murakami H., Goto D.B., Toda T., Chen ES, Grewal SI, Martienssen RA, Yanagida M. (2007) Ribonuclease activity of Dis3 is required for mitotic progression and provides a possible link between heterochromatin and kinetochore function. *PloS One,* 2(3), e317. doi: 10.1371/journal.pone.0000317.

Murata Y., Tsuji M., Tani M. (1984) Ultrastructure of glomus tumor *J. Cutan. Pathol.,* 11(1), 53-58.

Murphy S.M., Preble A.M., Patel U.K., O'Connell K.L., Dias D.P., Moritz M., Agard D., Stults J.T., Stearns T. (2001) GCP5 and GCP6: two new members of the human gamma-tubulin complex. *Mol. Biol. Cell.,* 12(11), 3340-3352.

Murray A.W. (2001) Cell cycle. Centrioles at the checkpoint. *Science,* 291(5508):1499-502. doi: 10.1126/science.291.5508.1499.

Murray R.G., Murray A.S., Pizzo A. (1965) The fine structure of mitosis in rat thymic lymphocytes. *J. Cell Biol.,* 26, 601-619.

Nachury M.V., Loktev A.V., Zhang Q, Westlake CJ, Peranen J, Merdes A., Slusarski D.C., Scheller R.H., Bazan J.F., Sheffield V.C., Jackson P.K. (2007) A core complex of BBS proteins cooperates with the GTPase Rab8 to promote ciliary membrane biogenesis. *Cell,* 129, 1201-1213.

Nadezhdina E.S., Fais D., Chentsov Yu.S. (1979) On the association of centrioles with the interphase nucleus. *Eur. J. Cell Biol.,* 19, 109-115.

Nakamura M., Masuda H., Horii J., Kuma K., Yokoyama N., Ohba T., Nishitani H., Miyata T., Tanaka M., Nishimoto T. (1998) When overexpressed, a novel centrosomal protein, RanBPM, causes ectopic microtubule nucleation similar to gamma-tubulin. *J. Cell Biol.,* 143(4), 1041-1052.

Nakanishi A., Han X., Saito H., Taguchi K., Ohta Y., Imajoh-Ohmi S., Miki Y. (2007) Interference with BRCA2, which localizes to the centrosome during S and early M phase, leads to abnormal nuclear division. *Biochem. Biophys. Res. Commun.,* 355(1), 34-40.

Nakazawa Y., Hiraki M., Kamiya R., Hirono M. (2007) SAS-6 is a cartwheel protein that establishes the 9-fold symmetry of the centriole. *Curr. Biol.,* 17(24), 2169-2174.

Nanjundappa R., Kong D., Shim K., Stearns T., Brody S.L., Loncarek J., Mahjoub M.R. (2019) Regulation of cilia abundance in multiciliated cells. *ELife,* 8, e44039.

Naro C., Barbagallo F., Chieffi P., Bourgeois C.F., Paronetto M.P., Sette C. (2014) The centrosomal kinase NEK2 is a novel splicing factor kinase involved in cell survival. *Nucleic Acids Res.,* 42(5), 3218-27. doi: 10.1093/nar/gkt1307.

Nasmyth K., Petersand J-M., Uhlmann F. (2000) Splitting the Chromosome: Cutting the Ties That Bind Sister Chromatids. *Science,* 288(5470), 1379-1384. doi: 10.1126/science.288.5470.1379.

Nauli S.M., Alenghat F.J., Luo Y., Williams E., Vassilev P., Li X., Elia A.E.H., Lu W., Brown E.M., Quinn S.J., Ingber D.E., Zhou J. (2003) Polycystins 1 and 2 mediate mechanosensation in the primary cilium of kidney cells. *Nat. Genet.,* 33, 129-37. doi: 10.1038/ng1076.

Nauli S.M., Jin X., Hierck B.P. (2011) The mechanosensory role of primary cilia in vascular hypertension. *Int. J. Vasc. Med.,* 2011, 376281.

Navone N.M., Labate M.E., Troncoso P., Pisters L.L., Conti C.J., von Eschenbach A.C., Logothetis C.J. (1999) p53 mutations in prostate cancer bone metastases suggest that selected p53 mutants in the primary site define foci with metastatic potential. *J. Urol.,* 161(1), 304-308. PMID: 10037428.

Navone N.M., Troncoso P., Pisters L.L., Goodrow T.L., Palmer J.L., Nichols W.W., von Eschenbach A.C., Conti C.J. (1993) p53 protein accumulation and gene mutation in the progression of human prostate carcinoma. *J. Natl. Cancer Inst.,* 85(20), 1657-1669. doi: 10.1093/jnci/85.20.1657.

Nechipurenko I.V., Berciu C., Sengupta P., Nicastro D. (2017) Centriolar remodeling underlies basal body maturation during ciliogenesis in caenorhabditis elegans. *ELife,* 6, e25686.

Negi S.S., Olson M.O. (2006) Effects of interphase and mitotic phosphorylation on the mobility and location of nucleolar protein B23. *J. Cell Sci.,* 119(17), 3676-3685.

Neveling K., Mensenkamp A.R., Derks R., Kwint M., Ouchene H., Steehouwer M., van Lier B., Bosgoed E., Rikken A., Tychon M., Zafeiropoulou D., Castelein S., Hehir-Kwa J., Tjwan Thung D., Hofste T., Lelieveld S.H., Bertens S.M., Adan I.B., Eijkelenboom A., Tops B.B., Yntema H., Stokowy T., Knappskog P.M., Høberg-Vetti H., Steen V.M., Boyle E., Martin B., Ligtenberg M.J., Shendure J., Nelen M.R., Hoischen A. (2017) BRCA Testing by Single-Molecule Molecular Inversion Probes. *Clin Chem.,* 63(2), 503-512.

Neverova A.L., Uzbekov R.E, Votchal M.S., Vorobjev I.A. (1996) Effect of UV microirradiation of the centrosome on cell behavior IV. Synthetic activity, spreading and growth of cells with the irradiated centrosome. *Tsytologiia*, 38(2) 145-154. In Russian.

Neverova A.L., Uzbekov R.E., Votchal M.S., Zatsepina O.V., Vorobjev I.A. (1999) Postmitotic reconstruction of nucleoli in culture cells with UV-microbeam photoinactivated centrosome. *Membr. Cell Biol.,* 12(6), 805-815.

Ng D.C.H., Richards D.K., Mills R.J., Ho U.Y., Perks H.L., Tucker C.R., Voges H.K., Pagan J.K., Hudson J.E. (2020) Centrosome Reduction Promotes Terminal

Differentiation of Human Cardiomyocytes. *Stem Cell Reports,* 15(4), 817-826. doi: 10.1016/j.stemcr.2020.08.007.

Nicastro D., McIntosh J.R., Baumeister W. (2005) 3D structure of eukaryotic flagella in a quiescent state revealed by cryo-electron tomography. *Proc. Natl. Acad. Sci. USA,* 102, 15889-15894.

Niethammer M., Smith D.S., Ayala R., Peng J., Ko J., Lee M.S., Morabito M., Tsai L.H. (2000) NUDEL is a novel Cdk5 substrate that associates with LIS1 and cytoplasmic dynein. *Neuron,* 28(3), 697-711.

Nigg E. (2002) Centrosome aberrations: cause or consequence of cancer progression? *Nature Rev. Cancer,* 2, 15-825.

Nigg E.A. (2007) Centrosome duplication: of rules and licenses. *Trends Cell Biol.*, 17(5), 215-221. doi: S0962-8924(07)00055-4 [pii] doi: 10.1016/j.tcb.2007.03.003

Nigg E.A., Holland A.J. (2018) Once and only once: Mechanisms of centriole duplication and their deregulation in disease. *Nat. Rev. Mol. Cell Biol.,* 19, 297-312.

Nigg E.A., Raff J.W. (2009) Centrioles, centrosomes, and cilia in health and disease. *Cell,* 139(4), 663-678.

Nigg E.A., Stearns T. (2011) The centrosome cycle: Centriole biogenesis, duplication and inherent asymmetries. *Nat. Cell Biol.,* 13, 1154-1160.

Nishimura T., Takahashi M., Kim H.S., Mukai H., Ono Y. (2005) Centrosome-targeting region of CG-NAP causes centrosome amplification by recruiting cyclin E-cdk2 complex. *Genes Cells,* 10(1), 75-86.

Noguchi E., Hayashi N., Azuma Y., Seki T., Nakamura M., Nakashima N., Yanagida M., He X., Mueller U., Sazer S., Nishimoto T. (1996) Dis3, implicated in mitotic control, binds directly to Ran and enhances the GEF activity of RCC1. *EMBO J.,* 15(20), 5595-5605. PMID: 8896453; PMCID: PMC452304.

Nolte F., Giehl M., Haass W., Nowak V., Schumann C., Nowak D., Mossner M, Popp H.D., Schulze T.J., Klein S., Seifarth W., Hofmann W.K., Fabarius A. (2013) Centrosome aberrations in bone marrow cells from patients with myelodysplastic syndromes correlate with chromosomal instability. *Ann. Hematol.,* 92(10),1325-1333. doi: 10.1007/s00277-013-1772-7

Nonaka S., Tanaka Y., Okada Y., Takeda S., Harada A., Kanai Y., Kido M., Hirokawa N. (1998) Randomization of left-right asymmetry due to loss of nodal cilia generating leftward flow of extraembryonic fluid in mice lacking KIF3B motor protein. *Cell,* 95, 829-837.

Nunes de Almeida F., Walther R.F., Pressé M.T., Vlassaks E., Pichaud F. (2019) Cdc42 defines apical identity and regulates epithelial morphogenesis by promoting apical recruitment of Par6-aPKC and Crumbs. *Development,* 146(15), dev175497. doi: 10.1242/dev.175497.

Nurse P. (1990) Universal control mechanism regulating onset of M-phase. *Nature,* 344, 503-508.

O'Connell K.F. (2002) The zyg-1 kinase, a mitotic and meiotic regulator of centriole replication. *Oncogene,* 21, 6201-6208.

O'Connell K.F., Caron C., Kopish K.R., Hurd D.D., Kemphues K.J., Li Y., White J.G. (2001) The c. Elegans zyg-1 gene encodes a regulator of centrosome duplication with distinct maternal and paternal roles in the embryo. *Cell,* 105, 547-558.

O'Connell, C.B., and Wang, Y.-l. (2000). Mammalian spindle orientation and position respond to changes in cell shape in a dynein-dependent fashion. *Mol. Biol. Cell,* 11, 1765-1774.

O'Hara P.T. (1970) Spiral Tilt of Triplet Fibers in Human Leukocyte Centrioles. *J. Ultrastr. Res.,* 31, 195-198.

O'Toole E.T., Giddings Jr T.H., Dutcher, S.K. (2007) Understanding microtubule organizing centers by comparing mutant and wild-type structures with electron tomography. *Methods Cell Biol.,* 79, 125-143.

Oakley C.E., Oakley B.R. (1989) Identification of gamma-tubulin, a new member of the tubulin superfamily encoded by mipA gene of *Aspergillus nidulans*. *Nature,* 338(6217), 662-664.

Obino D., Farina F., Malbec O., Sáez P.J., Maurin M., Gaillard J., Dingli F., Loew D., Gautreau A., Yuseff M-I., Blanchoin L., Théry M., Lennon-Duménil A-M. (2016). Actin nucleation at the centrosome controls lymphocyte polarity. *Nat. Comm.,* 7, 10969.

Odell J., Sikirzhytski V., Tikhonenko I., Cobani S., Khodjakov A., Koonce M. (2019). Force balances between interphase centrosomes as revealed by laser ablation. *Mol. Biol. Cell,* 30, 1705-1715.

Oegema K., Wiese C., Martin O.C., Milligan R.A., Iwamatsu A., Mitchison T.J., Zheng Y. (1999) Characterization of two related Drosophila gamma-tubulin complexes that differ in their ability to nuclrate microtubules. *J. Cell Biol.,* 144(4), 721-733.

Ohkura H., Adachi Y., Kinoshita N., Niwa O., Toda T., Yanagida M. (1988) Coldsensitive and caffeine-supersensitive mutants of the Schizosaccharomyces pombe dis genes implicated in sister chromatid separation during mitosis. *EMBO J.,* 7(5), 1465-1473. doi: 10.1002/j.1460-2075.1988.tb02964.x.

Ohta M., Ashikawa T., Nozaki Y., Kozuka-Hata H., Goto H., Inagaki M., Oyama M., Kitagawa D. (2014) Direct interaction of Plk4 with STIL ensures formation of a single procentriole per parental centriole. *Nat. Commun.,* 5 (1), 5267. 10.1038/ncomms6267.

Okuda M., Horn H. F., Tarapore P., Tokuyama Y., Smulian A.G., Chan P. K., Knudsen E. S., Hofmann I. A., Snyder J. D., Bove K. E., Fukasawa K. (2000) Nucleophosmin/B23 is a target of CDK2/cyclin E in centrosome duplication. *Cell,* 103 (1), 127-140.

O'Neill A.C., Uzbas F., Antognolli G., Merino F., Draganova K., Jäck A., Zhang S., Pedini G., Schessner J.P., Cramer K., Schepers A., Metzger F., Esgleas M., Smialowski P., Guerrini R., Falk S., Feederle R., Freytag S., Wang Z., Bahlo M., Jungmann R., Bagni C., Borner G.H.H., Robertson S.P., Hauck S.M., Götz M. (2022) Spatial centrosome proteome of human neural cells uncovers disease-relevant heterogeneity. *Science,* 376(6599):eabf9088.

Onishchenko G.E., Bystrevskaia V.B., Chentsov Iu.S. (1979) Multipolar mitosis in normal tissue culture cells and following experimental induction with 2-mercaptoethanol. *Dokl. Akad. Nauk SSSR*, 248(6), 1443-1446. In Russian.

Onishchenko G.E., Chentsov Iu.S. (1986) Structure of the centriolar complex in the hepatocytes of intact and regenerating mouse live. *Tsitologiia,* 28(4), 403-408. In Russian.

Onuchic L.F., Furu L., Nagasawa Y., Hou X., Eggermann T., Ren Z., Bergmann C., Senderek J., Esquivel E., Zeltner R., Rudnik-Schöneborn S., Mrug M., Sweeney W.,

Avner E.D., Zerres K., Guay-Woodford L.M., Somlo S., Germino G.G. (2002) PKHD1, the polycystic kidney and hepatic disease 1 gene, encodes a novel large protein containing multiple immunoglobulin-like plexin-transcription-factor domains and parallel beta-helix 1 repeats. *Am. J. Hum. Genet.,* 70, 1305-1317.

Osborn M., Weber K. (1976) Cytoplasmic microtubules in tissue culture cells appear to grow from an organizing structure towards the plasma membrane. *Proc. Natl. Acad. Sci. USA,* 73(3), 867-871.

Osman I., Drobnjak M., Fazzari M., Ferrara J., Scher H.I., Cordon-Cardo C. (1999) Inactivation of the p53 pathway in prostate cancer: impact on tumor progression. *Clin. Cancer Res.,* 5(8), 2082-2088. PMID: 10473090.

O'Toole E.T., Giddings T.H., Dutcher S.K. (2007) Understanding of microtubule organizing centers by comparing mutant and wild) type structures with elrctron tomography. *Methods Cell Biol.,* 79, 135-143.

O'Toole E.T., Giddings T.H., McIntosh J.R., Dutcher S.K. (2003) Three-dimensional organization of basal bodies from wild-type and δ-tubulin deletion strains of Chlamydomonas reinhardtii. *Mol. Biol. Cell,* 14 (7), 2999-3012.

Otto E., Loeys B., Khanna H., Hellemans J., Sudbrak R., Fan S., Muerb U., O'Toole J.F., Helou J., Attanasio M., Utsch B., Sayer J.A., Lillo C., Jimeno D., Coucke P., De Paepe A., Reinhardt R., Klages S., Tsuda M., Kawakami I., Kusakabe T., Omran H., Imm A., Tippens M., Raymond P.A., Hill J., Beales P., He S., Kispert A., Margolis B., Williams D.S., Swaroop A., Hildebrandt F. (2005) A novel ciliary IQ domain protein, NPHP5, is mutated in Senior-Loken syndrome (nephronophthisis with retinitis pigmentosa), and interacts with RPGR and calmodulin. *Nat. Genet.,* 37, 282-288.

Ou Y., Rattner J.B. (2000) A subset of centrosomal proteins are arranged in a tubular conformation that is reproduced during centrosome duplication. *Cell Motil. Cytoskeleton,* 47(1), 13-24.

Ou Y., Rattner J.B. (2004) The centrosome in higher organisms: structure, composition, and duplication. *Int. Rev. Cytol.,* 238, 119-182.

Ou Y.Y., Mack G.J., Zhang M., Rattner J.B. (2002) CEP110 and ninein are located in a specific domain of the centrosome associated with centrosome maturation. *J. Cell Sci.,* 115(9), 1825-1835.

Ou Y.Y., Zhang M., Chi S., Matyas J.R., Rattner J.B. (2003) Higher order structure of the pcm adjacent to the centriole. *Cell Motil. Cytoskeleton*, 55, 125-133.

Ouchi M., Fujiuchi N., Sasai K., Katayama H., Minamishima Y.A., Ongusaha P.P., Deng C., Sen S., Lee S.W., Ouchi T. (2004) BRCA1 phosphorylation by Aurora-A in the regulation of G2 to M transition. *J. Biol. Chem.,* 279(19), 19643-19648.

Ouyang X., Wang X., Xu K., Jin D.Y., Cheung A.L., Tsao S.W., Wong Y.C. (2001) Effect of p53 on centrosome amplification in prostate cancer cells. *Biochim. Biophys. Acta,* 1541(3), 212-20. doi: 10.1016/s0167-4889(01)00157-4.

Overy H.R., Priest R.E. (1966) Mitotic cell division in postnatal cardiac growth. *Lab. Invest.,* 15(6), 1100-1103.

Pabis K., Barardo D., Sirbu O., Selvarajoo K., Gruber J., Kennedy B.K. (2024) A concerted increase in readthrough and intron retention drives transposon expression during aging and senescence. *Elife,* 12, RP87811. doi: 10.7554/eLife.87811.

Paccagnini E., De Marzo L., Giusti F., Dallai R. (2006) The aberrant spermatogenesis of the *Haplothrips simplex* (Buffa) (*Thysanoptera*): ultrastructural study. *Tissue Cell,* 38, 177-186.

Paccagnini E., Mencarelli C., Mercati D., Afzelius B.A., Dallai R. (2007) Ultrastructural analysis of the aberrant axoneme morphogenesis in thrips (Thysanoptera, Insecta). *Cell Motil. Cytoskeleton,* 64(9), 645-661.

Paccagnini E., Mercati D., Giusti F., Conti B., Dallai R. (2010) The spermatogenesis and the sperm structure of *Terebrantia* (*Thysanoptera, Insecta*). *Tissue Cell,* 42(4), 247-258.

Paintrand M., Moudjou M., Delacroix H., Bornens M. (1992) Centrosome Organization and Centriole Architecture: Their Sensitivity to Divalent Cations. *J. Struct. Biol.,* 108, 107-128.

Palazzo A.F., Joseph H.L., Chen Y.J., Dujardin D.L., Alberts A.S., Pfister K.K., Vallee R.B., Gundersen G.G. (2001) Cdc42, dynein, and dynactin regulate MTOC reorientation independent of Rho-regulated microtubule stabilization. *Curr. Biology,* 11, 1536-1541.

Palazzo R.E., Vaisberg E., Cole R.W., Rieder C.L. (1992) Centriole duplication in lysates of Spisula solidissima oocytes. *Science,* 256, 219-221.

Palazzo R.E., Vogel J.M., Schnackenberg B.J., Hull D.R., Wu X. (2000) Centrosome maturation. *Curr. Top. Dev. Biol.,* 49, 449-470.

Palla A.R., Hilgendorf K.I., Yang A.V., Kerr J.P., Hinken A.C., Demeter J., Kraft P., Mooney N.A., Yucel N., Burns D.M., Wang Y.X., Jackson P.K., Blau H.M. (2022) Primary cilia on muscle stem cells are critical to maintain regenerative capacity and are lost during aging. *Nat. Commun.,* 13(1), 1439.

Pannu V., Mittal K., Cantuaria G., Reid M.D., Li X., Donthamsetty S., McBride M., Klimov S., Osan R., Gupta M.V., Rida P.C., Aneja R. (2015) Rampant centrosome amplification underlies more aggressive disease course of triple negative breast cancers. *Oncotarget,* 6(12), 10487-97. doi: 10.18632/oncotarget.3402.

Paoletti A., Bornens M. (1997) Organisation and functional regulation of the centrosome in animal cells. Prog. Cell Cycle Res., 3, 285-299. Review.

Paoli F., Gottardo M., Dallai R., Roversi P.F. (2013) Morphology of the male reproductive system and sperm ultrastructure of the egg parasitoid *Gryon pennsylvanicum* (*Ashmead*) (*Hymenoptera, Platygastridae*). *Arthropod Struct. Dev.,* 42(4), 297-308.

Paoli F., Roversi P.F., Gottardo M., Callaini G., Mercati D., Dallai R. (2015) A microtubule organizing centre (MTOC) is responsible for the production of the sperm flagellum in Matsucoccus feytaudi (*Hemiptera: Coccoidea*). *Arthropod Struct. Dev.,* 44(3), 237-242.

Park E.Y., Woo Y.M., Park J.H. (2011) Polycystic kidney disease and therapeutic approaches. *BMB Rep.,* 44(6), 359-368.

Parker J.D., Hilton L.K., Diener D.R., Rasi M.Q., Mahjoub M.R., Rosenbaum J.L., Quarmby L.M. (2010) Centrioles are freed from cilia by severing prior to mitosis. *Cytoskeleton*, 67, 425-430.

Parvin A., Hao S.L., Tan F.Q., Yang W.X. (2020) Inhibition of kinesin motor protein KIFC1 by AZ82 induces multipolar mitosis and apoptosis in prostate cancer cell. *Gene,* 760, 144989. doi: 10.1016/j.gene.2020.145232.

Paschal B.M., Holzbaur E.L., Pfister K.K., Clark S., Meyer D.I., Vallee R.B. (1993) Characterization of 50 kD polypeptide in cytoplasmic dynein preparations reveals a complex with p150 Glued and a novel actin. *J. Biol. Chem.*, 268, 15318-15323.

Patel S., Player M.R. (2008) Small-molecule inhibitors of the p53-HDM2 interaction for the treatment of cancer. *Expert. Opin. Investig. Drugs,* 17(12), 1865-1882.

Pazour G.J., Baker S.A., Deane J.A., Cole D.G., Dickert B.L., Rosenbaum J.L., Witman G.B., Besharse J.C. (2002a) The intraflagellar transport protein, IFT88, is essential for vertebrate photoreceptor assembly and maintenance. *J. Cell Biol.,* 157, 103-113.

Pazour G.J., Dickert B.L., Vucica Y., Seeley E.S., Rosenbaum J.L., Witman G.B., Cole D.G. (2000) Chlamydomonas IFT88 and its mouse homologue, polycystic kidney disease gene tg737, are required for assembly of cilia and flagella. *J. Cell Biol.,* 151(3), 709-718.

Pazour G.J., San Agustin J., Follit J.A., Rosenbaum J.L., Witman G.B. (2002b) Polycystin-2 localizes to kidney cilia and the ciliary level is elevated in orpk mice with polycystic kidney disease. *Curr. Biol.,* 12, R378-380.

Pazour G.J., Witman G.B. (2003) The vertebrate primary cilium is a sensory organelle. *Curr. Opin. Cell Biol.,* 15, 105-110.

Pearson C.G., Osborn D.P., Giddings T.H. Jr., Beales P.L., Winey M. (2009) Basal body stability and ciliogenesis requires the conserved component Poc1. *J. Cell Biol.,* 187(6), 905-920.

Pedersen L.B., Rosenbaum J.L. (2008) Intraflagellar transport (IFT) role in ciliary assembly, resorption and signalling. *Curr. Top. Dev. Biol.,* 85, 23-61.

Peel N., Stevens N.R., Basto R., Raff J.W. (2007) Overexpressing centriole-replication proteins *in vivo* induces centriole overduplication and *de novo* formation. *Curr. Biol.,* 17(10), 834-843. doi: 10.1016/j.cub.2007.04.036.

Pelechano V. (2017) From transcriptional complexity to cellular phenotypes: Lessons from yeast. *Yeast,* 34(12), 475-482. doi: 10.1002/yea.3277.

Pelletier L., O'Toole E., Schwager A., Hyman A.A., Muller-Reichert T. (2006) Centriole assembly in Caenorhabditis elegans. *Nature*, 444, 619-623.

Pelletier L., Ozlu N., Hannak E., Cowan C., Habermann B., Ruer M., Muller-Reichert T., Hyman A.A. (2004) The caenorhabditis elegans centrosomal protein spd-2 is required for both pericentriolar material recruitment and centriole duplication. *Curr. Biol.,* 14, 863-873.

Pelletier L., Ozlü N., Hannak E., Cowan C., Habermann B., Ruer M., Müller-Reichert T., Hyman A.A. (2004) The Caenorhabditis elegans centrosomal protein SPD-2 is required for both pericentriolar material recruitment and centriole duplication. *Curr Biol.,* 14(10), 863-873. Erratum in: *Curr Biol.,* 2006, 16(12), 1255.

Peng G.E., Wilson S.R., Weiner O.D. (2011) A pharmacological cocktail for arresting actin dynamics in living cells. *Mol. Biol. Cell,* 22, 3986-3994.

Persico V., Callaini G., Riparbelli M.G. (2019) The Microtubule-Depolymerizing Kinesin-13 Klp10A Is Enriched in the Transition Zone of the Ciliary Structures of Drosophila melanogaster. Front. *Cell Dev. Biol.,* 7, 173.

Persico V., Callaini G., Riparbelli M.G. (2021) Sas-4 Colocalizes with the Ciliary Rootlets of the Drosophila Sensory Organs. *J. Dev. Biol.*, 9(1), 1. doi: 10.3390/jdb9010001.

Persico V., Migliorini M., Callaini G., Riparbelli M.G. (2020) The Singularity of the Drosophila Male Germ Cell Centriole: The Asymmetric Distribution of Sas4 and Sas6. *Cells,* 9(1), 115.

Pessa H.K., Will C.L., Meng X., Schneider C., Watkins N.J., Perälä N., Nymark M., Turunen J.J., Lührmann R., Frilander M.J. (2008) Minor spliceosome components are predominantly localized in the nucleus. *Proc. Natl. Acad. Sci. USA,* 105(25), 8655-8660.

Peterson S.P., Berns M.W. (1980) The Centriolar Complex. *Int. Rev. Cytology,* 64, 81-106.

Petriman N.A., Lorentzen E. (2020) Structural insights into the architecture and assembly of eukaryotic flagella. *Microb. Cell,* 7(11), 289-299.

Phillips D.M. (1967) Giant centriole formation in Sciara. *J. Cell Biol.,* 33(1), 73-92.

Pickett-Heaps JD. (1971) The autonomy of the centriole: fact or fallacy? *Cytobios*, 3, 205-214.

Piehl M., Cassimeris L. (2003) Organization and dynamics of growing microtubule plus ends during early mitosis. *Mol. Biol. Cell,* 14(3), 916-925.

Piel M., Meyer P., Khodjakov A., Rieder C.L., Bornens M. (2000) The respective contributions of the mother and daughter centrioles to centrosome activity and behavior in vertebrate cells. *J. Cell Biol.,* 149(2), 317-330.

Piemonte K.M., Anstine L.J., Keri R.A. (2021) Centrosome Aberrations as Drivers of Chromosomal Instability in Breast Cancer. *Endocrinology*, 162(12), bqab208. doi: 10.1210/endocr/bqab208.

Pierson-Mullany L.K., Lange C.A. (2004) Phosphorylation of progesterone receptor serine 400 mediates ligand-independent transcriptional activity in response to activation of cyclin-dependent protein kinase 2. *Mol. Cell Biol.,* 24, 10542-10557. doi: 10.1128/MCB.24.24.10542-10557.2004.

Pietromonaco S.F., Seluja G.A., Elias L. (1995) Identification of enzymatically active Ca2+/calmodulin-dependent protein kinase in centrosomes of hemopoietic cells. *Blood Cells Mol Dis.,* 21(1), 34-41.

Pihan G.A., Doxsey S.J. (1999) The mitotic machinery as a source of genetic instability in cancer. *Semin Cancer Biol.,* 4, 289-302. doi: 10.1006/scbi.1999.0131.

Pihan G.A., Purohit A., Wallace J., Knecht H., Woda B., Quesenberry P., Doxsey S.J. (1998) Centrosome defects and genetic instability in malignant tumors. *Cancer Res.,* 58, 3974-3985.

Pihan G.A., Purohit A., Wallace J., Malhotra R., Liotta L., Doxsey S.J. (2001) Centrosome defects can account for cellular and genetic changes that characterize prostate cancer progression. *Cancer Res.,* 61, 2212-2219.

Pimenta-Marques A., Bento I., Lopes C.A., Duarte P., Jana S.C., and Bettencourt-Dias M. (2016) A mechanism for the elimination of the female gamete centrosome in *Drosophila melanogaster*. *Science,* 353, aaf4866.

Pinot M., Chesnel F., Kubiak J.Z., Arnal I., Nedelec F.J., Gueroui Z. (2009) Effects of confinement on the self-organization of microtubules and motors. *Curr. Biol.,* 19, 954-960.

Piperno G., Huang B., Ramanis Z., Luck D. J. (1981) «Radial spokes of Chlamydomonas flagella: polypeptide composition and phosphorylation of stalk components». *J. Cell Biol.,* 88, 73-79.

Pitaval A., Senger F., Letort G., Gidro,X., Guyon L., Sillibourne J., Théry M. (2017) Microtubule stabilization drives 3D centrosome migration to initiate primary ciliogenesis. *J. Cell Biol.,* 216, 3713-3728.

Pitaval A., Tseng Q., Bornens M., Théry M. (2010) Cell shape and contractility regulate ciliogenesis in cell cycle-arrested cells. *J. Cell Biol.,* 191, 303-312.

Pittenger M.F., Cleveland D.W. (1985) Retention of autoregulatory control of tubulin synthesis in cytoplasts: demonstration of a cytoplasmic mechanism that regulates the level of tubulin expression. *J. Cell Biol.,* 101(5 Pt 1), 1941-1952.

Plessmann U., Weber K. (1997) Mammalian sperm tubulin: an exceptionally large number of variants based on several posttranslational modifications. *J. Protein Chem.,* 16, 385-390.

Pockwinse S.M., Krockmalnic G., Doxsey S. J., Nickerson J., Lian J. B., van Wijnen A.J., Stein J.L., Stein G.S., Penman S. (1997) Cell cycle independent interaction of CDC2 with the centrosome, which is associated with the nuclear matrix-intermediate filament scaffold. *Proc. Natl. Acad. Sci. USA,* 94(7), 3022-3027.

Poole C.A., Flint M.H., Beaumont B.W. (1985) Analysis of the morphology and function of primary cilia in connective tissues: a cellular cybernetic probe? *Cell Motil. Cytoskeleton,* 5, 175-193.

Poole C.A., Jensen C.G., Snyder J.A., Gray C.G., Hermanutz V.L., Wheatley D.N. (1997) Confocal analysis of primary cilia structure and colocalization with the Golgi apparatus in chondrocytes and aortic smooth muscle cells. *Cell Biol. Int.,* 21, 483-494.

Porter K.R. (1966) Cytoplasmic microtubules and their functions. In "Principles of Biomolecular Organization" (eds. G.E.W. Wolstenholme and M. O'Connor). United Kingdom, London: Churchill (J. and A.) Ltd., 308-356.

Porter M.E., Sale, W.S. (2000) The 9+2 axoneme anchors multiple inner arm dyneins and a network of kinases and phosphatases that control motility. *J. Cell Biol.,* 151, F37-F42.

Portman R.W., LeCluyse E.L., Dentler W.L. (1987) Development of microtubule capping structures in ciliated epithelial cells. *J. Cell Sci.,* 87, 85-94.

Pouthas F., Girard P., Lecaudey V., Ly T.B., Gilmour D., Boulin C., Pepperkok R., Reynaud E.G. (2008) In migrating cells, the Golgi complex and the position of the centrosome depend on geometrical constraints of the substratum. *J. Cell Sci.,* 121, 2406-2414.

Praetorius H.A., Spring K.R. (2001) Bending the MDCK cell primary cilium increases intracellular calcium. *J. Membr. Biol.,* 184(1), 71-79.

Praetorius H.A., Spring K.R. (2003) The renal cell primary cilium functions as a flow sensor. *Curr. Opin. Nephrol. Hypertens.,* 12(5), 517-520.

Pratt M.B., Titlow J.S., Davis I., Barker A.R., Dawe H.R., Raff J.W., Roque H. (2016) Drosophila sensory cilia lacking MKS proteins exhibit striking defects in development but only subtle defects in adults. *J. Cell Sci.,* 129(20), 3732-3743.

Prescott D.M., Myerson D., Wallace J. (1972) Enucleation of mammalian cells with cytochalasin B. *Exp. Cell Res.,* 71(2), 480-485.

Prigent C., Uzbekov R. (2022) Duplication and Segregation of Centrosomes during Cell Division. *Cells,* 11(15), 2445. doi: 10.3390/cells11152445.

Prinzhorn W., Stehle M., Kleiner H., Ruppenthal S., Muller M.C., Hofmann W.K., Fabarius A, Seifarth W. (2016) c-MYB is a transcriptional regulator of ESPL1/Separase in BCR-ABL-positive chronic myeloid leukemia. *Biomark Res.,* 4, 5. doi: 10.1186/s 40364-016-0059-2

Prosser S.L., Pelletier L. (2020) Centriolar satellite biogenesis and function in vertebrate cells. *J. Cell Sci.,* 133, jcs239566.

Prudovskiĭ I.A., Zelenin A.V. (1978) Effect of enucleation on the structural and functional state of the cytoplasm in L cells. *Tsitologiia,* 20(8), 952-956.PMID: 726069 In Russian.

Przybylski R.J. (1971) Occurrence of centrioles during skeletal and cardiac myogenesis. *J. Cell Biol.,* 49, 214-221.

Pugacheva E.N., Jablonski S A., Hartman T.R., Henske E.P., Golemis E.A. (2007) HEF1-dependent Aurora A activation induces disassembly of the primary cilium. *Cell,* 129, 1351-1363.

Purohit A., Tynan S.H., Vallee R., Doxsey S.J. (1999) Direct interaction of pericentrin with cytoplasmic dynein light intermediate chain contributes to mitotic spindle organization. *J. Cell Biol.,* 147(3), 481-492.

Quarmby, L.M. (2004) Cellular deflagellation. *Int. Rev. Cytol.,* 233, 47-91.

Raff J.W. (2004) Centrosome in a developing organism: lessons from Drosophila. In "*Centrosomes in Development and Disease*" (ed. E.A. Nigg). Germany, Weinheim: Wiley-VCH Verlag Gmbh and Co, KGaA, 251-270.

Rafiq N.B.M., Nishimura Y., Plotnikov S.V., Thiagarajan V., Zhang Z., Shi S., Natarajan M., Viasnoff V., Kanchanawong P., Jones G.E., Bershadsky A.D. (2019) A mechano-signaling network linking microtubules, myosin IIA filaments and integrin-based adhesions. *Nature Materials,* 18, 638-649.

Rapley J., Baxter J.E., Blot J., Wattam S.L., Casenghi M., Meraldi P., Nigg E.A., Fry A.M. (2005) Coordinate regulation of the mother centriole component nlp by nek2 and plk1 protein kinases. *Mol. Cell Biol.,* 25(4), 1309-1324.

Rasi, M.Q., Parker, J.D., Feldman, J.L., Marshall, W.F., Quarmby, L.M. (2009) Katanin knockdown supports a role for microtubule severing in release of basal bodies before mitosis in Chlamydomonas. *Mol. Biol. Cell,* 20, 379-388.

Rath O., Kozielski F. (2012) Kinesins and cancer. *Nat. Rev. Cancer,* 12, 527-539. doi: 10. 1038/nrc3310.

Rattner J. B., Phillips S. G. (1973) Independence of centriole formation and DNA synthesis. *J. Cell Biol.,* 57(2), 359-372.

Rauch A., Thiel C.T., Schindler D., Wick U., Crow Y.J., Ekici A.B., van Essen A.J., Goecke T.O., Al-Gazali L., Chrzanowska K.H., Zweier C., Brunner H.G., Becker K., Curry C.J., Dallapiccola B., Devriendt K., Dörfler A., Kinning E., Megarbane A., Meinecke P., Semple R.K., Spranger S., Toutain A., Trembath R.C., Voss E, Wilson L., Hennekam R., de Zegher F., Dörr H-G., Reis A. (2008) Mutations in the Pericentrin (PCNT) Gene Cause Primordial Dwarfism. *Science,* 319, 816-819.

Reese T.S. (1965) Olfactory cilia in the frog. *J. Cell Biol.,* 25, 209-230.

Reilein A., Nelson W. J. (2005) APC is a component of an organizing template for cortical microtubule networks. *Nat. Cell Biol.,* 7, 463-473.

Reina J., Gottardo M., Riparbelli M.G., Llamazares S., Callaini G., Gonzalez C. (2018) Centrobin is essential for C-tubule assembly and flagellum development in Drosophila melanogaster spermatogenesis. *J. Cell Biol.,* 217(7), 2365-2372.

Reiter J.F., Leroux M.R. (2017) Genes and molecular pathways underpinning ciliopathies. *Nat. Rev. Mol. Cell Biol.*, 18, 533-547.

Reynolds E.S. (1963) The use of lead citrate at high pH as electron-opaque stain in electron microscopy. *J. Cell Biol.,* 17, 208.

Ried T., Heselmeyer-Haddad K., Blegen H., Schröck E., Auer G. (1999) Genomic changes defining the genesis, progression, and malignancy potential in solid human tumors: a phenotype/genotype correlation. *Genes Chromosomes Cancer,* 25(3), 195-204. doi: 10.1002/(sici)1098-2264(199907)25:3<195::aid-gcc1>3.0.co;2-8.

Rieder C. L. (1979) Ribonucleoprotein staining of centrioles and kinetochores in newt lung cell spindles. *J. Cell Biol.,* 80, 1-9.

Rieder C.L., Alexander S.P. (1990) Kinetochores are transported poleward along a single astral microtubule during chromosome attachment to the spindle in newt lung cells. *J. Cell Biol.,* 110, 81-96.

Rieder C.L., Borisy G.G. (1982) The centrosome cycle in PtK2 cells: asymmetric distribution and structural changes in pericentriolar material. *Biol. Cell,* 44, 117-132.

Rieder C.L., Jensen C.G., Jensen L.C. (1979) The resorption of primary cilia during mitosis in a vertebrate (PtK1) cell line. *J. Ultrastruct. Res.,* 68(2), 173-185.

Ringo D.L. (1967) Flagellar motion and fine structure of the flagellar apparatus in Chlamydomonas. *J. Cell Biol.,* 33, 543-571.

Ringo D.L. (1967) The arrangement of subunits in flagellar fibers. *J. Ultrastruct. Res.,* 17(3), 266-277.

Riparbelli M.G., Callaini G., Megraw T.L. (2012) Assembly and persistence of primary cilia in dividing Drosophila spermatocytes. *Dev. Cell,* 23(2), 425-432.

Riparbelli M.G., Callaini G., Mercati D., Hertel H., Dallai R. (2009) Centrioles to basal bodies in the spermiogenesis of Mastotermes darwiniensis (Insecta, Isopoda). *Cell Motil. Cytoskeleton,* 66(5), 1100-1105.

Riparbelli M.G., Dallai R., Callaini G. (2010) The insect centriole: A land of discovery. *Tissue Cell,* 42(2), 69-80.

Riparbelli M.G., Dallai R., Mercati D., Bu Y., Callaini G. (2009) Centriole symmetry: A big tale from small organisms. *Cell. Motil. Cytoskeleton,* 66(12), 1100-1105.

Riparbelli M.G., Persico V., Callaini G. (2018) A transient microtubule-based structure uncovers a new intrinsic asymmetry between the mother centrioles in the early *Drosophila* spermatocytes. *Cytoskeleton (Hoboken),* 75(11), 472-480.

Riparbelli M.G., Persico V., Callaini G. (2021) Early *Drosophila* Oogenesis: A Tale of Centriolar Asymmetry. *Cells,* 10(8), 1997.

Riparbelli M.G., Persico V., Dallai R., Callaini G. (2020) Centrioles and Ciliary Structures during Male Gametogenesis in *Hexapoda*: Discovery of New Models. *Cells,* 9(3), 744.

Riparbelli M.G., Pratelli A., Callaini G. (2024) The cilium like region of the *Drosophila bifurca* spermatocyte: Elongation of a giant axoneme without intraflagellar transport. *Cytoskeleton (Hoboken),* 81(11), 529-538.

Robbins E., Gonatas N.K. (1964) The ultrastructure of a mammalian cell during the mitotic cycle. *J. Cell Biol.,* 21, 429-463.

Robbins E., Jentzsch G., Micali A. (1968) The centriole cycle in synchronized HeLa cells. *J. Cell Biol.,* 36, 329-339.

Robinson S.R., Oliver A.W., Chevassut T.J., Newbury S.F. (2015) The 3' to 5' exoribonuclease DIS3: from structure and mechanisms to biological functions and role in human disease. *Biomolecules,* 5(3), 1515-1539.

Rodionov V., Nadezhdina E., Borisy G. (1999) Centrosomal control of microtubule dynamics. *Proc. Natl. Acad. Sci. USA,* 96(1), 115-120.

Rodionov V.I., Hope A.J., Svitkina T.M., Borisy G.G. (1998) Functional coordination of microtubule-based and actin-based motility in melanophores. *Curr. Biol.,* 8(3), 165-168.

Rodionov V.I., Lim S., Gelfand V.I., Borisy G.G. (1994) Microtubule dynamic in fish melanofores. *J. Cell Biol.,* 126, 1455-1464.

Rodrigues-Martins A., Riparbelli M., Callaini G., Glover D.M., Bettencourt-Dias M. (2007) Revisiting the role of the mother centriole in centriole biogenesis. *Science,* 316(5827), 1046-1050.

Rodrigues-Martins A., Riparbelli M., Callaini G., Glover D.M., Bettencourt-Dias M. (2008) From centriole biogenesis to cellular function: centrioles are essential for cell division at critical developmental stages. *Cell Cycle,* 7(1), 11-16.

Rogers G.C., Rusan N.M., Roberts D.M., Peifer M., Rogers S.L. (2009) The SCF Slimb ubiquitin ligase regulates Plk4/Sak levels to block centriole reduplication. *J. Cell Biol.,* 184(2), 225-239.

Rogers K.A., McKee N.H., Kalnins V.I. (1985) Preferential orientation of centrioles toward the heart in endothelial cells of major blood vessels is reestablished after reversal of a segment. *Proc. Nat. Acad. Sci. USA,* 82, 3272-3276.

Roghi C., Giet R., Uzbekov R., Morin N., Chartrain I., Le Guellec R., Couturier A., Doree M., Philippe M., Prigent C. (1998) The Xenopus Protein Kinase pEg2 Associated with the Centrosome in the Cell Cycle-Dependent Manner, bind to the Spindle Microtubules and is involved in Bipolar Mitotic Spindle Assembly. *J. Cell Sci.,* 111, 557-572.

Roignant J.-Y. Treisman J.E. (2009) Pattern formation in the Drosophila eye disc. *Int. J. Dev. Bio.,* 53(5-6), 795-804.

Roman A.C., Garrido-Jimenez S., Diaz-Chamorro S., Centeno F., Carvajal-Gonzalez J.M. (2019) Centriole Positioning: Not Just a Little Dot in the Cell. *Results and Problems in Cell Differentiation,* 67, 201-221.

Roos U.P. (1973). Light and electron microscopy of rat kangaroo cells in mitosis. I. Formation and breakdown of the mitotic apparatus. *Chromosoma,* 40 (1), 43-82.

Rosales J.L., Rattner J.B., Lee K-Y. (2010) Cdk5 in the centriolar appendages mediatescenexin1 localization and primary cilia formation. *Cell Cycle,* 9(10), 2037-2039.

Ross L., Normark B.B. (2015) Evolutionary problems in centrosome and centriole biology. *J. Evol. Biol.,* 28(5), 995-1004.

Roth K.E., Rieder C.L., Bowser S.S. (1988) Flexible-substratum technique for viewing cells from the side: some *in vivo* properties of primary (9+0) cilia in cultured kidney epithelia. *J. Cell Sci.*, 89 (Pt 4), 457-466. doi: 10.1242/jcs.89.4.457.

Roth L.E., Shigenaka Y. (1964) The structure and formation of cilia and filaments in rumen protozoa. *J. Cell Biol.,* 20, 249-270.

Ruffner H., Jiang W., Craig A.G., Hunter T., Verma I.M. (1999) BRCA1 is phosphorylatedat serine 1497 *in vivo* at a cyclin-dependent kinase 2 phosphorylation site. *Mol. Cell Biol.,* 19, 4843-4854.

Ruiz F., Krzywicka A., Klotz C., Keller A., Cohen J., Koll F., Balavoine G., Beisson J. (2000) The SM19 gene, required for duplication of basal bodies in Paramecium, encodes a novel tubulin, eta-tubulin. *Curr. Biol.,* 10(22), 1451-1454.

Rumyantsev P.P., Snigirevskaya E. S. (1968) The ultrastructure of differetiating cells of the heart muscle in the state of mitotic division. *Acta. Morphol. Acad. Sci. Hung.,* 16 (3), 271-283.

Ruppenthal S., Kleiner H., Nolte F., Fabarius A., Hofmann W.K., Nowak D., Seifarth W. (2018) Increased separase activity and occurrence of centrosome aberrations concur with transformation of MDS. *PLoS One,* 2513(1), e0191734. doi: 10.1371/journal.pone.0191734.

Ruthmann A. (1959) The fine structure of the meiotic spindle of the crayfish. J. *Biophys. Biochem. Cytol.,* 5, 177-180.

Saavedra H.I., Maiti B., Timmers C., Altura R., Tokuyama Y., Fukasawa K., Leone G. (2003) Inactivation of E2F3 results in centrosome amplification. *Cancer Cell,* 3, 333-346.

Sabatini D.D., Bensch K.G., Barnett R.J. (1963) Cytochemistry and electron microscopy. The preservation of cellular ultrastructure and enzymatic activity by aldehyde fixation. *J. Cell Biol.,* 17, 19-58.

Safieddine A., Coleno E., Salloum S., Imbert A., Traboulsi A-M., Kwon O.S., Lionneton F., Georget V., Robert M-C., Gostan T., Lecellier C-H., Chouaib R., Pichon X., Le Hir H., Zibara K., Mueller F., Walter T., Peter M., Bertrand E. (2021) A choreography of centrosomal mRNAs reveals a conserved localization mechanism involving active polysome transport. *Nat. Commun.,* 2(1), 1352. doi: 10.1038/s41467-021-21585-7.

Sager P.R., Rothfield N.L., Oliver J.M., Berlin R.D. (1986) A novel mitotic spindle pole component that originated from the cytoplasm during prophase. *J. Cell Biol.,* 103, 1863-1872.

Sakaguchi H. (1965) Pericentriolar filamentous bodies. *J. Ultrastr. Res.,* 12, 13-21.

Sakharov V.N., Voronkova L.N. (1993) The kinetics of the transition to DNA synthesis in the cell cycle of sister cells of the ESK line in culture. *Tsitologiia,* 35(6-7), 79-85. In Russian.

Salim A., Werther P., Hatzopoulos G.N., Reymond L., Wombacher R., Gönczy P., Johnsson K. (2023) Chemical Probe for Imaging of Polo-like Kinase 4 and Centrioles. *JACS Au.,* 3(8), 2247-2256. doi: 10.1021/jacsau.3c00271.

Salisbury J.L. (2003) Centrosome size is controlled by centriolar SAS-4. *Trends Cell Biol.,* 13(7), 340-343. Review.

Salisbury J.L., Floyd G.L. (1978) Calcium-induced contraction of the rhizoplast of a quadriflagellate green alga. *Science,* 202, 975-977.

Salmon W.C., Adams M.C., Waterman-Storer C.M. (2002) Dual-wavelength fluorescent speckle microscopy reveals coupling of microtubule and actin movements in migrating cells. *J. Cell Biol.,* 158, 31-37.

Salpingidou G., Smertenko A., Hausmanowa-Petrucewicz I., Hussey P.J., Hutchison C.J. (2007) A novel role for the nuclear membrane protein emerin in association of the centrosome to the outer nuclear membrane. *J. Cell Biol.,* 178, 897-904.

Salwini-Plawen L. (1982) On the phylogenetic origin of photoreceptors. In *" Visual cells in evolution"* (ed. J.A. Westffall). USA, New York: Raven Press, 137-154.

Sanchez E., Liu X., Huse M. (2019) Actin clearance promotes polarized dynein accumulation at the immunological synapse. *PLoS ONE,* 14, e0210377.

Sánchez-Martínez C., Lallena M.J., Sanfeliciano S.G., de Dios A. (2019) Cyclin dependent kinase (CDK) inhibitors as anticancer drugs: recent advances (2015-2019). *Bioorg. Med. Chem. Lett.,* 29, 126637. doi: 10.1016/j.bmcl.2019.126637.

Sancho D., Vicente-Manzanares M., Mittelbrunn M., Montoya M.C., Gordón-Alonso M., Serrador J.M., Sánchez-Madrid F. (2002) Regulation of microtubule-organizing center orientation and actomyosin cytoskeleton rearrangement during immune interactions. *Immunological Reviews,* 189, 84-97.

Sankaran S., Starita L.M., Simons A.M., Parvin J.D. (2006) Identification of domains of BRCA1 critical for the ubiquitin-dependent inhibition of centrosome function. *Cancer Res.* 66(8), 4100-4107. doi: 10.1158/0008-5472.CAN-05-4430.

Sarkar A., Rieger H. (2019) Search and Capture Efficiency of Dynamic Microtubules for Centrosome Relocation during IS Formation. *Biophysical J.* 116, 2079-2091.

Sasai K., Parant J.M., Brandt M.E., Carter J., Adams H.P., Stass S.A., Killary A.M., Katayama H., Sen S. (2008) Targeted disruption of Aurora A causes abnormal mitotic spindle assembly, chromosome misalignment and embryonic lethality. *Oncogene,* 27(29), 4122-4127.

Sasaki S., Shionoya A., Ishida M., Gambello M.J., Yingling J., Wynshaw-Boris A., Hirotsune S. (2000) A LIS1/NUDEL/cytoplasmic dynein heavy chain complex in the developing and adult nervous system. *Neuron,* 28(3), 681-696.

Sathananthan A.H. (1997) Ultrastructure of the human egg. *Hum. Cell,* 10(1), 21-38. Review.

Sathananthan A.H. (1998) Paternal centrosomal dynamics in early human development and infertility. *J. Assist. Reprod. Genet.,* 15(3), 129-139. Review.

Sathananthan A.H., Kola I., Osborne J., Trounson A., Ng S.C., Bongso A., Ratnam S.S. (1991) Centrioles in the beginning of human development. *Proc. Natl. Acad. Sci. USA,* 88(11), 4806-4810.

Sathananthan A.H., Ratnam S.S., Ng S.C., Tarín J.J., Gianaroli L., Trounson A. (1996) The sperm centriole: its inheritance, replication and perpetuation in early human embryos. *Hum. Reprod.,* 11(2), 345-356.

Sato N., Mizumoto K., Nakamura M., Nakamura K., Kusumoto M., Niiyama H., Ogawa T., Tanaka M. (1999) Centrosome abnormalities in pancreatic ductal carcinoma. *Clin. Cancer Res.,* 5, 963-970. PMID:10353727.

Sauer C.M., Hall J.A., Couturier D.L., Bradley T., Piskorz A.M., Griffiths J., Sawle A., Eldridge M.D., Smith P., Hosking K., Reinius M.A.V., Morrill Gavarró L., Mes-Masson A.M., Ennis D., Millan D., Hoyle A., McNeish I.A., Jimenez-Linan M., Martins F.C., Tischer J., Vias M., Brenton J.D. (2023) Molecular landscape and functional characterization of centrosome amplification in ovarian cancer. *Nat. Commun.,* 14(1), 6505. doi: 10.1038/s41467-023-41840-3.

Sawin K.E., Le Guellec K., Philippe M., Mitchison T.J. (1992) Mitotic spindle organization by a plus-end-directed microtubule motor. *Nature,* 359, 540-543.

Schaeffer A., Buracco S., Gazzola M., Gelin M., Vianay B., de Pascalis C., Blanchoin L., Théry T. (2025) Microtubule-driven cell shape changes and actomyosin flow synergize to position the centrosome. *J. Cell Biol.,* 224(7), e202405126.

Schafer D.A., Gill S.R., Cooper J.A., Heuser J.E., Schroer T.A. (1994) Ultrastructural analysis of the dynactin complex: an actin-related protein is a component of a filament that resembles F-actin. *J. Cell Biol.,* 126(2), 403-412.

Schatten G., Simerly C., Schatten H. (1985). Microtubule configuration during fertilization, mitosis and early development in the mouse. *Proc. Nat. Acad. Sci. USA,* 82, 4152-4156.

Schatten H. (2013) The Impact of Centrosome Abnormalities on Breast Cancer Development and Progression with a Focus on Targeting Centrosomes for Breast Cancer Therapy. In *"Cell and Molecular Biology of Breast Cancer"* (ed. H. Schatten). USA, New Jersey: Humana Press, 261-287.

Schatten H., Sun QY. (2009). The role of centrosomes in mammalian fertilization and its significance for ICSI. *Mol. Hum. Reprod.,* 5(9), 531-538.

Schatten H., Walter M., Biessmann H., Schatten G. (1988) Microtubules are required for centrosome separation in sea urchin eggs during fertilization and mitosis. *Cell Motil. Cytoskeleton,* 11, 248-259.

Scheer U. (2014) Historical roots of centrosome research: Discovery of Boveri's microscope slides in Würzburg. *Philos. Trans. of the Royal Society B,* 369, 20130469.

Schleiden M.J. (1838) Beitrage zur Phytogenesis [Contributions to phytogenesis]. *Muller's Archiv*. Translation *in Sydenham Soc., XII. London, 1847.* In German. Citation from Wilson (1900).

Schmidt T.I., Kleylein-Sohn J., Westendorf J., Le Clech M., Lavoie S.B., Stierhof Y.D., Nigg E.A. (2009) Control of centriole length by CPAP and CP110. *Curr Biol.,* 19(12), 1005-1011.

Scholey J.M. (2003). Intraflagellar transport. *Annu. Rev. Cell Dev. Biol.,* 19, 423-443.

Schreiner A., Schreiner K.E. (1905) Űber die Entwicklung der männlichen Geschlechtszellen von Myxine glutinosa (L.) [On the development of male sex cells of Myxine glutinosa (L.)]. *Arch. Biol.,* 21, 183-357. In German.

Schrøder J.M., Schneider L., Christensen S.T., Pedersen L.B. (2007) EB1 is required for primary cilia assembly in fibroblasts. *Curr. Biol.,* 17(13), 1134-1139.

Schulze E, Kirschner M. (1987) Dynamic and stable populations of microtubules in cells. *J. Cell Biol.,* 104(2), 277-288.

Schwann Th. (1839) Mikroscopische Untersuchungen uber die Uebereinstimmung in der Structur und dem Wachsthum der Thiere und Pflanzen [Microscopic investigations into the similarity in structure and growth of animals and plants]. Germany, Berlin. Translation in Sydenham Soc., XII. London, 1847. In German. Citation from Wilson (1900).

Schwartz E.A., Leonard M.L., Bizios R., Bowser S.S. (1997) Analysis and modeling of the primary cilium bending response to fluid shear. *Am. J. Physiol.,* 272(1 Pt 2), F132-138. doi: 10.1152/ajprenal.1997.272.1.F132.

Sedjaï F., Acquaviva C., Chevrier V., Chauvin J.P., Coppin E., Aouane A., Coulier F., Tolun A., Pierres M., Birnbaum D., Rosnet O. (2010) Control of ciliogenesis by FOR20, a novel centrosome and pericentriolar satellite protein. *J. Cell Sci.,* 123(14), 2391-2401.

Seeger-Nukpezah T., Golemis E.A. (2012) The extracellular matrix and ciliary signaling. *Curr. Opin. Cell Biol.,* 24(5), 652-661. doi: 10.1016/j.ceb.2012.06.002.

Seetharaman S., Etienne-Manneville S. (2020) Cytoskeletal Crosstalk in Cell Migration. *Trends in Cell Biol.,* 30, 720-735.

Seetharaman S., Vianay B., Roca V., Farrugia A.J., De Pascalis C., Boëda B., Dingli F., Loew D., Vassilopoulos S., Bershadsky A.D., Théry M., Etienne-Manneville S. (2022) Microtubules tune mechanosensitive cell responses. *Nat. Mater.,* 21(3), 366-377. doi: 10.1038/s41563-021-01108-x.

Sekar T., Sebire N.J. (2024) Renal Pathology of Ciliopathies. *Pediatr. Dev. Pathol.,* 27(5), 411-425. doi: 10.1177/10935266241242173

Selby C.C. (1953) Electron micrographs of mitotic cells of the Ehrluch mouse ascites tumor in thin section. *Exp. Cell Res.,* 5(2), 386-393.

Seo S., Zhang Q., Bugge K., Breslow D.K., Searby C.C., Nachury M.V., Sheffield V.C. (2011) A novel protein LZTFL1 regulates ciliary trafficking of the BBSome and Smoothened. *PLoS Genet.,* 7, e1002358.

Shao S., Wang Y., Jin S., Song Y., Wang X., Fan W., Zhao Z., Fu M., Tong T., Dong L., Fan F., Xu N., Zhan Q. (2006) Gadd45a interacts with aurora-A and inhibits its kinase activity. *J. Biol. Chem.,* 281(39), 28943-28950.

Sharma A., Markey M., Torres-Munoz K., Varia S., Kadakia M., Bubulya A., Bubulya P.A. (2011) Son maintains accurate splicing for a subset of human pre-mRNAs. *J. Cell Sci.,* 124, 4286-4298.

Sharp D.J., McDonald K.L., Brown H.M., Matthies H.J., Walczak C., Vale R.D., Mitchison T.J., Scholey J.M. (1999) The bipolar kinesin, KLP61F, cross-links microtubules within interpolar microtubule bundles of Drosophila embryonic mitotic spindles. *J. Cell Biol.,* 44(1), 125-38. doi: 10.1083/jcb.144.1.125.

Shastri A., Will B., Steidl U., Verma A. (2017) Stem and progenitor cell alterations in myelodysplastic syndromes. *Blood,* 129(12), 1586-1594. doi: 10.1182/blood-2016-10-696062

Shay J.W., Porter K.R., Prescott D.M. (1974) The surface morphology and fine structure of CHO (Chinese hamster ovary) cells following enucleation. *Proc. Natl. Acad. Sci. USA,* 71(8), 3059-3063. doi: 10.1073/pnas.71.8.3059.

Shen H., Yang M., Li S., Zhang J., Peng B., Wang C., Chang Z., Ong J., Du P. (2021) Mouse totipotent stem cells captured and maintained through spliceosomal repression. *Cell,* 184(11), 2843-2859. e20. doi: 10.1016/j.cell.2021.04.020.

Shin I., Yakes F.M., Rojo F., Shin N.Y., Bakin A.V., Baselga J., Arteaga C.L. (2002) PKB/Akt mediates cell-cycle progression by phosphorylation of p27(Kip1) at threonine 157 and modulation of its cellular localization. *Nat. Med.,* 8, 1145-1152. doi: 10.1038/nm759.

Shinar T., Mana M., Piano F., Shelley M.J. (2011) A model of cytoplasmically driven microtubule-based motion in the single-celled Caenorhabditis elegans embryo. *Proc. Nat. Acad. Sci. USA,* 108, 10508-10513.

Shinmura K., Bennett R.A., Tarapore P., Fukasawa K. (2007) Direct evidence for the role of centrosomally localized p53 in the regulation of centrosome duplication. *Oncogene,* 26(20), 2939-2944.

Shukla A., Kong D., Sharma M., Magidson V. Loncarek J. (2015) Plk1 relieves centriole block to reduplication by promoting daughter centriole maturation. *Nat. Commun.,* 6, 8077.

Sibon O.C., Kelkar A., Lemstra W., Theurkauf W.E. (2000) DNA-replication/DNA-damage-dependent centrosome inactivation in Drosophila embryos. *Nat. Cell Biol.,* 2, 90-95.

Silflow C.D., Liu B., LaVoie M., Richardson E.A., Palevitz, B.A. (1999) Gamma-tubulin in Chlamydomonas: characterization of the gene and localization of the gene product in cells. *Cell Motil. Cytoskeleton,* 42, 285-297.

Siller S.S., Sharma H., Li S., Yang J., Zhang Y., Holtzman M.J., Winuthayanon W., Colognato H., Holdener B.C., Li F.Q., Takemaru K-I. (2017) Conditional knockout mice for the distal appendage protein cep164 reveal its essential roles in airway multiciliated cell differentiation. *PLoS Genet.,* 13, e1007128.

Sillibourne J.E., Milne D.M., Takahashi M., Ono Y., Meek D.W. (2002) Centrosomal anchoring of the protein kinase CK1delta mediated by attachment to the large, coiled-coil scaffolding protein CG-NAP/AKAP450. *J. Mol. Biol.,* 322(4), 785-797.

Silver D.L., Watkins-Chow D. E., Schreck K. C., Pierfelice T. J., Larson D. M., Burnetti A. J., Liaw H.-J., Myung K., Walsh C. A., Gaiano N., Pavan W.J. (2010). The exon junction complex component Magoh controls brain size by regulating neural stem cell division. *Nat. Neurosci.,* 13(5), 551-558. doi: 10.1038/nn.2527.

Silverman-Gavrila, R.V., Forer, A. (2000) Evidence that actin and myosin are involved in the poleward flux of tubulin in metaphase kinetochore microtubules of crane-fly spermatocytes. *J. Cell Sci.,* 113, 597-609.

Singh C.K., Denu R.A., Nihal M., Shabbir M., Garvey D.R., Huang W., Iczkowski K.A., Ahmad N. (2022) PLK4 is upregulated in prostate cancer and its inhibition reduces centrosome amplification and causes senescence. *Prostate,* 82(9), 957-969. doi: 10.1002/pros.24342.

Singh G., Kucukural A., Cenik C., Leszyk J.D., Shaffer S.A., Weng Z., Moore M.J. (2012) The cellular EJC interactome reveals higher-order mRNP structure and an EJC-SR protein nexus. *Cell,* 151, 750-764.

Sir J. H., Barr A.R., Nicholas A.K., Carvalho O.P., Khurshid M., Sossick A., Reichelt S., D'Santos C., Woods C.G., Gergely F. (2011) A Primary Microcephaly Protein Complex Forms a Ring around Parental Centrioles. *Nat. Genet.,* 43, 1147-1153.

Sir J.H., Barr A.R., Nicholas A.K., Carvalho O.P., Khurshid M., Sossick A., Reichelt S., D'Santos C., Woods C.G., Gergely F. (2011) A primary microcephaly protein complex forms a ring around parental centrioles. *Nat. Genet.,* 43(11), 1147-1153.

Skop A.R., White J.G. (1998) The dynactin complex is required for cleavage plane specification in early Caenorhabditis elegans embryos. *Curr. Biol.,* 8, 1110-1116.

Slankis T., Gibbs S.P. (1972) The fine structure of mitosis and cell division in the chrysophycean alga Ochromonas danica. *J. Phycol.,* 8, 213-256.

Slautterback D.B. (1963) Cytoplasmic microtubules. I. Hydra. *J. Cell Biol.,* 18, 367-388.

Sluder G. (1989) Centrosomes and the cell cycle. *J. Cell Sci. Suppl.,* 12, 253-275.

Sluder G. (2016) Using sea urchin gametes and zygotes to investigate centrosome duplication. *Cilia,* 5, 20.

Sluder G., Begg D.A. (1985) Experimental analysis of the reproduction of spindle poles. *J. Cell Sci.,* 76, 35-51.

Smith S.B., Kiss D.L., Turk E., Tartakoff A.M., Andrulis E.D. (2011) Pronounced and extensive microtubule defects in a Saccharomyces cerevisiae DIS3 mutant. *Yeast Chichester Engl.,* 28(11), 755-769. doi: 10.1002/yea.1899.

Snee M.J., Wilson W.C., Zhu Y., Chen S.Y., Wilson B.A., Kseib C., O'Neal J., Mahajan N., Tomasson M.H., Arur S., Skeath J.B. (2016) Collaborative control of cell cycle progression by the RNA exonuclease Dis3 and Ras is conserved across species. *Genetics,* 203(2), 749-762.

Snell W.J., Dentler W.L., Haimo L.T., Binder L.I., Rosenbaum J.L. (1974) Assembly of chick brain tubulin onto isolated basal bodies of Chlamydomonas reinhardi. *Science,* 185(148), 357-360.

Snow J.J., Ou G., Gunnarson A.L., Walker M.R., Zhou H.M., Brust-Mascher I., Scholey J.M. (2004) Two anterograde intraflagellar transport motors cooperate to build sensory cilia on C. elegans neurons. *Nat. Cell Biol.,* 6, 1109-1113.

Snyder J.A., McIntosh J.R. (1975) Initiation and growth of microtubules from mitotic centers in lysed mammalian cells. *J. Cell Biol.,* 67, 744-760.

Somlo S., Ehrlich B. (2001) Human disease: calcium signaling in polycystic kidney disease. *Curr Biol.,* 11, R356-360.

Song M.H., Aravind L., Müller-Reichert T., O'Connell K.F. (2008) The conserved protein SZY-20 opposes the Plk4-related kinase ZYG-1 to limit centrosome size. *Dev. Cell.,* 15(6), 901-912.

Song S., Jung S., Kwon M. (2023) Expanding roles of centrosome abnormalities in cancers. *BMB Rep.,* 56(4), 216-224.

Sonn S., Oh G.T., Rhee K. (2011) Nek2 and its substrate, centrobin/Nip2, are required for proper meiotic spindle formation of the mouse oocytes. *Zygote,* 19(1), 15-20.

Sorokin S. (1962) Centrioles and the formation of rudimentary cilia by fibroblasts and smooth muscle cells. *J. Cell Biol.,* 15, 363-377.

Sorokin S.P. (1968) Reconstructions of centriole formation and ciliogenesis in mammalian lungs. *J. Cell Sci.,* 3, 207-230.

Sotelo J.R., Trujillo-Cenoz O. (1958) Electron microscope study on the development of ciliary components of the neural epithelium of the chick embryo. *Z. Zellforsch. Mikrosk. Anat.,* 49, 1-12.

Soung N.K., Park J.E., Yu L.R., Lee K.H., Lee J.M., Bang J.K., Veenstra T.D., Rhee K., Lee K. S. (2009). Plk1-dependent and -independent roles of an ODF2 splice variant, hCenexin1, at the centrosome of somatic cells. *Dev. Cell,* 16(4), 539-550.

Spektor A., Tsang W.Y., Khoo D., Dynlacht B.D. (2007) Cep97 and CP110 suppress a cilia assembly program. *Cell,* 130(4), 678-690.

Sperling A.S, Gibson C.J, Ebert B.L. (2017) The genetics of myelodysplastic syndrome: from clonal haematopoiesis to secondary leukaemia. *Nat. Rev. Cancer,* 17(1), 5-19. doi: 10.1038/nrc.2016.112.

Spring L.M., Wander S.A., Andre F., Moy B., Turner N.C., Bardia A. (2020) Cyclin-dependent kinase 4 and 6 inhibitors for hormone receptor-positive breast cancer: past, present, and future. *Lancet,* 395, 817-827. doi: 10.1016/S0140-6736(20)30165-3.

Spurck T.P., Stonington O.G., Snyder J.A., Pickett-Heaps J.D., Bajer A., Mole-Bajer J. (1990) UV microbeam irradiations of the mitotic spindle. II. Spindle fiber dynamics and force production. *J. Cell Biol.,* 111(4), 1505-1518.

Sredni S.T, Bailey A.W., Suri A., Hashizume R., He X., Louis N., Gokirmak T., Piper D.R., Watterson D.M., Tomita T. (2017b). Inhibition of polo-like kinase 4 (PLK4): a new therapeutic option for rhabdoid tumors and pediatric medulloblastoma. *Oncotarget.,* 8 (67), 111190-111212. doi:10.18632/oncotarget.22704.

Sredni S.T., Suzuki M., Yang J.P., Topczewski J., Bailey A.W., Gokirmak T., Gross J.N., de Andrade A., Kondo A., Piper D.R., Tomita T. (2017a) A functional screening of the kinome identifies the Polo-like kinase 4 as a potential therapeutic target for malignant rhabdoid tumors, and possibly, other embryonal tumors of the brain. *Pediatric Blood and Cancer,* 64 (11), e 26551. doi:10.1002/pbc.26551.

Sreng L., Quennedey A. (1976) Role of a temporary ciliary structure in the morphogenesis of insect glands. An electron microscope study of the tergal glands of male *Blattella germanica L.* (*Dictyoptera, Blattellidae*). *J. Ultrastruct. Res.,* 56(1), 78-95.

St Johnston D., Sanson B. (2011) Epithelial polarity and morphogenesis. *Curr. Opin. Cell Biol.,* 23, 540-546.

Starostina N.G., Kipreos E.T. (2012) Multiple degradation pathways regulate versatile CIP/KIP CDK inhibitors. *Trends Cell Biol.,* 22, 33-41. doi: 10.1016/j.tcb.2011.10.004.

Steffen W., Linck R.W. (1988). Evidence for tektins in centrioles and axonemal microtubules. *Proc. Natl. Acad. Sci. USA,* 85(8), 2643-2647.

Steib E., Gambarotto D., Laporte M.H., Olieric N., Zheng C., Borgers S., Olieric V., Le Guennec M., Koll F., Tassin A.-M., Steinmetz M.O., Guichard P., Hamel V. (2020) WDR90 is a centriolar microtubule wall protein important for centriole architecture integrity. *Elife,* 9, e57205. doi: 10.7554/eLife.57205.

Steinman J.B., Santarossa C.C., Miller R.M., Yu L.S., Serpinskaya A.S., Furukawa H., Morimoto S., Tanaka,Y., Nishitan M., Asano M., Zalyte R., Ondrus A.E., Johnson A.G., Ye F., Nachury M.V., Fukase Y., Aso K., Foley M.A., Gelfand V.I., Chen J.K., Carter A.P., Kapoor T.M. (2017) Chemical structure-guided design of dynapyrazoles, cell-permeable dynein inhibitors with a unique mode of action. *eLife,* 6, 1-36.

Stemm-Wolf A. J., O'Toole E. T., Sheridan R. M., Morgan J. T., Pearson C. G. (2021). The SON RNA splicing factor is required for intracellular trafficking structures that promote centriole assembly and ciliogenesis. *Mol. Biol. Cell,* 32(20):ar4. doi: 10.1091/mbc.E21-06-0305.

Stephens R.E., Lamieux N.A. (1998) Tektins as structural determinants in basal bodies. *Cell Motil. Cytoskeleton,* 40, 379-392.

Stevens N.R., Roque H., Ra J.W. (2010) Sas-6 and Ana2 coassemble into tubules to promote centriole duplication and engagement. *Dev. Cell,* 19(6), 913-919.

Stinchcombe J.C., Majorovits E., Bossi G., Fuller S., Griffiths G.M. (2006) Centrosome polarization delivers secretory granules to the immunological synapse. *Nature,* 443, 462-465.

Stingele J., Bellelli R., Boulton S.J. (2017) Mechanisms of DNA-protein crosslink repair. *Nat. Rev. Mol. Cell Biol.,* 18, 563-573.

Stockinger L., Cireli E. (1965) Eine bisher unbekannte art der zentriolenvermehrung [A previously unknown type of centriole proliferation]. *Zeitschrift fur Zellforschung und mikroskopische anatomie*, 68, 733-740. In German.

Strnad P., Gonczy P. (2008) Mechanisms of procentriole formation. *Trends in Cell Biology,* 18(8), 389-396.

Strnad P., Leidel S., Vinogradova T., Euteneuer U., Khodjakov A., Gönczy P. (2007) Regulated HsSAS-6 levels ensure formation of a single procentriole per centriole during the centrosome duplication cycle. *Dev. Cell,* 13(2), 203-213.

Stubblefield E. (1968). Centriole Replication in a Mammalian Cell. In *"The Proliferation and Spread of Neoplastic Cells"*. USA, Baltimore: The Williams and Wilkins Co. 175-193.

Stubblefield E., Brinkley B.R. (1967) Architecture and Function of the Mammalian Centriole. In *"Formation and Fate of Cell organelles"* (ed. K.B. Warren), USA, New York: Academic Press Inc. 175-218.

Subbiah A., Caswell D.L., Turner K., Jaiswal A., Avidor-Reiss T. (2024) CP110 and CEP135 Localize Near the Proximal Centriolar Remnants of Mice Spermatozoa. *MicroPubl Bio.,* 10.17912/micropub.biology.001083. doi: 10.17912/micropub.biology.001083.

Such J. (1969) Concerning the presence of structures resembling rudimentary abortive cilia in the retinal cells of the young embryo of *Carausius morosus. Br. C R Acad. Sci. Hebd Seances Acad. Sci. D.,* 268 (6), 948-949.

Sullenberger C., Vasquez-Limeta A., Kong D., Loncarek J. (2020) With Age Comes Maturity: Biochemical and Structural Transformation of a Human Centriole in the Making. *Cells,* 9(6), 1429. doi: 10.3390/cells9061429.

Sullivan W. (2015) The institute for the study of non-model organisms and other fantasies. *Mol. Biol. Cell,* 26, 387-389.

Sulston J.E., Albertson D.G., Thomson J.N. (1980) The caenorhabditis elegans male: Postembryonic development of nongonadal structures. *Dev. Biol.,* 78, 542-576.

Sun Y. (2006) p53 and its downstream proteins as molecular targets of cancer. *Mol. Carcinog.,* 45(6), 409-415.

Suri A., Bailey A.W., Tavares M.T., Gunosewoyo H., Dyer C.P., Grupenmacher A.T., Piper D.R., Horton R.A., Tomita T., Kozikowski A.P., Roy S.M., Sredni S.T. (2019). Evaluation of Protein Kinase Inhibitors with PLK4 Cross-Over Potential in a Pre-Clinical Model of Cancer. *Int. J. Mol. Sci.,* 20 (9), 2112. doi:10.3390/ijms20092112.

Susalka S.J., Nikulina K., Salata M.W., Vaughan P.S., King S.M. Vaughan K.T., Pfister K.K. (2002) The roadblock light chain binds a novel region of the cytoplasmic Dynein intermediate chain. *J. Biol. Chem.,* 277(36), 32939-32946.

Susalka S.J., Nikulina K., Salata M.W., Vaughan P.S., King S.M. Vaughan K.T., Pfister K.K. (2002) The roadblock light chain binds a novel region of the cytoplasmic Dynein intermediate chain. *J. Biol. Chem.,* 277(36), 32939-32946. doi: 10.1074/jbc. M205510200.

Sutrisno S., Marlina D., Tjandraprawira K.D., Adinda Adriansyah P.N. (2024) The impact of pathogenic BRCA1/2 tumor mutation status on high grade serous epithelial ovarian cancer survival outcome: A multicenter study from Indonesia. *SAGE Open Med.,* 12, 20503121241299849. doi: 10.1177/20503121241299849.

Sydor A.M., Coyaud E., Rovelli C., Laurent E., Liu H., Raught B., Mennella V. (2018) PPP1R35 is a novel centrosomal protein that regulates centriole length in concert with the microcephaly protein rttn. *eLife*, 7, e37846.

Szaraz D., Danek Z., Lipovy B., Krivanek J., Buchtova M., Putnova Moldovan B., Putnova I., Stembirek J., Andrasina T., Divacka P., Holla L.I., Linhartova P.B (2024) Primary cilia and hypoxia-associated signaling in developmental odontogenic cysts in relation to autosomal dominant polycystic kidney disease - A novel insight. *Heliyon*, 9(6), e17130. doi: 10.1016/j.heliyon.2023.e17130.

Szebenyi G., Hall B., Yu R., Hashim A.I., Krämer H. (2007) Hook2 localizes to the centrosome, binds directly to centriolin/CEP110 and contributes to centrosomal function. *Traffic*, 8(1), 32-46.

Szöllösi A., Ris H., Szöllösi D., Debec A. (1986) A centriole-free Drosophila cell line. A high voltage EM study. *Eur. J. Cell Biol.,* 40(1), 100-104.

Szollosi D. (1964) The structure and function of centrioles and their satellites in the jellyfish Phialidium gregarium. *J. Cell Biol.,* 21, 465-479.

Szollosi D., Calarco P., Donahue R.P. (1972) Absence of centrioles in the first and second meiotic spindles of mouse oocytes. *J. Cell Sci.,* 11(2), 521-541.

Szollosi D., Ozil J.P. (1991) *De novo* formation of centrioles in parthenogenetically activated, diploidized rabbit embryos. *Biol. Cell,* 72(1-2), 61-66.

Tadesse S., Anshabo A.T., Portman N., Lim E., Tilley W., Caldon C.E., Wang S. (2020) Targeting CDK2 in cancer: challenges and opportunities for therapy. *Drug Discov. Today,* 25, 406-413. doi: 10.1016/j.drudis.2019.12.001.

Tadesse S., Caldon E.C., Tilley W., Wang S. (2019) Cyclin-dependent kinase 2 inhibitors in cancer therapy: an update. *J. Med. Chem.,* 62, 4233-4251. doi: 10.1021/acs.jmedchem.8b01469.

Takahashi M., Mukai H., Oishi K., Isagawa T., Ono Y. (2000) Association of immature hypophosphorylated protein kinase cepsilon with an anchoring protein CG-NAP. *J. Biol. Chem.,* 275(44), 34592-34596.

Takahashi M., Shibata H., Shimakawa M., Miyamoto M., Mukai H., Ono Y. (1999) Characterization of a novel giant scaffolding protein, CG-NAP, that anchors multiple signaling enzymes to centrosome and the golgi apparatus. *J. Biol. Chem.,* 274(24), 17267-17274.

Takahashi M., Yamagiwa A., Nishimura T., Mukai H., Ono Y. (2002) Centrosomal proteins CG-NAP and kendrin provide microtubule nucleation sites by anchoring gamma-tubulin ring complex. *Mol. Biol. Cell,* 13(9), 3235-3245.

Tamm S.L. (1972) Free kinetosomes in Austrialian flagellates. I. Types and spatial arrangement. *J. Cell Biol.,* 54(1), 39-55.

Tang C.J., Fu R.H., Wu K.S., Hsu W.B., Tang T.K. (2009) CPAP is a cell-cycle regulated protein that controls centriole length. *Nat. Cell Biol.,* 11(7), 825-831.

Tang N., Marshall W.F. (2012) Centrosome positioning in vertebrate development. *J. Cell Biol.,* 125, 51-61.

Tang S.L., Gao Y.L., Wen-Zhong H. (2018) Knockdown of TRIM37 suppresses the proliferation, migration and invasion of glioma cells through the inactivation of

PI3K/Akt signaling pathway. *Biomed. Pharmacother,* 99, 59-64. doi: 10.1016/j. biopha.2018.01.054.

Tang X., Cao J., Zhang L., Huang Y., Zhang Q., Rong Y.S. (2017) Maternal Haploid, a Metalloprotease Enrichedat the Largest Satellite Repeat and Essential for Genome Integrity in Drosophila Embryos. *Genetics,* 206, 1829-1839.

Tanimoto H., Kimura A. (2016) Shape-motion relationships of centering microtubule asters. *J. Cell Biol.,* 212, 777-787.

Tanimoto H., Sallé J., Dodin L., Minc N. (2018) Physical Forces Determining the Persistency and Centering Precision of Microtubule Asters. *Nature Physics,* 14, 848-854.

Tanos B.E., Yang H.J., Soni R., Wang W.J., Macaluso F.P., Asara J.M., Tsou M.F. (2013) Centriole distal appendages promote membrane docking, leading to cilia initiation. *Genes Dev.,* 27, 163-168.

Tassin A.M., Maro B., Bornens M. (1985) Fate of microtubule-organizing centers during myogenesis *in vitro. J. Cell Biol.,* 100, 35-46.

Tatsuka M., Sato S., Kanda A., Miki T., Kamata N., Kitajima S., Kudo Y., Takata T. (2009) Oncogenic role of nuclear accumulated Aurora-A. *Mol. Carcinog.,* 48(9), 810-820.

Tay J.Y., Ho J.X., Cheo F.F., Iqbal J. (2024) The Tumour Microenvironment and Epigenetic Regulation in BRCA1 Pathogenic Variant-Associated Breast Cancers. *Cancers (Basel).* 16(23), 3910. doi: 10.3390/cancers16233910.

Taylor J.H., Woods P.S., Hughes W.L. (1957) The organization and duplication of chromosomes as revealed by autoradiographic studies using tritium-labeled thymidine. *Proc. Nat. Acad. Sci. USA,* 43, 122-128.

Tchakhotin S. (1959) Study of the localized effects of ultraviolet rays on the living cell by micro-prick method. *Tsitologiia,* 1(6), 614-626. In Russian.

Telentschak S., Soliwoda M., Nohroudi K., Addicks K., Klinz F-J. (2015) Cytokinesis failure and successful multipolar mitoses drive aneuploidy in glioblastoma cells. *Oncol. Rep.,* 33(4), 2001-2008. doi: 10.3892/or.2015.3751.

Terada Y., Uetake Y., Kuriyama R. (2003) Interaction of Aurora-A and centrosomin at the microtubule-nucleating site in Drosophila and mammalian cells. *J. Cell Biol.,* 162(5), 757-763.

Theesfeld C.L., Irazoqui J.E., Bloom K., Lew D.J. (1999) The role of actin in the spindle orientation changes during the Saccaromyces cerevisiae cell cycle. *J. Cell Biol.,* 146, 1019-1031.

Thein K.H., Kleylein-Sohn J., Nigg EA, Gruneberg U. (2007) Astrin is required for the maintenance of sister chromatid cohesion and centrosome integrity. *J. Cell Biol.,* 178(3), 345-354.

Théry M. (2010) Micropatterning as a tool to decipher cell morphology and functions. *J. Cell Sci.,* 123, 4201-4213.

Théry M., Racine V., Piel M., Pépin A., Dimitrov A., Chen Y., Sibarita J.B., Bornens M. (2006). Anisotropy of cell adhesive microenvironment governs cell internal organization and orientation of polarity. *Proc. Natl. Acad. Sci. USA,* 103, 19771.

Tian X., Zhou D., Chen L., Tian Y., Zhong B., Cao Y., Dong Q., Zhou M., Yan J., Wang Y., Qiu Y., Zhang L., Li Z., Wang H., Wang D., Ying G., Zhao Q. (2018) Polo-like kinase 4 mediates epithelial-mesenchymal transition in neuroblastoma via PI3K/Akt signaling pathway. *Cell Death. Dis.,* 9, 54. doi: 10.1038/s41419-017-0088-2.

Tian X-x., Rai D., Li J., Zou C., Bai Y., Wazer D., Band V., Gao Q. (2005) BRCA2 suppresses cell proliferation via stabilizing MAGE-D1. *Cancer Res.,* 65(11), 4747-4753. doi: 10.1158/0008-5472.CAN-05-0018.

Tilney L.G., Bryan J., Bush D.J., Fujiwara K., Mooseker M.S., Murphy D.B., Snyder D.H. (1973) Microtubules: evidence for 13 protofilaments. *J. Cell Biol.,* 59, 267-275.

Tilney L.G., Goddard J., (1970) Nucleating sites for the assembly of cytoplasmic microtubules in the ectodermal cells of blastulae of Arbacia punctulata. *J. Cell Biol.,* 46, 564-575.

Todoerti K., Ronchetti D., Favasuli V., Maura F., Morabito F., Bolli N., Taiana E., Neri A. (2022) DIS3 mutations in multiple myeloma impact the transcriptional signature and clinical outcome. *Haematologica.* 107(4), 921-932. doi: 10.3324/haematol.2021.278342.

Tomlinson A., Ready D.F. (1987) Cell fate in the Drosophila ommatidium. *Dev. Biol.,* 123(1), 264-275.

Torchia E.C., Chen Y., Sheng H., Katayama H., Fitzpatrick J., Brinkley W.R., Sen S., Roop D.R. (2009) A genetic variant of Aurora Kinase A promotes genomic instability leading to highly malignant skin tumors. *Cancer Res.,* 69(18), 7207-7215.

Toh B., Jones C., Kavanaugh G., Hamill M., Allen S., Choi S., Garrett A., Valganon-Petrizan M., MacMahon S., Yuan L., Way R., Harder H., Gold R., Taylor A., Gabe R., Lucassen A., Manchanda R., Fallowfield L., Jenkins V., Gandhi A., Evans D. G., George A., Hubank M., Kemp Z., Bremner S., Turnbull C. (2024) BRCA-DIRECT digital pathway for diagnostic germline genetic testing within a UK breast oncology setting: a randomised, non-inferiority trial. *Br. J. Cancer,* 131(9), 1506-1515.

Tournier F., Bornens M. (1994) Cell Cycle regulation of Centrosome Function. In "Microtubules". USA, New York: Wiley-Liss, Inc. 303-324.

Tournier F., Karsenti E., Bornens M. (1989) Parthenogenesis in Xenopus eggs injected with centrosomes from synchronized human lymphoid cells. *Dev. Biol.,* 136, 321-329.

Tournier F., Komesli S., Paintrand M., Job D., Bornens M. (1991) The intercentriolar linkage is critical for the ability of heterologous centrosomes to induce parthenogenesis in Xenopus. *J. Cell Biol.,* 113(6), 1361-1369.

Toya M., Terasawa M., Nagata K., Iida Y., Sugimoto A. (2011) A kinase-independent role for aurora a in the assembly of mitotic spindle microtubules in caenorhabditis elegans embryos. *Nat. Cell Biol.,* 13, 708-714.

Trachana V., van Wely K.H., Guerrero A.A., Fütterer A., Martínez-A C. (2007) Dido disruption leads to centrosome amplification and mitotic checkpoint defects compromising chromosome stability. *Proc. Natl. Acad. Sci. U S A,* 104(8), 2691-2696. doi: 10.1073/pnas.0611132104.

Tsai J.J., Hsu W.B., Liu J.H., Chang C.W., Tang T.K. (2020) Author correction: Cep120 interacts with c2cd3 and talpid3 and is required for centriole appendage assembly and ciliogenesis. *Sci. Rep.,* 10, 1265.

Tsang W.Y., Spektor A., Luciano D.J., Indjeian V.B., Chen Z., Salisbury J.L., Sanchez I., Dynlacht B.D. (2006) CP110 cooperates with two calcium-binding proteins to regulate cytokinesis and genome stability. *Mol. Biol. Cell,* 17, 3423-3434.

Tschachotin S. (1912) Die mikroskopische strahlen-stichmethode eine zelloperationsmethode. [The microscopic radiation puncture method is a cell surgery method]. *Biol. Zentralblatt,* 32, 623-630 In German.

Tschachotin S.S. (1912) Eine Microoperationsvorrichtung [A microsurgery device]. *Zs. Wiss. Mikrosk.,* 29, 188. In German.

Tsou M.F., Stearns T. (2006) Mechanism limiting centrosome duplication to once per cell cycle. *Nature,* 442(7105), 947-951.

Tucker J.B., Paton C.C., Richardson G.P., Mogensen M.M., Russell I.J. (1992) A cell surface-associated centrosomal layer of microtubule-organizing material in the inner pillar cell of the mouse cochlea. *J. Cell Sci.,* 102 (Pt 2), 215-226. doi: 10.1242/jcs.102.2.215.

Tucker R.W., Pardee A.B., Fujiwara K. (1979a) Centriole ciliation is related to quiescence and DNA synthesis in 3T3 cells. *Cell,* 17(3), 527-535. doi: 10.1016/0092-8674(79)90261-7.

Tucker R.W., Scher C.D., Stiles C.D. (1979b) Centriole deciliation associated with the early response of 3T3 cells to growth factors but not to SV40. *Cell,* 18(4), 1065-72. doi: 10.1016/0092-8674(79)90219-8.

Turner K., Solanki N., Salouha H.O., Avidor-Reiss T. (2022) Atypical Centriolar Composition Correlates with Internal Fertilization in Fish. *Cells,* 11(5), 758. doi: 10.3390/cells11050758.

Turner K.A., Caswell D.L., McGrady B.M., Pietras-Allen A., Sedlak J., Nathan C., Parasuraman S., McGann A.P., Fazili F.M., Bell J.R., El Smail K.N., Pillai S.B., Parry K.R., Richardson K.P., Ruble K., Jaiswal A., Shah T.A., Sindhwani P., Avidor-Reiss T. (2023) CP110 and CEP135 localize near the proximal and distal centrioles of cattle and human spermatozoa. *MicroPubl Biol.*, 10.17912/micropub.biology.000951. doi: 10.17912/micropub.biology.000951.

Tyzack G.E., Neeves J., Crerar H., Klein P., Ziff O., Taha D.M., Luisier R., Luscombe N.M., Patani R. (2021) Aberrant cytoplasmic intron retention is a blueprint for RNA binding protein mislocalization in VCP-related amyotrophic lateral sclerosis. *Brain,* 144(7), 1985-1993. doi: 10.1093/brain/awab078.

Tzivion G., Shen Y.H., Zhu J. (2001) 14-3-3 proteins; bringing new definitions to scaffolding. *Oncogene,* 20(44), 6331-6338.

Uetake Y., Loncarek J., Nordberg J.J., English C.N., La Terra S., Khodjakov A., Sluder G. (2007) Cell cycle progression and *de novo* centriole assembly after centrosomal removal in untransformed human cells. *J. Cell Biol.,* 176(2), 173-182.

Uetake Y., Terada Y., Matuliene J., Kuriyama R. (2004) Interaction of Cep135 with a p50 dynactin subunit in mammalian centrosomes. *Cell Motil. Cytoskeleton.,* 58, 53-66.

Uhlmann F., Wernic D., Poupart M.A., Koonin E.V., Nasmyth K. (2000) Cleavage of cohesin by the CD clan protease separin triggers anaphase in yeast. *Cell,* 103(3), 375-386.

Unal S., Alanay Y., Cetin M., Boduroglu K., Utine E., Cormier-Daire V., Huber C., Ozsurekci, Y., Kilic E., Simsek Kiper, O.P., Gumruk F. (2014) Striking Hematological Abnormalities in Patients with Microcephalic Osteodysplastic Primordial Dwarfism Type II (MOPD II): A Potential Role of Pericentrin in Hematopoiesis: Hematological Abnormalities in MOPD II. *Pediatr. Blood Cancer,* 61, 302-305.

Uzbekov R., Alieva I. (2018) Who are you, subdistal appendages of centriole? *Open Biol.,* 8(7). pii: 180062. doi: 10.1098/rsob.180062. Review.

Uzbekov R., Avidor-Reiss T. (2024) Comment on: 'The proximal centriole age in spermatozoa is a potential reason for its different fate in the zygote after fertilization. *Open Biol.,* 14(3):230458. doi: 10.1098/rsob.230458.

Uzbekov R., Burlaud-Gaillard J., Garanina AS, Bressac C. (2017) The length of a short sperm: elongation and shortening during spermiogenesis *in Cotesia congregata* (*Hymenoptera, Braconidae*). *Arthropod Structure and Development,* 46(2), 265-273.

Uzbekov R., Garanina A, Bressac C. (2018a) Centrioles without microtubules: a new morphological type of centriole. *Biol. Open.,* 7(8). pii: bio036012. doi: 10.1242/bio. 036012.

Uzbekov R., Garanina A.S., Burlaud-Gaillard J., Bressac C. (2018b) The flagellum of the shortest spermatozoon in the animal kingdom. Elongation and shortening of the axoneme in the process of spermiogenesis of the parasitic wasp cotesia congregata. In *"Flagella and Cilia: Types, Structure and Functions"* (ed. R. Uzbekov). USA, New York: Nova Science Publishers, Inc. pp. 83-108.

Uzbekov R., Kireyev I., Prigent C. (2002) Centrosome separation: Respective role of microtubules and actin filaments. *Biol. Cell,* 94, 275-288. doi: 10.1016/s0248-4900(02)01202-9.

Uzbekov R., N'goran SMO, Garanina A., Pancione M., Chaika S.Y., Bressac C. (2024a) The Centrosomes of Haploid and Diploid Cells Have an Equal Number of Centrioles in the Parasitoid Wasp Anisopteromalus Calandrae. *Front. Biosci. (Landmark Ed),* 29(1), 28. doi: 10.31083/j.fbl2901028.

Uzbekov R., Prigent C. (2007) Clockwise or anticlockwise? Turning the centriole triplets in the right direction! *FEBS Lett.,* 581(7), 1251-1254.

Uzbekov R., Prigent C., Arlot-Bonnemains Y. (1999) Cell Cycle Analysis and Synchronisation of the Xenopus laevis XL2 Cell Line: Study of the Kinesin Related Protein XlEg5. *Micr. Res Tech.,* 45, 31-42.

Uzbekov R., Singina G.N., Shedova E.N., Banliat C., Avidor-Reiss T., Uzbekova S. (2023) Centrosome Formation in the Bovine Early Embryo. *Cells,* 12(9), 1335. doi: 10.3390/ cells12091335.

Uzbekov R., Uzbekova S., Severin F., Prigent C., Arlot-Bonnemains Y. (2024b) Aurora A Kinase Begins to Localize to the Centrosome in the S-phase of the Cell Cycle in the XL2 Cell Line. *Front. Biosci (Landmark Ed),* 29(9), 317. doi:

Uzbekov R.E. (2004) Dynamics of functional activity of the centrosome during cell cycle. Doctor of Sci. Thesis. Moscow State University, Moscow. In Russian.

Uzbekov R.E. (2007) Centriole duplication in PE (SPEV) cells starts before beginning of DNA replication. Biol.Membranes, 24(4), 292-298. In Russian. In English - *Biochemistry (Moscow) Suppl. Series A: Membrane and Cell Biology,* 1(3), 206-211.

Uzbekov R.E. and Alieva I.B. (2013*) Centrosome - history of investigations and new discoveries. From cytoplasmic granule to the centre of intracellular regulation.* (ed. E.S. Nadezhdina). Moscow, Russia: Moscow University Press. 320 pages.

Uzbekov R.E., Alieva I.B. (2010) Centriole duplication or DNA replication - what starts earlier? In "Cytoskeleton: Cell Movement, Cytokinesis and Organelles Organization" (eds. S. Lansing and T. Rousseau). USA, New York: Nova Science Publishers, Inc. Chapter 5.

Uzbekov R.E., Arlot-Bonnemains Y., Giet R., Prigent C. (2001) Level of expression protein kinase pEg2 and localization in the centrosome depend from cell cycle stage. *Tsitologiia,* 43(4), 404-405. In Russian.

Uzbekov R.E., Avidor-Reiss T. (2020) Principal postulates of centrosomal biology. Version 2020. *Cells,* 9(2156). doi: 10.3390/cells91021569.

Uzbekov R.E., Giet R., Arlot-Bonnemains Y., Prigent C. (2001) Role of pEg2 kinase in mitotic spindle construction process. *Biol. Cell,* 93, 337.

Uzbekov R.E., Maurel D.B., Aveline P.C., Pallu S., Benhamou C.L., Rochefort G.Y. (2012) Fine ultrastructure of the centrosome and morphological aspects of *in vivo* realization of mechanotransduction primary cilia function in rat osteocytes. *Microsc. Microanal.,* 18(6), 1430-1441.

Uzbekov R.E., Vorob'ev I.A., Drachev V.A. (1989) The effect of the laser microirradiation of the cell center on neutrophil motility. *Tsitologia,* 31, 874-881.

Uzbekov R.E., Vorobjev I.A. (1991a) Effect of UV microirradiation of the centrosome on cell behavior. 1. Microirradiation in metaphase: disorganization of the spindle and disturbance of cell division. *Tsytologiia,* 33(2), 15-22. In Russian.

Uzbekov R.E., Vorobjev I.A. (1991b) Effect of UV microirradiation of the centrosome on cell behavior. 2. Microirradiation in anaphase: completion of cell division and destination of the interfase cell. *Tsytologiia,* 33(10), 79-84. In Russian.

Uzbekov R.E., Vorobjev I.A. (1992) Effect of UV microirradiation of the centrosome on cell behavior 3. Ultrastructure of the centrosome after ultraviolet microirradiation. *Tsytologiia,* 34 (2), 62-67. In Russian.

Uzbekov R.E., Votchal M.S., Vorobjev I.A. (1995) Role of the centrosome in mitosis: UV microirradiation study. *J. Photochem. Photobiol.,* 29, 163-170.

Uzbekova S., Arlot-Bonnemains Y., Dupont J., Dalbiès-Tran R., Papillier, P., Pennetier S., Thelie A., Perreau C., Mermillod, P., Prigent, C., Uzbekov, R. (2008) Spatio-temporal expression patterns of aurora kinases A, B, and C and cytoplasmic polyadenylation-element-binding protein in bovine oocytes during meiotic maturation. *Biol. Repr.,* 78(2), 218-233.

Vaisberg E.A., Koonce M.P., McIntosh J.R. (1993) Cytoplasmic dynein plays a role in mammalian mitotic spindle formation. *J. Cell Biol.,* 123, 849-858.

Valent P., Horny H.P., Bennett J.M., Fonatsch C., Germing U., Greenberg P., Haferlach T., Haase D., Kolb H.J., Krieger O., Loken M., van de Loosdrecht A., Ogata K., Orfao A., Pfeilstöcker M., Rüter B., Sperr W.R., Stauder R., Wells D.A. (2007) Definitions and standards in the diagnosis and treatment of the myelodysplastic syndromes: Consensus statements and report from a working conference. *Leuk. Res.,* 31(6), 727-36. doi: 10.1016/j.leukres.2006.11.009

Vallee R.B. (1992) Dynamin: motor protein or regulatory GTPase. *J. Muscle Res. Cell Motil.,* 13(5), 493-496.

Vallee R.B., Gee M.A. (1998) Make room for dynein. *Trends Cell Biol.,* 8(12), 490-494.

Van Beneden E. (1876) Recherches sur les Dicyémides, survivants actuels d'un embranchement des mésozoaires. [Research on Dicyemids, current survivors of a branch of mesozoa]. *Bull. Acad. Roy. Méd. Belg.,* 41 (sér. II), 1160-1205. In French.

Van Beneden E., Neyt A. (1887) Nouvelles recherches sur la fecondation et la division mitoshes l'Ascaride megalocephale [New research on fertilization and mitotic division in the *Ascaris megalocephalus*]. *Bull. Acad. Roy. Méd. Belg.,* 14 (sér. III) (8). In French.

Van Breugel M., Hirono M., Andreeva A., Yanagisawa H.A., Yamaguchi S., Nakazawa Y., Morgner N., Petrovich M., Ebong I.O., Robinson C.V., Johnson C.M., Veprintsev D., Zuber B. (2011) Structures of SAS-6 suggest its organization in centrioles. *Science,* 331(6021), 1196-1199.

Vandre D.D., Davis F.M., Rao P.N., Borisy G.G. (1984a) Phosphoproteins are components of mitotic microtubule organizing centers. *Proc. Natl. Acad. Sci. USA,* 81, 4439-4443.

Vandre D.D., Kronebusch P., Borisy G.G. (1984b) Interphase-mitosis transition: microtubule rearrangements in cultured cells and sea urchin eggs. In "*Molecular biology of the cytoskeleton*" (eds. G.G. Borisy, D.M. Cleveland, D.B. Murphy). USA, Cold Spring Harbor, NY: Cold Spring Harbor Laboratory. 3-16.

Vasquez R.J., Gard D.L., Cassimeris L. (1999) Phosphorylation by CDK1 regulates XMAP215 function *in vitro*. *Cell Motil. Cytoskeleton,* 43(4), 310-321.

Vaughan K.T., Vallee R.B. (1995) Cytoplasmic dynein binds dynactin through a direct interaction between the intermediate chains and p150Glued. *J. Cell Biol.,* 131(6 pt 1), 1507-1516.

Venkei Z.G., Yamashita Y.M. (2015) The centrosome orientation checkpoint is germline stem cell specific and operates prior to the spindle assembly checkpoint in Drosophila testis. *Development,* 142, 62-69.

Verde F., Labbé J.C., Dorée M., Karsenti, E. (1990) Regulation of microtubule dynamics by cdc2 protein kinase in cell-free extracts of Xenopus eggs. *Nature,* 343, 233-238.

Vernos I., Karsenti E. (1996) Motors involved in spindle assembly and chromosome segregation. *Curr. Opin. Cell Biol.,* 8, 4-9.

Vidwans S.J., Wong M.L., O'Farrell P.H. (2003) Anomalous centriole configurations are detected in Drosophila wing disc cells upon Cdk1 inactivation. *J. Cell Sci.,* 116, 137-143.

Vieillard J., Paschak M., Duteyrat J.L., Augière C., Cortier E., Lapart J.A., Thomas J., Durand B. (2016) Transition zone assembly and its contribution to axoneme formation in Drosophila male germ cells. *J. Cell Biol.,* 214(7), 875-889.

Virchow R. (1855) Cellular-Pathologie [Cellular pathology]. *Arch. Fur Path. Anat.* VIII, 23. In German. Citation from Wilson (1900).

Vladimirov Y.A., Potapenko A.Ja. (1989) *Physico-chemical basis of photo-biological processes.* USSR, Moscow: High School Press, 199 p. In Russian.

Von Lenhossék M. (1898) Uber Flimmerzellen. [About ciliated cells]. *Verh. anat. Ges., Kiel,* 12, 106-128. In German.

Vorobjev I.A., Alieva I.B., Grigoriev I.S., Borisy G.G. (2003) Microtubule dynamics in living cells: direct analysis in the internal cytoplasm. *Cell Biol. Int.,* 27 (3), 293-294.

Vorobjev I.A., Chentsov Ju.S. (1977) The ultrastructure of the centrosome in the cells of hematopoietic tissues of axolotl. *Tsitologiia,* 19, 598-603. In Russian.

Vorobjev I.A., Chentsov Ju.S. (1987) Ultrastructure of centrosomes in L-cells. *Ontogenez,* 18(4), 354-359. In Russian.

Vorobjev I.A., Chentsov Yu.S. (1980) The ultrastructure of centriole in mammalian tissue culture cells. *Cell Biol. Int. Rep.,* 4(11), 1037-1044.

Vorobjev I.A., Chentsov Yu.S. (1982) Centrioles in the Cell Cycle. I Epithelial Cells. *J. Cell Biol.,* 98, 938-949.

Vorobjev I.A., Drachev V.A., Chentsov Ju.S. (1988) Centrosome inactivation in mitosis by laser microirradiation. *Biopolymers and Cell,* 4(N6), 313-321. In Russian.

Vorobjev I.A., Nadezhdina E.S. (1986) Centriolar apparatus and its role in the organization of microtubules. In *"Common Problems of Physico-Chemical Biology"* (ed. Ju.S. Chentsov), USSR, Moscow: VINITI Press, 164p. In Russian.

Vorobjev I.A., Nadezhdina, E.S. (1987) The centrosome and its role in the organization of microtubules. *Int. Rev. Cytol.,* 106, 227-293.

Vorobjev I.A., Svitkina T.M., Borisy G.G. (1997) Cytoplasmic assembly of microtubules in cultured cells *J. Cell Sci.,* 110, 2635-2645.

Vorobjev I.A., Uzbekov R.E., Komarova Yu.A., Alieva I.B. (2000) Gamma-tubulin distribution in interphase and mitotic cells upon stabilization and depolymerization of microtubules. *Biol. Membranes,* 17(2), 173-187. In Russian. In English - *Membr. Cell Biol.,* 14(2), 219-235.

Wachmann E., Hennig, A. (1974) Centriolen in der Entwicklung der Retinulazellen von *Megachile rotundata (F.)* (*Hymenoptera, Apidae*) [Centrioles in the development of retinula cells of *Megachile rotundata (F.) (Hymenoptera, Apidae)*]. *Zoomorphology,* 77, 337-344. In German.

Wachmann E., Pfannenstiel H. D., Wellmann H., Shelton P.M.J. (1983) Morphogenesis of open rhabdoms in ommatidia of *Leptinotarsa decemlineata* and *Crioceris asparagi* (*Coleoptera: Chrysomelidae*). *Zoomorphology,* 103, 165-176.

Waizenegger I.C., Hauf S., Meinke A., Peters J.M. (2000) Two distinct pathways remove mammalian cohesin from chromosome arms in prophase and from centromeres in anaphase. *Cell,* 103(3), 399-410.

Walczak C.E., Verma S., Mitchison T.J. (1997) XCTK2: a kinesin-related protein that promotes mitotic spindle assembly in *Xenopus laevis* egg extracts. *J. Cell Biol.,* 136, 859-870.

Wang H., Shao N., Ding Q.M., Cui J., Reddy E.S., Rao V.N. (1997) BRCA1 proteinsare transported to the nucleus in the absence of serum and splice variants BRCA1a, BRCA1b are tyrosine phosphoproteins that associate with E2F, cyclins and cyclin dependent kinases. *Oncogene,* 15, 143-157.

Wang H.F., Takenaka K., Nakanishi A., Miki Y. (2011) BRCA2 and nucleophosmin coregulate centrosome amplification and form a complex with the Rho effector kinase ROCK2. *Cancer Res.,* 71(1), 68-77.

Wang L., Failler M., Fu W., Dynlacht B.D. (2018) A distal centriolar protein network controls organelle maturation and asymmetry. *Nat. Commun.,* 9, 3938.

Wang L.H., Mayer B., Stemmann O., Nigg E.A. (2010) Centromere DNA decatenation depends on cohesin removal and is required for mammalian cell division. *J. Cell Sci.,* 123, 806-813.

Wang X., Yang Y., Duan Q., Jiang N., Huang Y., Darzynkiewicz Z., Dai W. (2008) sSgo1, a major splice variant of Sgo1, functions in centriole cohesion where it is regulated by Plk1. *Dev Cell.,* 14(3), 331-41. doi: 10.1016/j.devcel.2007.12.007. PMID: 18331714.

Wang X., Zhou Y.X., Qiao W., Tominaga Y., Ouchi M., Ouchi T., Deng C.X. (2006) Overexpression of aurora kinase A in mouse mammary epithelium induces genetic instability preceding mammary tumor formation. *Oncogene,* 25(54), 7148-7158.

Ward C.J., Hogan M.C., Rossetti S., Walker D., Sneddon T., Wang X., Kubly V., Cunningham J.M., Bacallao R., Ishibashi M., Milliner D.S., Torres V.E., Harris P.C. (2002) The gene mutated in autosomal recessive polycystic kidney disease encodes a large, receptor-like protein. *Nat. Genet.,* 30, 259-269.

Watanabe K., Takao D., Ito K.K., Takahashi M., Kitagawa D. (2019) The cep57-pericentrin module organizes pcm expansion and centriole engagement. *Nat. Commun.,* 10, 931.

Waterman-Storer C.M., Salmon E.D. (1997) Actomyosin-based retrograde flow of microtubules in the lamella of migrating epithelial cells influences microtubule dynamic instability and turnover and is associated with microtubule breakage and treadmilling. *J. Cell Biol.,* 139(2), 417-434.

Watson M.L. (1958) Staining of tissue sections for electron microscopy with heavy metals. *J. Biophys. Biochem. Cytol.,* 4(4), 475-478.

Wei Q., Xu Q., Zhang Y., Li Y., Zhang Q., Hu Z., Harris P.C., Torres V.E., Ling K., Hu J. (2013) Transition fibre protein fbf1 is required for the ciliary entry of assembled intraflagellar transport complexes. *Nat. Commun.,* 4, 2750.

Wei Z., Kim T.S., Ahn J.I., Meng L., Chen Y., Ryu E.K., Ku B., Zhou M., Kim S.J., Bang J.K., van Deursen J.M., Park J-E., Lee K.S. (2020) Requirement of the cep57-cep63 interaction for proper cep152 recruitment and centriole duplication. *Mol. Cell. Biol.,* 40, doi:10.1128/MCB.00535-19.

Went H.A. (1977) Can a reverse transcriptase by involved in centriole duplication? *J. Theor. Biol.,* 68(1), 95-100.

Westlake C.J., Baye L.M., Nachury M.V., Wright K.J., Ervin K.E., Phu L., Chalouni C., Beck J.S., Kirkpatrick D.S., Slusarski D.C., Sheffield V.C., Scheller R.H., Jackson P.K. (2011) Primary cilia membrane assembly is initiated by Rab11 and transport protein particle II (TRAPPII) complex-dependent trafficking of Rabin8 to the centrosome. *Proc. Natl. Acad. Sci. USA,* 108(7), 2759-2764.

Wheatley D.N. (1969) Cilia in cell-cultured fibroblasts. I. On their occurrence and relative frequencies in primary cultures and established cell lines. *J. Anat.,* 105(2), 351-362.

Wheatley D.N. (1971) Cilia in cell-cultured fibroblasts. 3. Relationship between mitotic activity and cilium frequency in mouse 3T6 fibroblasts. *J. Anat.,* 110(Pt 3), 367-82. PMID: 5169621

Wheatley D.N. (1974) Pericentriolar virus-like particles in Chinese hamster ovary cells. *J. Gen. Virol.,* 24(2), 395-399.

Wheatley D.N. (1982) *The Centriole: a central enigma of cell biology.* Niederlande/USA, Amsterdam/New York: Elsevier Biomedical Press, 232 p.

Wheatley D.N., Feilen E.M., Yin Z., Wheatley S.P. (1994) Primary cilia in cultured mammalian cells: detection with an antibody against detyrosinated alpha-tubulin (ID5) and by electron microscopy. *J. Submicrosc. Cytol. Pathol.*, 26(1), 91-102.

Wheatley D.N., Wang A.M., Strugnell G.E. (1996) Expression of primary cilia in mammalian cells. *Cell Biol. Int.,* 20(1), 73-81. doi: 10.1006/cbir.1996.0011.

White R.A., Pan Z., Salisbury J.L. (2000) GFP-centrin as a marker for centriole dynamics in living cells. *Microsc. Res. Tech.,* 49 (5), 451-457.

Whitehead C.M., Winkfein R.J., Rattner J.B. (1996) The relationship of HsEg5 and the actin cytoskeleton to centrosome separation. *Cell Motil. Cytoskeleton,* 35(4), 298-308.

Widdowson E.M. (1970) Harmony of Growth. *Lancet,* 1, 902-905.

Williamson R.A, Weiner C.P., Yuh W.T., Abu-Yousef M.M. (1989) Magnetic resonance imaging of anomalous fetuses. *Obstet. Gynecol.* 73(6), 952-956.

Wilson E.B. (1900) *The Cell in Development and Inheritance.* 2nd ed. New York, NY, USA: The Macmillan Company,

Wilson E.B. (1925) *The Cell in Development and Inheritance.* 3rd ed. New York, NY, USA: The Macmillan Company.

Wilson P.G., Fuller M.T., Borisy G.G. (1997) Monastral bipolar spindles: implications for dynamic centrosome organization. *J. Cell Sci.,* 110, 451-464.

Witman G.B. (2003) Cell motility: Deaf Drosophila keep the beat. *Curr. Biol.,* 13(20), R796-R798.

Wolf K.W., Kyburg J. (1989) The restructuring of the flagellar base and the flagellar necklace during spermatogenesis of Ephestia kuehniella Z. (Pyralidae, Lepidoptera). *Cell and Tissue Research,* 256, 77-86.

Wolf M.T., Hildebrandt F. (2011) Nephronophthisis. *Pediatr. Nephrol.,* 26, 181-194.

Wolf N., Hirsh, D., McIntosh J.R. (1978) Spermatogenesis in males of the free-living nematode, caenorhabditis elegans. *J. Ultrastruct. Res.,* 63, 155-169.

Wolfe J. (1972) Basal body fine structure and chemistry. *Adv. Cell Mol. Biol.,* 2, 151-192.

Wong C., Stearns T. (2003) Centrosome number is controlled by a centrosome-intrinsic block to reduplication. *Nat. Cell Biol.,* 5(6), 539-544. doi: 10.1038/ncb993.

Wong Y. L., Anzola J. V., Davis R. L., Yoon M., Motamedi A., Kroll A., Seo C. P., Hsia J. E., Kim S.K., Mitchell J.W., Mitchell B.J., Desai A., Gahman T.C., Shiau A.K., Oegema K. (2015) Reversible centriole depletion with an inhibitor of Polo-like kinase 4. *Science,* 348 (6239), 1155. 10.1126/science.aaa5111.

Woodruff J.B., Wueseke O., Hyman A.A. (2014) Pericentriolar material structure and dynamics. *Philosophical Transactions of the Royal Society B: Biological Sciences.* 369, 20130459. doi: 10.1098/rstb.2013.0459.

Woolley D.M., Nickels, S.N. (1985) Microtubule termination patterns in mammalian sperm flagella. *J. Ultrastruct. Res.,* 90, 221-234.

Wu H.-Y., Kabacaoğlu G., Nazockdast E., Chang H.-C., Shelley M.J., Needleman D.J. (2023) Laser ablation and fluid flows reveal the mechanism behind spindle and centrosome positioning. *Nature Physics,* 20, 157-168.

Wu J., Misra G., Russell R.J., Ladd A.J., Lele T.P., Dickinson R.B. (2011) Effects of dynein on microtubule mechanics and centrosome positioning. *Mol. Biol. Cell,* 22, 4834-4841.

Wu L., Timmers C., Maiti B., Saavedra H.I., Sang L., Chong G.T., Nuckolls F., Giangrande P., Wright F.A., Field S.J., Greenberg M.E., Orkin S., Nevins J.R., Robinson M.L., Leone G. (2001) The E2F1-3 transcription factors are essential for cellular proliferation. *Nature,* 414 (6862), 457-462.

Wu Z., Chen H., Zhang Y., Wang Y., Wang Q., Augière C., Hou Y., Fu Y., Peng Y., Durand B., Wei Q. (2024) Cep131-Cep162 and Cby-Fam92 complexes cooperatively maintain Cep290 at the basal body and contribute to ciliogenesis initiation. *PLoS Biol.,* 22(3), e3002330.

Wühr M., Dumont S., Groen A.C., Needleman D.J., Mitchison T.J. (2009) How does a millimeter-sized cell find its center? *Cell Cycle,* 8, 1115-1121.

Xue P., Zheng J., Li R., Yan L., Wang Z., Jia Q., Zhang L., Li X. (2024) High Expression of KIFC1 in Glioma Correlates with Poor Prognosis. 67(3), *J. Korean Neurosurg. Soc.,* 364-375. doi: 10.3340/jkns.2023.0155

Yaffe D., Gershon D. (1967) Multinucleated muscle fibers: induction of DNA synthesis and mitosis by polyoma virus infection. *Nature (London),* 215, 421.

Yaguchi K., Yamamoto T., Matsui R., Tsukada Y., Shibanuma A., Kamimura K., Koda T., Uehara R. (2018) Uncoordinated centrosome cycle underlies the instability of non-diploid somatic cells in mammals. *J. Cell Biol.,* 217, 2463-2483.

Yamada E. (1956) The fine structure of centriole in some animal cells. *Proc. 1st Conf. Asia Oceania Tokyo*, 247-250.

Yamamoto S., Gaillard J., Vianay B., Guerin C., Orhant-Prioux M., Blanchoin L., Théry M. (2022) Actin network architecture can ensure robust centering or sensitive decentering of the centrosome. *EMBO J.,* 41(20), e111631.

Yamashiki, N., Kawamura N. (1998) Behavior of centrioles during meiosis in the male silkworm, *Bombyx mori* (*Lepidoptera*). *Dev. Growth Differ.,* 40(6), 619-630.

Yamashita Y.M., Jones D.L. (2003) Orientation of asymmetric stem cell division by the APC tumor suppressor and centrosome. *Science,* 301, 1547-1550.

Yan L.Y., Huang J.C., Zhu Z.Y., Lei Z.L., Shi L.H., Nan C.L., Zhao Z.J., Ouyang Y.C., Song X.F., Sun Q.Y., Chen D.Y. (2006) NuMA distribution and microtubule configuration in rabbit oocytes and cloned embryos. *Reproduction,* 132(6), 869-876.

Yan, X., Habedanck, R., Nigg, E. A. (2006) A complex of two centrosomal proteins, CAP350 and FOP, cooperates with EB1 in microtubule anchoring. *Mol. Biol. Cell,* 17(2), 634-644.

Yang J, Liu X, Yue G, Adamian M, Bulgakov O, Li T. (2002) Rootletin, a novel coiled-coil protein, is a structural component of the ciliary rootlet. *J. Cell Biol.,* 159(3), 431-440. doi: 10.1083/jcb.200207153.

Yang K., Zhu L., Liu C., Zhou D., Zhu Z., Xu N., Li W. (2024) Current status and prospect of the DNA double-strand break repair pathway in colorectal cancer development and treatment. *Biochim. Biophys. Acta Mol. Basis Dis.,* 1870(7), 167438.

Yang P., Diener D.R., Yang C., Kohno T., Pazour G.J., Dienes J.M., Agrin N.S., King S.M., Sale W.S., Kamiya R., Rosenbaum J.L., Witman G.B. (2006) Radial spoke proteins of Chlamydomonas flagella. *J. Cell Sci.,* 119(6), 1165-1174.

Yao J., Fu C., Ding X., Guo Z., Zenreski A., Chen Y., Ahmed K., Liao J., Dou Z., Yao X. (2004) Nek2A kinase regulates the localization of numatrin to centrosome in mitosis *FEBS Lett.,* 575(1-3), 112-118.

Yarden R.I., Brody L.C. (1999) BRCA1 interacts with components of the histone deacetylase complex. *Proc. Natl. Acad. Sci. USA,* 96, 4983-4988.

Ye X., Zeng H., Ning G., Reiter J.F., Liu A. (2014) C2cd3 is critical for centriolar distal appendage assembly and ciliary vesicle docking in mammals. *Proc. Natl. Acad. Sci. USA,* 111, 2164-2169.

Yi J., Wu X., Chung A.H., Chen J.K., Kapoor T.M., Hammer J.A. (2013) Centrosome repositioning in T cells is biphasic and driven by microtubule end-on capture-shrinkage. *J. Cell Biol.,* 202, 779-792.

Yigit G., Brown K.E., Kayserili H., Pohl E., Caliebe A., Zahnleiter D., Rosser E., Bögershausen, N., Uyguner Z.O., Altunoglu U., Nürnberg G., Nürnberg P., Rauch A., Li Y., Thiel C.T., Wollnik B. (2015) Mutations in CDK5RAP2 Cause Seckel Syndrome. *Mol. Genet. Genomic Med.,* 3, 467-480.

Yin X., Yu J., Zhou Y., Wang C., Jiao Z., Qian Z., Sun H., Chen B. (2018) Identification of CDK2 as a novel target in treatment of prostate cancer. *Future Oncol.,* 14, 709-718. doi: 10.2217/fon-2017-0561.

Yoder B.K., Hou X., Guay-Woodford L.M. (2002) The polycystic kidney disease proteins, polycystin-1, polycystin-2, polaris, and cystin, are co-localized in renal cilia. *J. Am. Soc. Nephrol.,* 13, 2508-2516.

Yu S., Jiang T., Jia D., Han Y., Liu F., Huang Y., Qu Z., Zhao Y., Tu J., Lv Y., Li J., Hu X., Lu Z., Han S., Qin Y., Liu X., Xie S., Wang Q.K., Tang Z., Luo D., Liu M. (2019) BCAS2 is essential for hematopoietic stem and progenitor cell maintenance during zebrafish embryogenesis. *Blood,* 133(8), 805-815. doi: 10.1182/blood-2018-09-876599.

Yvon A.-M., Wadsworth P. (2000) Region-specific transport of microtubules in motile cells. *J. Cell Biol.,* 151, 1003-1012.

Zach F., Grassmann F., Langmann T., Sorusch N., Wolfrum U., Stohr H. (2012) The retinitis pigmentosa 28 protein fam161a is a novel ciliary protein involved in intermolecular protein interaction and microtubule association. *Hum. Mol. Genet.,* 21, 4573-4586.

Zalokar M., Audit C., Erk I. (1975) Developmental defects of female-sterile mutants of Drosophila melanogaster. *Dev. Biol.,* 47, 419-432.

Zanetti A., Dujardin G., Fares-Taie L., Amiel J., Roger J.E., Audo I., Robert M.P., David P., Jung V., Goudin N., Guerrera I.C., Moriceau S., Amana D., Assia Batzir N., Bachar-Zipori A., Basel Salmon L., Boddaert N., Briault S., Bruel A.L., Costet-Fighiera C., Coutinho Santos L., Gitiaux C., Kaminska K., Kuentz P., Orenstein N., Philip-Sarles N., Plutino M., Quinodoz M., Santos C., Sigaudy S., Soeiro E Sá M., Sofrin E., Sousa A.B., Sousa-Luis R., Thauvin-Robinet C., van Dijk E.L., Zaafrane-Khachnaoui K., Zur D., Kaplan J., Rivolta C., Rozet J.M., Perrault I. (2024) GPATCH11 variants cause mis-splicing and early-onset retinal dystrophy with neurological impairment. *Nat. Commun.,* 15(1), 10096. doi: 10.1038/s41467-024-54549-8.

Zariwala M., Noone P.G., Sannuti A., Minnix S., Zhou Z., Leigh M.W., Hazucha M., Carson J. L., Knowles M.R. (2001) Germline mutations in an intermediate chain dynein cause primary ciliary dyskinesia. *Am. J. Respir. Cell Mol. Biol.,* 25(5), 577-583.

Zatsepina O.V., Rousselet A., Chan P.K., Olson M.O., Jordan E.G., Bornens M. (1999) The nucleolar phosphoprotein B23 redistributes in part to the spindle poles during mitosis. *J. Cell Sci.,* 112 (4), 455-466.

Zatsepina O.V., Zhelev N.Zh., Jordan G.J. (1995) Immunolocalization of nucleolar protein B23 in mitotic centrosomes. *Mol. Biol. (Moskow),* 29(6), 1359-1367. In Russian.

Zein-Sabatto H., Lerit D.A. (2021) The Identification and Functional Analysis of mRNA Localizing to Centrosomes. *Front. Cell Dev. Biol.,* 9, 782802. doi: 10.3389/fcell.2021.782802.

Zetterberg A., Larsson, O. (1995). Cell Cycle Progression and Cell Growth in Mammalian Cells: Kinetic Aspects of Transition Events. In *"Cell Cycle Control"* (eds. C. Hutchison, and D. M. Glover). United Kingdom, Oxford: IRL Press at Oxford Univpress. 206-227.

Zhang C., Fang Z., Xiong Y., Li J., Liu L., Li M., Zhang W., Wan J. (2010) Copy number increase of aurora kinase A in colorectal cancers: a correlation with tumor progression. *Acta Biochim. Biophys. Sin (Shanghai),* 42(11), 834-838. doi: 10.1093/abbs/gmq088.

Zhang H., Somasundaram K., Peng Y., Tian H., Zhang H., Bi D., Weber B.L., El-Deiry W.S. (1998) BRCA1 physically associates with p53 and stimulates itstranscriptional activity. *Oncogene,* 16, 1713-1721.

Zhang J., Megraw T.L. (2007) Proper recruitment of gamma-tubulin and D-TACC/Msps to embryonic Drosophila centrosomes requires Centrosomin Motif 1. *Mol. Biol. Cell,* 18(10), 4037-4049.

Zhang J., Wang Y.L. (2017) Centrosome defines the rear of cells during mesenchymal migration. *Mol. Biol.Cell,* 28, 3240-3251.

Zhang M., Chang, Z., Tian, Y., Wang, L., Lu, Y. (2020). Two novel TCTN2 mutations cause Meckel-gruber syndrome. *J. Hum. Genetics,* 65(11), 1039-1043. doi.org/10.3389/fped.2017.00244.

Zhang N., Pati D. (2017) Biology and insights into the role of cohesin protease separase in human malignancies. *Biol. Rev. Camb. Philos. Soc.,* 92(4), 2070-2083. doi: 10.1111/brv.12321.

Zhang P., Wu B., Wang Y., Ren Y., Li G., Qan Y., Lei C., Wang H. (2022) Identification of Pathogenic Variants in RPGRIP1L with Meckel Syndrome and Preimplantation Genetic Testing in a Chinese Family. *Reprod Sci.,* 29(8), 2200-2207. doi: 10.1007/s43032-022-00898-y.

Zhang W., Fletcher L., Muschel R.J. (2005) The role of Polo-like kinase 1 in the inhibition of centrosome separation after ionizing radiation. *J. Biol. Chem.,* 280(52), 42994-42999.

Zhang X., Yan C., Hang J., Finci L.I., Lei J., Shi Y. (2017) An Atomic Structure of the Human Spliceosome. *Cell,* 169(5), 918-929.e14. doi: 10.1016/j.cell.2017.04.033.

Zhao H., Chen Q., Fang C., Huang Q., Zhou J., Yan X., Zhu X. (2019) Parental centrioles are dispensable for deuterosome formation and function during basal body amplification. *EMBO Rep.,* 20, e46735.

Zhao X., Jin S., Song Y., Zhan Q. (2010) Cdc2/cyclin B1 regulates centrosomal Nlp proteolysis and subcellular localization. *Cancer Biol. Ther.,* 10(9), 945-952.

Zheng X., Ramani A., Soni K., Gottardo M., Zheng S., Ming Gooi L., Li W., Feng S., Mariappan A., Wason A., Widlund P., Pozniakovsky A., Poser I., Deng H., Ou G., Riparbelli M.G., Giuliano C., Hyman A.A., Sattler M., Gopalakrishnan J., Li H. (2016) Molecular basis for CPAP-tubulin interaction in controlling centriolar and ciliary length. *Nat. Commun.,* 7, 11874.

Zheng Y., Wong M.L., Alberts B., Mitchison T. (1995) Nucleation of microtubule assembly by a gamma-tubulin-containing ring complex. *Nature,* 378(6557), 578-583.

Zhou K., He Y., Lin X., Zhou H., Xu X., Xu J. (2024) KIFC1 depends on TRIM37-mediated ubiquitination of PLK4 to promote centrosome amplification in endometrial cancer. *Cell Death. Discov.,* 10(1), 419. doi: 10.1038/s41420-024-02190-1.

Zhu H., Chen D., Tang J., Huang C., Lv S., Wang D., Li G. (2018) Overexpression of centrosomal protein 55 regulates the proliferation of glioma cell and mediates proliferation promoted by EGFRvIII in glioblastoma U251 cells. *Oncol. Lett.,* 15(2), 2700-2706. doi: 10.3892/ol.2017.7573.

Zhu J., Abbruzzese J.L., Izzo J., Hittelman W.N., Li D. (2005) AURKA amplification, chromosome instability, and centrosome abnormality in human pancreatic carcinoma cells. *Cancer Genet. Cytogenet.,* 159, 10-17; PMID:15860351.

Zhu J., Burakov A., Rodionov V., Mogilner A. (2010) Finding the cell center by a balance of dynein and myosin pulling and microtubule pushing: A computational study. *Mol. Biol. Cell,* 21, 4418-4427.

Zhu R., Gundersen G.G. (2017) Centrifugal displacement of nuclei reveals multiple LINC complex mechanisms for homeostatic nuclear positioning. *Curr. Biol.,* 27, 3097-3110.

Zieske K., Schwille P. (2014) Reconstitution of self-organizing protein gradients as spatial cues in cell-free systems. *eLife*, 3, e03949.

Zimmerman K.W. (1898) Beitrage zur kenntniss cinigen drusen und epithelien [Contributions to the knowledge of glands and epithelia]. *Arch. Mikrosk. Anat. EntwMech.*, 52, 552-706. In German.

Zimmerman W.C., Sillibourne J., Rosa J., Doxsey S.J. (2004) Mitosis-specific anchoring of gamma tubulin complexes by pericentrin controls spindle organization and mitotic entry. *Mol. Biol. Cell,* 15(8), 3642-3657.

Zitouni S., Nabais C., Jana S.C., Guerrero A., Bettencourt-Dias M. (2014) Polo-like kinases: structural variations lead to multiple functions. *Nat. Rev. Mol. Cell Biol.,* 15, 433. 10.1038/nrm3819.

Zorn G.A., Lucas J.J., Kates J.R. (1979) Purification and characterization of regenerating mouse L929 karyoplasts. *Cell,* 18(3), 659-672. doi: 10.1016/0092-8674(79)90121-1.

About The Editor

Rustem E. Uzbekov was born in 1964 in Kazan, USSR. In 1987 he graduated from the Faculty of Biology of the Lomonosov Moscow State University, where he received PhD in Cell Biology in 1992 and Dr. of Science degree in 2004. Since 2006 he is an Associated Professor at Moscow State University and since 2008 he has been working at the University of Tours. He is the author of 6 books and over 100 scientific articles.

Index

H

I

K

L

M